THE LOGICAL
FOUNDATIONS
OF MATHEMATICS

FOUNDATIONS AND PHILOSOPHY OF SCIENCE AND TECHNOLOGY SERIES

General Editor: MARIO BUNGE, *McGill University, Montreal, Canada*

SOME TITLES IN THE SERIES

AGASSI, J.
The Philosophy of Technology

ALCOCK, J.
Parapsychology: Science or Magic?

ANGEL, R.
Relativity: The Theory and its Philosophy

BUNGE, M.
The Mind - Body Problem

GIEDYMIN, J.
Science and Convention

SIMPSON, G.
Why and How: Some Problems and Methods in Historical Biology

WILDER, R.
Mathematics as a Cultural System

*PERGAMON JOURNALS OF RELATED INTEREST

ANALYSIS MATHEMATICA

FUNDAMENTA SCIENTIAE
STUDIES IN HISTORY AND PHILOSOPHY OF SCIENCE

* *Free specimen copies available on request*

THE LOGICAL FOUNDATIONS OF MATHEMATICS

by

WILLIAM S. HATCHER

Département de Mathématiques, Université Laval, Québec, Canada

PERGAMON PRESS

OXFORD · NEW YORK · TORONTO · SYDNEY · PARIS · FRANKFURT

U.K.	Pergamon Press Ltd., Headington Hill Hall, Oxford OX3 0BW, England
U.S.A.	Pergamon Press Inc., Maxwell House, Fairview Park, Elmsford, New York 10523, U.S.A.
CANADA	Pergamon Press Canada Ltd., Suite 104, 150 Consumers Road, Willowdale, Ontario M2J 1P9, Canada
AUSTRALIA	Pergamon Press (Aust.) Pty. Ltd., P.O. Box 544, Potts Point, N.S.W. 2011, Australia
FRANCE	Pergamon Press SARL, 24 rue des Ecoles, 75240 Paris, Cedex 05, France
FEDERAL REPUBLIC OF GERMANY	Pergamon Press GmbH, 6242 Kronberg-Taunus, Hammerweg 6, Federal Republic of Germany

First edition 1982

British Library Cataloguing in Publication Data

Hatcher, William S

The logical foundations of mathematics. – (Foundations and philosophy of science and technology series).

1. Mathematics – Philosophy
I. Title   II. Series
510′.1   QA8.4   80–41253

ISBN 0–08–025800–X

Printed in Hungary by Franklin Printing House

To Judith, Sharon, Carmel, and Benjamin
for teaching me
that logic isn't everything

Preface

THE present volume offers a study of the foundations of mathematics stressing comparisons between and critical analyses of the major nonconstructive foundational systems. Constructivism is discussed, however, and it is treated sufficiently to give the reader an accurate idea of the position of constructivism within the spectrum of foundational philosophies.

The focus of the work is on the modern systems of foundations developed during the last 100 years, but some attempt is made to give historical perspective to these developments. Our treatment of type theory, especially, makes a serious attempt to present a technically correct system as close to the original as possible.

The first chapter spends a good deal of time setting up a general framework for the rest of the study. In particular, two complete systems of axioms and rules for the first-order predicate calculus are given, one for efficiency in proving metatheorems, and the other, in a "natural deduction" style, for presenting detailed formal proofs. A somewhat novel feature of this framework is a full semantic and syntactic treatment of variable-binding term operators as primitive symbols of logic. Not only is this interesting in itself, but it greatly facilitates the presentation of certain systems. In any event, once established, this general framework enables us to present a considerable variety of foundational systems in a way which is correct both in spirit and in detail.

This book is essentially a considerably expanded and revised version of Hatcher [3] which was received in a generally positive and favourable way by logicians, mathematicians, and philosophers interested in foundations. Most of the new material in the present volume represents work done during the 12-year period since Hatcher [3] was published. In particular, topos theory appears to me to be the most significant single foundational development during this period, and I have accordingly tried to include some nontrivial parts of it while maintaining the spirit and approach of Hatcher [3]. I have not, however, found it possible to develop the functorial methods so widely used in topos theory since this would have clearly necessitated an even more extensive treatment. As it is, the chapter on categorical algebra is now the longest in the book. Yet, these developments are exciting and the extended treatment we have given of them seems warranted.

In fact, since it turns out that topos theory is, essentially, local, intuitionistic type theory, the internal language of a topos is quite similar to the language of type theory of Chapter 4. The exact relationship between topos theory and set theory is also established. There is, consequently, a much greater feeling of unity and cohesiveness to the present work than was possible to achieve in 1968 when the clearly important categorical approach was just getting off the ground.

I want to express here my sincere thanks to all those friends and colleagues who have, at one time or another, said a kind and encouraging word about this project. I am sure that I would not have had the courage to take pen in hand again and push through to completion without the cumulative emotional effect of this positive reinforcement. There are, of course, those to whom I owe an intellectual as well as an emotional debt and whose critical comments and suggestions have been important in writing *The Logical Foundations of Mathematics*. I would like to mention in particular John Corcoran, Phil Scott, F. W. Lawvere, Newton da Costa, Stephen Whitney, J. Lambek, André Boileau, and André Joyal. Also my thanks to my students Michel Gagnon and Guy Jacob for their capable research assistance at various stages of the project. Thanks are also due to Mario Bunge for his kind invitation to include this work in his series on the foundations and philosophy of science and to Pergamon Press for their expert handling of the editing and publishing.

This project was completed during my sabbatical year (1979–80) spent in the Mathematics Department at McGill University in Montreal, and I would like to express my thanks to those members of the Department and its administrative staff who contributed to making my stay there both pleasant and productive. Finally, sincere thanks are also due to my home university, Laval, for granting the sabbatical year that was so crucial to completing this book as well as to the National Science and Engineering Research Council of Canada for its continued grant support over the years which has enabled me to maintain the study and research necessary to the success of this and other projects.

W. S. HATCHER
Département de Mathématiques
Université Laval
Québec, Canada

Contents

Chapter 1. First-order Logic   1

    Section 1. The sentential calculus   1
    Section 2. Formalization   9
    Section 3. The statement calculus as a formal system   13
    Section 4. First-order theories   19
    Section 5. Models of first-order theories   32
    Section 6. Rules of logic; natural deduction   39
    Section 7. First-order theories with equality; variable-binding term operators   57
    Section 8. Completeness with *vbto*s   61
    Section 9. An example of a first-order theory   63

Chapter 2. The Origin of Modern Foundational Studies   68

    Section 1. Mathematics as an independent science   68
    Section 2. The arithmetization of analysis   69
    Section 3. Constructivism   71
    Section 4. Frege and the notion of a formal system   72
    Section 5. Criteria for foundations   73

Chapter 3. Frege's System and the Paradoxes   76

    Section 1. The intuitive basis of Frege's system   76
    Section 2. Frege's system   78
    Section 3. The theorem of infinity   90
    Section 4. Criticisms of Frege's system   95
    Section 5. The paradoxes   97
    Section 6. Brouwer and intuitionism   98
    Section 7. Poincaré's notion of impredicative definition   100
    Section 8. Russell's principle of vicious circle   101
    Section 9. The logical paradoxes and the semantic paradoxes   102

Chapter 4. The Theory of Types   103

    Section 1. Quantifying predicate letters   103
    Section 2. Predicative type theory   105
    Section 3. The development of mathematics in **PT**   116
    Section 4. The system **TT**   120
    Section 5. Criticisms of type theory as a foundation for mathematics   122
    Section 6. The system **ST**   127
    Section 7. Type theory and first-order logic   130

Chapter 5. Zermelo–Fraenkel Set Theory   135

   Section 1. Formalization of **ZF**   138
   Section 2. The completing axioms   154
   Section 3. Relations, functions, and simple recursion   159
   Section 4. The axiom of choice   167
   Section 5. The continuum hypothesis; descriptive set theory   170
   Section 6. The systems of von Neumann–Bernays–Gödel and Mostowski–Kelley–Morse 171
   Section 7. Number systems; ordinal recursion   178
   Section 8. Conway's numbers   186

Chapter 6. Hilbert's Program and Gödel's Incompleteness Theorems   190

   Section 1. Hilbert's program   193
   Section 2. Gödel's theorems and their import   193
   Section 3. The method of proof of Gödel's theorems; recursive functions   195
   Section 4. Nonstandard models of **S**   207

Chapter 7. The Foundational Systems of W. V. Quine   213

   Section 1. The system **NF**   213
   Section 2. Cantor's theorem in **NF**   223
   Section 3. The axiom of choice in **NF** and the theorem of infinity   226
   Section 4. **NF** and **ST**; typical ambiguity   229
   Section 5. Quine's system **ML**   233
   Section 6. Further results on **NF**; variant systems   235
   Section 7. Conclusions   236

Chapter 8. Categorical Algebra   237

   Section 1. The notion of a category   238
   Section 2. The first-order language of categories   243
   Section 3. Category theory and set theory   253
   Section 4. Functors and large categories   255
   Section 5. Formal development of the language and theory **CS**   260
   Section 6. Topos theory   279
   Section 7. Global elements in toposes   284
   Section 8. Image factorizations and the axiom of choice   291
   Section 9. A last look at **CS**   293
   Section 10. **ZF** and **WT**   297
   Section 11. The internal logic of toposes   302
   Section 12. The internal language of a topos   306
   Section 13. Conclusions   311

Selected Bibliography   313

Index   317

First-order Logic

IN ORDER to understand clearly the formal languages to be presented in later chapters of this book, it will be necessary for the reader to have some knowledge of first-order logic. This chapter serves to furnish the necessary tools. The reader who is already familiar with these questions can easily treat this chapter as a review, though some attention should be given to our particular form of the rules for the predicate calculus, which will be used in the remainder of this study.

1.1. The sentential calculus

By a *statement* or *sentence* of some language, we mean an expression of that language which is either true or false in the language. Other expressions of a language may be meaningful without being sentences in this sense. Commands, for example, are meaningful expressions of English, but they are not sentences in our sense. "Go help your brother", "Thou shalt not kill", and "Stop!" are commands in English. Though they are correctly structured English expressions, they do not qualify as sentences in our restrictive definition. It makes little sense to ask, "Is the command to 'stop' true or false?" The reader may consider that by "statement" or "sentence" we shall mean roughly what a grammarian might designate as an "English sentence in the indicative mood".

The *sentential* or *statement calculus* considers certain locutions by which sentences are combined to form more complicated sentences. The basic sentence connectives are: "not", symbolized by " $\sim$ "; "and", symbolized by " $\wedge$ "; "or", symbolized by " $\vee$ "; "if ..., then ---", symbolized by " $\supset$ "; and "... if and only if ---", symbolized by " $\equiv$ ". In a natural language such as English, these locutions undoubtedly vary in meaning according to certain contexts. We shall now proceed to fix their meaning by way of explicit conventions, and this will be our first step toward what is called "formalization". These conventions should not be construed as asserting that the above-mentioned locutions are always used in ordinary discourse in a manner consistent with our conventional meanings. They are rather to be regarded as an explicit statement of how we shall, in fact, agree to use these same locutions. This point has been frequently misunderstood by philosophers of "ordinary language".

In fixing our conventions, we shall be concerned only with the truth or falsity of sentences (as opposed to other aspects of sentences such as meaning or length). For this reason, the logic we obtain is often called *truth-functional*. We express our conventional meanings by way of diagrammatic tables which tell us the truth or falsity of a compound statement (formed by means of one of our connectives) relative to the truth or falsity of its component parts. These tables are called *truth tables*.

For example, the effect of the operation of *negation* ("not") on a given sentence X is given by the following table:

X	$\sim X$
T	F
F	T

The truth table tells us that if the sentence is true, then its negation is false and if the sentence is false, its negation is true. Notice that all possible cases of truth and falsity have been considered. Thus, we have completely described the (truth-functional) meaning of negation.

Associated with each of our connectives is an *operation*, which is just a function or mapping in the usual mathematical sense of the term. Negation is a mapping from sentences into sentences. Our other connectives are *binary* and thus mappings from pairs of sentences into sentences. As we have done with the truth table for "not", we shall introduce with each truth table the name we give to the operation associated with the particular connective.

The truth table for the *conjunction* (associated with "and") of two sentences is the following:

X	Y	$X \wedge Y$
T	T	T
T	F	F
F	T	F
F	F	F

Since conjunction is a binary operation, the number of possible cases of truth and falsity is greater than that of the negation operation. Generally speaking, a sentence with n component sentences will give rise to 2^n different possibilities of truth and falsity. We shall speak of "truth" and "falsity" as our two *truth values*. A given line of the truth table of a compound sentence X assigns a unique truth value to each component sentence of X and indicates the truth value of X for that assignment of truth values to components. For example, the first line of the truth table for the conjunction tells us that the compound $X \wedge Y$ has the value T for the assignment of values $\langle T, T \rangle$ (i.e. the assignment T to the first conjunct and T to the second).

We have not yet considered truth tables with more than two component sentences since we are still defining our basic connectives. However, we shall be able to iterate our connectives and thus build compound sentences of increasing complexity. We will therefore have truth tables for compounds with any finite number of component sentences. This will be clearer once we have completed the task of defining our basic connectives.

The truth table for the operation of *disjunction* ("or") is given as follows:

X	Y	$X \vee Y$
T	T	T
T	F	T
F	T	T
F	F	F

Notice that the conjunction of X and Y is true when and only when the conjuncts are both true, whereas the disjunction is false when and only when the two disjuncts are false. The relation between these two connectives will become clearer in the light of later examples.

Because we interpret $X \lor Y$ to be true when both disjuncts are true, it is sometimes said that we have adopted the *inclusive* meaning of "or".

The *conditional* ("if ..., then – – –") of two sentences is defined by the following table:

X	Y	$X \supset Y$
T	T	T
T	F	F
F	T	T
F	F	T

In a conditional statement $X \supset Y$, we call X the *antecedent* or *hypothesis* and Y the *consequent* or *conclusion*. This table probably departs most radically from ordinary usage. Ordinarily a statement of the form "If X, then Y" is thought of as asserting Y as true conditional upon the truth of X. For this reason, one does not even consider the cases in which X is false. Logicians have extended the common usage by giving truth values in these two further cases. Failure to do so would leave the conditional undefined since we would not have exhausted all possible cases of truth and falsity. We have, therefore, decided to consider the conditional as false only when the antecedent is true and the consequent is false.

Let us consider some examples to motivate our particular choice. If a statement Y is true, then it is true whether anything else is true or false. "If X, then $1 = 1$" is true for any statement X, true or false, since $1 = 1$. We thus give the conditional the value T in the first and third lines of the table.

As for the fourth line, which involves the falsity of both antecedent and consequent, consider such statements as the following, which occur frequently in mathematical exposition: "For any x whatever, if x is a prime number greater than 2, then x is odd." This statement is true. Now, let x be 6. Both antecedent and consequent are false in the resulting conditional statement "If 6 is a prime number greater than 2, then 6 is odd". Yet we still wish to count the original statement as true (even though this statement is not, strictly speaking, a conditional one). In fact, we desire things to be arranged so that a statement such as the original one is true only when *every* conditional statement obtained by substituting particular values for x is true. But this state of affairs will obtain only if we require conditionals to be true when both the antecedent and consequent are false; hence the fourth line of our truth table.

The reader may feel that, in our treatment, too much weight is being given to mathematical considerations and that a different treatment of the conditional would be possible and even preferable. Many philosophers have indeed refused to accept this treatment and have developed their own theory of the conditional connective. Suffice it to say that experience seems to have shown that our truth-functional treatment of the conditional is entirely adequate for mathematics and this usage is the principal application that we envisage for our logic.

The *biconditional* ("$\equiv$") of two sentences X and Y is true when X and Y agree in truth value:

X	Y	$X \equiv Y$
T	T	T
T	F	F
F	T	F
F	F	T

Now that we have defined our truth-functional connectives, it is clear that any sentence built up from other sentences by means of these connectives will have a well-determined truth value for each assignment of values to its components. To make this more precise, let us introduce several definitions. Let us call a sentence *atomic* if it is not built up from other sentences by means of our sentential connectives. For example, "It is raining today" and "I am sick" are atomic, whereas "It is not raining today" or "If it is raining today, then I am sick" are not atomic. Notice that an atomic sentence X may contain occurrences of some of our sentential connectives, but such occurrences of our connectives will not serve to build X from other *sentences*. Our example of three paragraphs ago is a case in point.

Now clearly we can build a compound sentence by means of our logical connectives by starting with a given number of atomic sentences. Moreover, any such sentence will have a truth table, which will give the truth value of the compound sentence for each assignment of values to its atomic components. For example:

$$[[\sim\ X]\ \lor\ Y]\ \supset\ Z$$

$[[\sim$	$X]$	$\lor$	$Y]$	$\supset$	Z
F	T	T	T	T	T
F	T	T	T	F	F
F	T	F	F	T	T
F	T	F	F	T	F
T	F	T	T	T	T
T	F	T	T	F	F
T	F	T	F	T	T
T	F	T	F	F	F

Notice here that we have used a more compact way of giving the truth table: we have put the values for the component parts under each atomic component and put the resulting value for each compound statement under the connective that forms the compound. This procedure is simpler than stating each component separately as we did in introducing the connectives.

Involved in the generalized construction of compound sentences from simpler ones is the question of grouping. Where X, Y, and Z are sentences, the expression $X \land Y \supset Z$ is ambiguous as between $[X \land Y] \supset Z$ and $X \land [Y \supset Z]$. In the vernacular, punctuation serves to indicate the intended grouping. For instance, let X stand for "I am sick", Y for "I will stay home", and Z for "The game is lost". The two translations of these groupings would then be (1) "If I am sick and if I stay home, then the game is lost", and (2) "I am sick, and if I stay home then the game is lost". These two sentences do not have the same meaning.

Often the grouping of compound sentences in the vernacular is not clear or it is left to contextual interpretation. For this reason, let us introduce a convention for the use of brackets; it will remove all possible ambiguity. This convention represents a further step toward formalization.

Definition 1. We say that an expression W is *well formed according to the sentential calculus* if W is an atomic sentence or if there are well-formed expressions X and Y such that W is of the form $[\sim X]$, or $[X \land Y]$, or $[X \lor Y]$, or $[X \supset Y]$, or $[X \equiv Y]$. We further suppose that atomic sentences contain no brackets. No other expressions are well formed (according to the sentential calculus).

This definition is our first example of a *recursive* definition. This is a definition involving an inductive or iterative process (such as building up sentences from simpler ones).

As we have defined them, well-formed expressions are sentences. We further suppose that, from now on, all sentences with which we deal are well formed.

From the way we have defined well-formed expressions, any such expression will have the same number of left and right brackets. We use this fact to define precisely what we mean by the *principal connective* of a sentence. The principal connective is thought of as the last connective used in constructing the sentence from its component parts. More precisely:

Definition 2. Given a sentence X, let us count the brackets from left to right, counting $+1$ for all left brackets and -1 for all right brackets. The sentence connective we reach while on the count of "one" will be the *principal connective*.

Example. $[[X \supset Y] \supset Z]$. Counting brackets, we find that we cross the second "$\supset$" on the count of one. This "$\supset$" is the principal connective. Notice that the final count is always zero. We leave it as an exercise to the reader to prove that there is always one and only one principal connective for any nonatomic sentence. (*Hint:* The proof is by mathematical induction on the number of brackets occurring in the sentence.)

Definition 3. We say that a sentence is a *tautology* or is *tautologically true* if and only if the truth table of the sentence exhibits only Ts under its principal connective.

In other words, a tautology is a sentence that has the truth value T for every assignment of values to its atomic components. The reader can verify that $[X \supset X]$, $[[X \lor Y] \lor [\sim X]]$ are tautologies, where X and Y are any sentences.

From now on, we shall relax our convention on brackets by omitting them wherever there is no ambiguity possible. In particular, we shall often omit the outside set of brackets.

Definition 4. Given two sentences X and Y, we say that X *tautologically implies* Y if the conditional $X \supset Y$ is a tautology.

THEOREM 1. *If X and Y are sentences and if the conditional $X \supset Y$ is a tautology and if, further, X is a tautology, then Y is a tautology.*

Proof. Suppose there is some assignment of values which makes Y false. Then, for that assignment of values, the conditional $X \supset Y$ will be false since every assignment of values makes X true. But this is impossible since the conditional $X \supset Y$ is a tautology. Thus, there is no assignment of values making Y false, and Y is tautology. (A complete proof of the above theorem can be given by mathematical induction.)

Theorem 1 embodies what we call a *rule of inference*. A rule of inference is an operation which allows us to pass from certain given sentences to other sentences. We say that the latter are *inferred* from the former. The traditional name of the rule of Theorem 1 is *modus ponens*. *Modus ponens* allows us to infer Y from $X \supset Y$ and X.

We may sometimes refer to tautologically true statements simply as *logically true* statements, meaning "logically true according to the sentential calculus". The reason for the latter qualification is that we shall subsequently study a class of logical truths that is broader than the tautologies. Tautologies can be thought of as those logical truths which are true strictly by virtue of their structure in terms of our five sentential connectives. Our broader class will include other types of logical operations that we have not yet considered.

Definition 5. Two statements will be called *tautologically equivalent* if their biconditional is tautologically true.

Exercise. Show that two statements which are built up from the same atomic components $X_1, X_2, \ldots, X_n$ are tautologically equivalent if and only if they have the same truth value for the same given assignment of values to the atomic components.

Notice that two statements may be tautologically equivalent even where they do not have the same atomic components. Thus, X is equivalent to $X \wedge [Y \vee [\sim Y]]$ where X and Y are any sentences.

THEOREM 2. *Let X be any sentence which is a tautology and whose atomic components are sentences $a_1, a_2, \ldots, a_n$. If any sentences whatever are substituted for the atomic sentences a_i, where the same sentence is substituted for each occurrence in X of a given a_i, the resulting sentence X' is also a tautology.*

Proof. A rigorous proof of this is done by induction, but the basic idea can be easily expressed. Since X is a tautology, it is true for every assignment of values to its atomic components. Now consider X'. Its atomic components are the atomic components of the sentences Y_i that have been substituted for the a_i. Now, for any assignment of values to the atomic components of X', we obtain an assignment of values to the Y_i. But this assignment of values to the Y_i is the same as some assignment of values to the a_i of X and thus yields the value T as before.

Theorem 2 tells us that substitution in a tautology always yields a tautology. This again shows us the sense in which our analysis is independent of the meaning of the sentences making up a given compound sentence.

THEOREM 3. *If A is tautologically equivalent to B, and if A is replaced by B in some sentence X (at one or more places), then the resulting sentence X' will be tautologically equivalent to X.*

Proof. The student will prove Theorem 3 as an exercise (*Hint:* Show that the conditional $[[A \equiv B] \supset [X \equiv X']]$ is tautologically true.)

Exercise 1. Establish the following equivalences where the variables represent sentences:
(a) $[X \vee Y] \equiv [Y \vee X]$; (b) $[X \wedge Y] \equiv [Y \wedge X]$; (c) $[X \vee [Y \vee Z]] \equiv [[X \vee Y] \vee Z]$; (d) $[X \wedge [Y \wedge Z]] \equiv [[X \wedge Y] \wedge Z]$; (e) $[\sim[\sim X]] \equiv X$.

These equivalences establish the commutative and associative laws for disjunction and conjunction with respect to the equivalence relation of tautological equivalence. Furthermore, negation is involutory.

It is well known that we really have a Boolean algebra with respect to tautological equivalence, where negation represents complementation, disjunction represents supremum, and conjunction represents infimum. In other words, the substitutivity of equivalence (Theorem 3) shows that we have a congruence relation, and the quotient algebra is a Boolean algebra. This shows the precise sense in which two operations of conjunction and disjunction are dual to each other. They are dual in the Boolean algebraic sense. The equivalence class of sentences determined by the tautologies will be the maximal element of our Boolean algebra. The minimal or zero element of our Boolean algebra will be determined by the equivalence class of the negations of tautologies. These are the *refutable or tautologically false* sentences. Obviously the tautologically false sentences will be those whose truth table has all *F*s under the principal connective. They are dual to tautologies.

In view of the associative and commutative laws of conjunction and disjunction, we can speak unambiguously about the conjunction (or disjunction) of any finite number of sentences without regard to grouping or order. Where several different connectives are involved, grouping and order again become relevant.

Exercise 2. By an *argument* in English, we mean a finite collection of statements called *premisses* followed by a statement called the *conclusion.*[†] We say that an argument is *valid* if the conjunction of the premisses logically implies the conclusion. In each of the following arguments, test validity by using letters to represent atomic sentences, forming the conditional with the conjunction of the premisses as antecedent and the conclusion as consequent, and testing for tautology (thus, a tautological implication).

(a) If the team wins, then everyone is happy. If the team does not win, then the coach loses his job. If everyone is happy, then the coach gets a raise in pay. Everyone is happy if and only if the team makes money. Hence, either the coach gets a raise in pay, or the team does not win and the coach loses his job.

(*Hint:* When dealing with an implication involving many variables, it becomes prohibitive and unnecessary to give all possible cases. Observe that by the truth table for the conditional, a conditional statement is false only in one case, when antecedent is true and consequent is false. The best method, then, is a truth-value analysis, working from the outside in. Try to make the antecedent true and the consequent false. This gives necessary conditions on the variables. If these are satisfied, then we have a counterexample and implication does not hold. Otherwise, we establish the impossibility of the false case and implication holds.)

(b) If the sun shines today, then I am glad. If a high-pressure zone moves in, then the sun will shine. Either my friend will visit and the sun will shine, or a high-pressure zone moves in. Hence, I am glad.

(c) If I study, then I shall pass the test. If I pass the test, then I shall be surprised. If I fail the test, it is because the teacher is too difficult and grades unfairly. If I fail the test, then I do not blame the teacher. Thus, either I shall not be surprised, or I shall fail the test and blame the teacher.

Exercise 3. Verify that the following are tautologies where the variables represent sentences:
(a) $[X \equiv Y] \equiv [[Y \supset Y] \wedge [Y \supset X]]$; (b) $[X \wedge Y] \equiv [\sim[[\sim X] \vee [\sim Y]]]$; (c) $[X \supset Y] \equiv [[\sim X] \vee Y]$.

[†] In using the double s in "premiss", we follow Church [3].

The equivalences of Exercise 3 show that we can express some of our basic connectives in terms of others. In particular, we can see that negation and disjunction suffice to define the others. A more precise understanding of what such definability of some connectives in terms of others really means involves the notion of a truth function which we shall now briefly sketch.

Given a set E with exactly two elements, say $E = \{T, F\}$, the set of our two truth values, a *truth function* is defined as a function from E^n into E, where E^n is the set of all n-tuples of elements of E. Let some compound sentence X be given, and suppose that X has exactly n atomic components. Then, for a given ordering of the atomic components of X (there are $n!$ such orderings) there is a uniquely defined truth function f from E^n to E associated with X. For a given n-tuple $\langle a_1, \ldots, a_n \rangle$ of truth values, $f(\langle a_1, \ldots, a_n \rangle)$ is determined as follows: We assign the truth value a_1 to the first atomic component in our ordering, a_2 to the second, and generally a_i to the ith atomic component. When the assignment of truth values is complete, we have simply a line of the truth table of X and $f(\langle a_1, \ldots, a_n \rangle)$ is defined to be the truth value a_{n+1} of X for the given assignment of values to atomic components.

Suppose now that we agree on some fixed ordering of atomic components for all sentences X, say the order of first occurrence from left to right. Then every sentence X has a uniquely associated truth function f as defined above. We call f the *truth function expressed by* X. Now, we say that a set K of sentential connectives is *adequate* if and only if it is possible to express every truth function by means of some compound sentence X constructed from atomic sentences using only connectives from K.

It is easy to see that the set $K = \{\sim, \vee, \wedge\}$ is adequate, for let some truth function f from E^n to E be given. Let n atomic sentences $X_1, \ldots, X_n$ be chosen. We now construct a sentence in the following way: We consider only those n-tuples of truth values $\langle a_1, \ldots, a_n \rangle$ for which

$$f(\langle a_1, \ldots, a_n \rangle) = T.$$

Suppose first that there is at least one n-tuple whose image under f is T. For each such n-tuple we form the following conjunction $X_1^* \wedge \ldots \wedge X_n^*$, where X_i^* is X if a_i is T and X_i^* is $[\sim X_i]$ if a_i is F. This compound obviously yields the value T when the values a_i are assigned to the X_i. We now take the disjunction $W_1 \vee \ldots \vee W_m$ of all the conjunctions W_j of the X_i^*s. This is a compound sentence whose expressed truth function is obviously f. On the other hand, if $f\langle a_1, \ldots, a_n \rangle = F$ for all n-tuples of truth values, then any contradictory sentence with n atomic components will express f. The sentence $X_1 \wedge [\sim X_1] \wedge W$, where W is the conjunction of the rest of the X_i, will do fine.

If two sentences have the same atomic components in the same left-to-right order of first occurrence, then they will be tautologically equivalent if and only if they express the same truth function. The equivalences of Exercise 3 thus show that we can express all truth functions by means of $\sim$ and $\vee$, since $\wedge$ is definable by means of $\sim$ and $\vee$, and we have shown $K = \{\sim, \vee, \wedge\}$ to be adequate. In other words, the set $\{\sim, \vee\}$ is adequate.

Exercise 4. Show that each of $\{\sim, \supset\}$ and $\{\sim, \wedge\}$ is adequate. Show that $\{\sim\}$, $\{\wedge\}$, and $\{\supset\}$ are not adequate.

Although it may not appear possible at first, there are single binary connectives which are alone adequate to express all truth functions. If we define a new connective " $|$ " by the table

X	Y	$[X\,\vert\,Y]$
T	T	F
T	F	T
F	T	T
F	F	T

then $[\sim X]$ is definable by $[X\,\vert\,X]$, and $[X\vee Y]$ is definable as

$$[[X\,\vert\,X]\,\vert\,[Y\,\vert\,Y]].$$

Since $\{\sim,\ \vee\}$ is adequate, the adequacy of $\{\,\vert\,\}$ follows immediately.

The binary connective "↓" defined by

X	Y	$[X\!\downarrow\! Y]$
T	T	F
T	F	F
F	T	F
F	F	T

is also adequate as the reader can check by defining $\sim$ and $\vee$ (or $\sim$ and $\wedge$) purely in terms of ↓. It is easy to show that | and ↓ are the only two adequate binary connectives and we leave this as an exercise to the reader. Church [3] contains a detailed discussion of the definability and expressibility of various connectives for the sentential calculus, as well as detailed references to early work on the question.

1.2. Formalization

In the preceding section, we have considered the English language and subjected it to a certain logical analysis having mathematical overtones. The reader may have noticed that, as we progressed, we moved further and further away from the necessity of giving examples involving specific sentences of English such as "I am sick". It was sufficient simply to consider certain forms involving only letters representing sentences together with our logical connectives. The reason for this should be clear if one reflects on our point of departure. Because our analysis was truth-functional, we were concerned only with sentences as objects capable of being considered true or false. Not much else, not even meaning, except on a very rudimentary level, was relevant to our analysis.

The whole process is typical of mathematics. One starts with a particular concrete situation, and then subjects it to an analysis which ignores some aspects while considering others important. In the empirical sciences, this process leads eventually to certain "laws" or general statements of relationship, which are afterwards capable of various degrees of verification. In mathematics, the process leads to the definition of abstract structures independent of any concrete situation. These abstract structures are then studied in their own right. Because they supposedly carry certain important features shared by many concrete cases, applications of these abstract structures are often found (sometimes in surprising ways), and one gains in economy by finding general results true, within a single given structure, for many different cases. Finally, any particular structure having all of the properties of a given abstract structure is called a "model" of the abstract structure.

Mathematical logic has given rise to the study of abstract structures called *formal systems* or *formal languages*. We define these as follows:

Definition 1. A *formal system* F is a quadruple $\langle A, S, P, R \rangle$ of sets which satisfies the following conditions: A is a nonempty set called the *alphabet* of F and whose elements are called *signs*, S is a nonempty subset of the set $\mathcal{M}(A)$ of all finite sequences of elements of A, P is a subset of S, and R is a set of finitary relations over S (a finitary relation over S being some subset of S^n, where $n \geqslant 1$ is some positive integer). Elements of $\mathcal{M}(A)$ are called *expressions* of F, elements of S are called *well-formed formulas*, abbreviated as *wffs*, or just *formulas* of F, elements of P are called *axioms* of F, and elements of R are called *primitive rules of inference* or just *rules of inference* of F. By the *degree* of a rule of inference r in R we mean the integer n such that r is a subset of S^n. Finally, the triple $\langle A, \mathcal{M}(A), S \rangle$ is called the *language* of F, and the pair $\langle P, R \rangle$ is called the *primitive deductive structure* of F relative to the language of F.

The above definition is a bit more general than we really need. In most cases, the alphabet will be countable. Moreover, we shall often require that the alphabet as well as the sets S and P be "effective" or "decidable" in the sense that there is some procedure allowing us to decide whether a given object is or is not a sign of the system, or whether a given expression is or is not a wff, or whether a given wff is or is not an axiom. We may similarly require that a given rule of inference R_n be effective in the sense that, for any n-tuple $\langle x_1, \ldots, x_n \rangle$ of elements of S, there is some effective procedure which allows us to determine whether or not the n-tuple is in the relation R_n. Also, the set R of rules of inference may usually be assumed finite.

The intuitive idea underlying our informal notion of "effectiveness" in the preceding paragraph is that there exists a set of rules or operations which furnish a mechanical test allowing us to decide a question in a finite length of time. Elementary arithmetic calculation is a good example of an effective procedure. Our bracket-counting method for determining the principal connective of a sentence is another. In Chapter 6, we give a precise mathematical definition of the notion of effectiveness by means of the so-called recursive functions.

The primitive deductive structure of a formal system F allows us to develop a fairly rich *derived* deductive structure which we now describe.

Definition 2. Given a formal system F and a set X of wffs of F, we say that a wff y is *immediately inferred* from the set X of wffs if there exists a primitive rule of inference R_n of degree n and a finite sequence $b_1, \ldots, b_{n-1}$ of elements of X such that the relation $R_n(b_1, \ldots, b_{n-1}, y)$ holds (i.e. the n-tuple $\langle b_1, \ldots, b_{n-1}, y \rangle$ is an element of the relation R_n).

Definition 3. Given a set $X \subset S$ of wffs, we say that a wff $y \in S$ is *deducible from the hypotheses* X if there exists a finite sequence $b_1, \ldots, b_n$ of wffs such that y is b_n and such that every member of the sequence is either (i) an element of X, or (ii) an axiom of F, or (iii) immediately inferred from a set of prior members of the sequence by some primitive rule of inference. The finite sequence $b_1, \ldots, b_n$ itself is called a *formal proof* (or *formal deduction*) *from the hypotheses* X. If X is empty (thus only axioms are used in the deduction together with rules of inference), then the sequence $b_1, \ldots, b_n$ is called simply a *formal deduction* or *formal proof in F*. In this case, y is said to be a *provable* wff or a *theorem* of F. A formal proof $b_1, \ldots, b_n$ is a *proof of b_n*.

We denote the set of all wffs which can be deduced from a given set of hypotheses X by $K(X)$, the set of consequences of X. The theorems are thus the set $K(P)$ (which is the same as the set $K(\Lambda)$).[†] The theorems are those wffs which can be obtained as the last line of a formal deduction. Notice that any axiom is a one-line proof of itself, and every formal proof must begin with an axiom.

A formal deduction in a formal system is an abstract analogue of our usual informal notion of a proof in which we "prove" an assertion by showing that it "follows from" other previously proved statements by successive application of logical laws or principles. These previously proved statements are ultimately based on axioms, and our logical principles or laws (such as the rule of *modus ponens*) allow us to pass from a given statement or statements to other statements.

We write $X \vdash y$ for $y \in K(X)$. When X is empty, we write simply $\vdash y$ instead of $\Lambda \vdash y$. Since $K(\Lambda)$ is the set of theorems, $\vdash y$ is a short way of asserting that "y is a theorem" or "y is provable". Sometimes we talk about provability with respect to different systems F and F', and we indicate this by writing "$\vdash_F y$", "$\vdash_{F'} y$", and so on. Also, if $\{z\} \vdash y$, for some one-element set $\{z\}$, we prefer to write simply "$z \vdash y$". This latter is properly read "y is deducible from the hypothesis z". Finally, we may sometimes wish to emphasize the fact that certain wffs occur as hypotheses and so we display them by writing $X, z_1, z_2, \ldots, z_n \vdash y$ to mean $X \cup \{z_1, z_2, \ldots, z_n\} \vdash y$.

Let us note in passing that a formal deduction from hypotheses X in a system F is the same thing as a formal deduction in the system which is the same as F except that its set of axioms is $P_F \cup X$. It is useful to keep this simple fact in mind since it means that some metatheorems about formal deduction have greater generality than may appear at first glance.

Given a formal proof $b_1, \ldots, b_n$ from the hypotheses X in a formal system F, each member of the sequence is *justified* in that it satisfies at least one of the conditions (i) to (iii) of Definition 3. It is quite possible that a given member of the proof could satisfy more than one of these conditions, even all three of them. By a *justification* for a member b_i of the proof, we mean any one of the three conditions (i) to (iii) which is true of b_i. Later on, when we deal with the languages known as first-order theories, we shall countenance a slightly different notion of proof, one in which it is necessary that every line in the proof be accompanied by an explicit justification.

As has already been mentioned, we generally suppose that the set S of wffs and the set P of axioms are decidable sets, and that our rules of inference are effective. These requirements have the result that the notion of formal proof is effective; given any finite sequence of wffs, we can decide whether or not it is a formal proof (from no hypotheses). It will not follow, however, that the set $K(\Lambda)$ of theorems is decidable even though the notion of deduction from no hypotheses is. A system F for which the theorems $K(\Lambda)$ are a decidable set is called a *decidable* system. Most interesting systems will not be decidable, though there are some exceptions.

In practically all systems which logicians consider, the set of wffs is decidable and the rules of inference are effective. However, certain branches of logic, such as model theory, do consider systems whose set P of axioms is not decidable (and even some for which the set S of wffs is not decidable). A system F whose set of axioms is decidable is called *axiomatized*.

[†] "Λ" is our symbol for the null set.

A system F for which there is a decidable set of axioms yielding the same theorems (keeping the rules and wffs the same) is said to be *axiomatizable*. There are many interesting axiomatizable systems.

Exercise 1. Prove our assertion that $K(P) = K(\Lambda)$.

Exercise 2. Show that the K function has the following properties in any formal system F: $X \subset K(X)$; $K(K(X)) = K(X)$; if $X \subset Y$, then $K(X) \subset K(Y)$; $K(X) = \bigcup_{Y \in \mathcal{F}} K(Y)$ where $\mathcal{F}$ is the class of finite subsets of X.

Exercise 3. Let two wffs, x and y, be *deductively equivalent* if each is deducible from the other. Show that any two theorems are deductively equivalent for any formal system F.

Exercise 4. Let some formal system F be given and let $X = a_1, a_2, \ldots, a_n$ and $Y = b_1, b_2, \ldots, b_m$ each be sequences of wffs which constitute formal proofs in F. Prove that the juxtaposition XY of these two deductions is again a deduction. Explain how this fact justifies the usual practice in mathematics of citing previously proved theorems, as well as axioms, in a proof.

Obviously, a provable wff of a formal system F will have an infinite number of different proofs. This is because any given proof of a wff b can be arbitrarily extended by adding new lines which are justified but which are not essential to obtaining b. For example, we can uselessly repeat hypotheses, add extraneous axioms and inferences obtained from them, or even repeat b itself any number of times. It is useful to make all of this precise by defining clearly when a formula in a deduction *depends* on another formula. Intuitively, a given formula will depend on another if the other formula has been used in obtaining the given formula.

Definition 4. Let $b_1, \ldots, b_n$ be a formal deduction from hypotheses X in some formal system F. Then we say that an occurrence b_i of a formula *depends on* an occurrence b_j of a formula, $j \le i$, if $i = j$ or else if $j < i$ and b_i is immediately inferred from a set of prior formulas at least one of whose occurrences depends on b_j. We say that a formula y occurring in the deduction *depends on* a formula z occurring in the deduction if at least one occurrence of y depends on an occurrence of z.

Notice that every formula occurring in the deduction depends on itself, and no formula occurring in the deduction depends on any formula not occurring in the deduction.

Using these concepts, we now prove:

THEOREM 1. *If $b_1, b_2, \ldots, b_n$ is a formal deduction from hypotheses X, in a formal system F, then there exists a subsequence $b_{i_1}, \ldots, b_{i_k}$ which is a formal deduction in F of b_n from hypotheses X and such that b_{i_k} depends on every occurrence of every formula in the new deduction.*

Proof. We prove this by induction on the length n of the original deduction. We assume the proposition is true for all deductions of length less than n and consider the case of length n. If b_n is an hypothesis or an axiom, then the sequence of length 1 consisting of the formula b_n

alone is a proof from the hypotheses X of the formula b_n. Moreover, this is a subsequence of the appropriate kind, chosen by putting $i_1 = n$.

Now, suppose b_n is immediately inferred from prior wffs $b_{j_1}, \ldots, b_{j_m}$ in order of occurrence in the deduction. Since each of these wffs is the last line of a formal deduction from hypotheses X and of length less than n, we can apply the induction hypothesis to each of them and obtain subsequences of our original deduction which are proofs of each of them consisting only of formulas on which they each depend. We now choose the smallest subsequence Y containing all of these (which will, in fact, just be their set-theoretic union). Y will be a formal deduction from hypotheses X because each line is justified (if it was justified as a part of one of the subsequences proving one of the b_{j_s}, then it remains justified as a part of the union of all of these subsequences). Moreover, b_{j_m} is the last line of the formal deduction Y in which all the b_{j_s} appear. Now, applying the rule originally used to infer b_n, we obtain b_n as the next line of a formal deduction Yb_n. Yb_n is a subsequence of the original deduction as is clear. Finally, b_n depends on every member of Yb_n since it depends on each of $b_{j_1}, \ldots, b_{j_m}$, and every other occurrence of a formula in Y is depended upon by at least one of the b_{j_s}. This completes the proof of the theorem.

Several consequences of Theorem 1 deserve to be stated explicitly as corollaries.

COROLLARY 1. *In any formal system F, every theorem has a proof involving only formulas on which it depends.*

Proof. Any theorem has a formal deduction and hence, by Theorem 1, a formal deduction involving only formulas on which it depends.

COROLLARY 2. *In any formal system F, if $y \vdash x$ and if x does not depend on y, then $\vdash x$.*

Proof. Since $y \vdash x$, there is a proof of x from the hypothesis y involving only formulas on which x depends. Since x does not depend on y, y does not appear in the new proof. This new deduction thus involves only axioms and rules of inference and is therefore a proof of x from no hypotheses.

We shall sometimes speak of the *interpretation* of a formal system F. This notion will be made quite precise for a large class of formal systems, the so-called first-order systems, which will be treated later in this chapter. For the moment we will not attempt a precise definition, but we will have in mind some language whose sentences can be interpreted as wffs of F and whose true sentences (or some of whose true sentences) can be interpreted as theorems of F. By way of example, we will now formulate a formal system whose intuitive interpretation will be precisely the statement calculus of Section 1.

1.3. The statement calculus as a formal system

We define a formal system **P** whose alphabet consists of the signs "[" ,"]", "*" (called *star*), " ~", " $\lor$ ", and the small italic letter "x". An expression of **P** is, of course, any finite sequence of occurrences of these signs. By a *statement letter* of **P**, we mean the letter "x"

followed by any nonzero finite number of occurrences of star. x_n will stand for the letter "x" followed by n occurrences of star. Thus, "x_5" stands for "x^{*****}", for example.

We define recursively the set S of wffs of **P**: (i) Any statement letter is a wff. (ii) If X and Y are wffs, then expressions of the form $[\sim X]$ and $[X \lor Y]$ are wffs. (iii) S has no other members except as given by (i) and (ii).

The reader will notice that the wffs, as we have defined them, are completely analogous to the informal wffs of Section 1.1, as we here simply replace the informal notion of "atomic sentence" by the formal notion of a statement letter. The introduction of the star into our formal system is simply a device to enable us to obtain a countably infinite number of distinct statement letters from a finite alphabet. There is no particular virtue in this and we shall countenance countably infinite alphabets many times in this book.

The reader will also notice that we have not included some of our basic connectives as elements of our alphabet. This is because we can economize by *defining* some connectives in terms of others. The exact meaning of this sort of definition is best illustrated by an example.

Definition 1. $[X \supset Y]$ for $[[\sim X] \lor Y]$.

In this definition the letters "X" and "Y" are not signs of our system but rather dummy letters representing arbitrary wffs. The sign "$\supset$" is not part of our system either, and it is not intended that our definition introduces the sign into the system, but rather that, in every case, the expression on the left is an abbreviation for the expression on the right.

More precisely, the meaning of such a definition is as follows: When any two wffs of **P** are substituted for the letters "X" and "Y" in the above two forms of Definition 1, then the expression resulting from substitution in the left-hand form is an abbreviation of the expression resulting from the same substitution in the right-hand form. The expression resulting from such substitution in the right-hand form will be a wff of **P** while the expression resulting from substitution in the left-hand form will not. Such forms, involving letters together with brackets and other special signs, are called *schemes*. Particular expressions resulting from substitution of formal expressions for the letters of a scheme are called *instances* of the scheme. Thus, in the foregoing example, an instance of the left-hand scheme arising from substitution of wffs of **P** for the letters "X" and "Y" is an abbreviation of the *corresponding* instance (resulting from the same substitution for the letters "X" and "Y") of the right-hand scheme.

The distinction between *use* and *mention* of signs is highly important when dealing with formal systems. The alphabet of a given formal system may contain signs that are used in ordinary English. In order to avoid confusion, we must carefully distinguish between the roles such signs play within the formal system and the informal use made of them in the vernacular. We speak of the formal system we are studying as the *object language*, and the language, such as English, we use in studying the formal system is called the *metalanguage*. In the system **P**, we have, for instance, the sign "x", which is part of our formal system but which also has a usage in ordinary English, our metalanguage. Our abbreviative definitions of certain expressions of a formal system are technical parts of the metalanguage and not part of the formal system itself. To avoid confusion, logicians make the following rule: In order to talk about an object (to *mention* it) we must *use* a name of the object. Where the objects of a discourse are nonlinguistic, there is little danger of confusion. We would not, for instance, use New York in order to mention New York. But where the objects are themselves linguistic ones (i.e. signs) we must be more careful.

The introduction of such metalinguistic definitions is practically necessary for any sort of manipulation of a formal language. Because we deal so extensively with such abbreviations, we speak of the original, formal notation as the *primitive* notation, and the abbreviated notation as *defined*. Thus, in the case of the formalization of **P** presently at hand, the symbols " ∼ " and " ∨ " are primitive while " ⊃ " is defined.

We often form the name of a linguistic object (a sign) by putting it into quotation marks, but it is possible to use signs as names of themselves. For example, we can say that "*x*" is a sign of **P**. Here, ' "*x*" ' is being used as a name for the sign which is the italic twenty-fourth letter of the Latin alphabet. But we often write that "*x* is a sign of **P**", thus using *x* as a name of itself. This is called an *autonomous* (self-naming) use of a sign. We can also use other symbols explicitly designated as names for our signs if it suits our purpose.

Similarly, expressions of a formal system, formed by juxtaposing the signs of the alphabet (writing them one after the other) must also be distinguished from expressions of English. We can form the names of formal expressions by juxtaposing in the metalanguage the names for the signs making up the finite sequence in the formal system. Again, we shall often use formal expressions as names of themselves.

There should be no confusion in this autonomous use of signs and expressions of a formal system, since those expressions and signs which are part of the system will always be explicitly designated for every formal system with which we deal in this study. Autonomous use avoids the myriad quotation marks which result from rigorous adherence to the use of quotation marks for name-forming.

Names, as we have here understood them, are simply *constants* in the sense commonly understood in mathematics. A constant is a symbol (linguistic object) which names or designates a particular object. A *variable* is a symbol which is thought of not as designating a particular object but rather as designating ambiguously any one of a given collection of objects. The collection of objects thus associated with a given variable is called the *domain of values* of the variable. It is, of course, we who decide what the domain of values of a given variable is, either by explicit designation or through some convention or contextual understanding. A given variable may obviously have different domains in different contexts, but its domain must be fixed and unambiguous in any particular context.

When substitutions are made for variables, it is usually understood that constants, names of elements of the domain of values, are actually substituted for the variables, rather than the values themselves. For example, in the phrase "*x* is prime", where the domain of *x* is the natural numbers, we can substitute "7" for "*x*" and obtain the sentence "7 is prime". We did not substitute the number seven (which is an abstract entity) but the numeral "7" for *x*.

However, when we are talking about linguistic entities, it does become possible to directly substitute the thing about which we are talking (a value) for another symbol. Thus, in dealing with formal systems, we can make use of dummy letters like the "*X*" and "*Y*" in the schemes of Definition 1. These dummy letters are not variables in the sense we have just defined, but rather *stand for* arbitrary wffs, and we can substitute wffs directly for them when our wffs are linguistic objects as is the case for our system **P**. (There is nothing in our definition of a formal system which requires that the set *S* of wffs must consist of linguistic objects.) Of course, if we want to consider such dummy letters as variables, we are free to do so, for we can regard the wffs as names of themselves and thus as constants which can be substituted for variables. In this case, a particular instance of a scheme is technically a name of a wff (itself) rather than a wff.

A variable in the metalanguage whose domain of values consists of signs or expressions of an object language is called a *metavariable* or *syntactical variable*. For example, our use of the letters "X" and "Y" in our recursive definition of the wffs of **P** is as metavariables rather than as dummy letters. We will not usually bother to distinguish explicitly between the use of letters of the metalanguage as metavariables and as dummy letters, since in the final analysis, the main difference in the two uses lies more in the way we regard what is going on than in what is going on. What we are doing is talking about formal expressions by means of letters and schemes, and we are allowing substitution of formal expressions for the letters of a scheme. In every case the class of formal expressions will be clearly designated. Hence, the main distinction resides in whether we regard a particular instance of a scheme as the autonym of a wff or as a wff, or whether, in the case of an abbreviative expression, we regard an instance of a scheme as standing for the wff it abbreviates or as naming it.

In any case, juxtaposition of letters in the metalanguage with other special signs will always represent the operation of juxtaposing the formal expressions they represent. Also, as follows from our discussion of abbreviations, we prefer to regard an abbreviation of a formal expression, such as "$[x^* \supset x^{**}]$" for "$[[\sim x^*] \vee x^{**}]$", as a technical part of the metalanguage which stands for the formal expression it abbreviates, rather than naming it. All these semantic matters will not concern us much in this book and we enter into a brief discussion here only to dispose of the matter for the rest of our study.

In the spirit of this last remark, let us point out one further complication that may have already occurred to the semantically precocious reader: When we speak of a *sign* such as the star "*", we cannot really mean the particular blob of ink on the particular part of this page. Such an ink blob is rather a *token* of the sign. The sign itself is the equivalence class of ink blobs under the relation "sameness of shape". A token is thus a representative of the equivalence class (i.e. the sign) in the usual mathematical sense. Similarly, sequences of signs are represented by linear strings of tokens.

We now return to our consideration of **P** and state several more definitions preparatory to designating its axioms.

Definition 2. $[X \wedge Y]$ for $[\sim[[\sim X] \vee [\sim Y]]]$.

Definition 3. $[X \equiv Y]$ for $[[X \supset Y] \wedge [Y \supset X]]$.

The intuitive justification for these definitions is clear in the light of Exercise 2 following Theorem 3 of Section 1 of this chapter.

We now use schemes and dummy letters to describe the set P of axioms of **P**. P consists of all wffs that are instances of the following schemes:

$$[[X \vee X] \supset X]; \quad [X \supset [X \vee X]]; \quad [[X \vee Y] \supset [Y \vee X]];$$
$$[[X \supset Z] \supset [[Y \vee X] \supset [Y \vee Z]]]$$

where X, Y, and Z stand for any wffs of **P**.

Notice that our axiom set is infinite.

The set R of rules of inference of **P** consists of one relation of degree 3, which is as follows: $R_3(X, Y, Z)$ if and only if there are wffs A and B such that X is $[A \supset B]$, Y is A, and Z is B. This is the formal analogue of *modus ponens*, and we will apply the name *modus ponens* to it. This completes the description of **P**.

Exercise. Determine which, if any, of the following are instances of our axiom schemes:

$$[[[x_1 \supset x_2] \vee [x_1 \supset x_2]] \supset [x_1 \supset x_2]];$$
$$[x_1 \supset [x_1 \vee x_2]]; \qquad [[x_1 \supset x_1] \supset [[x_1 \vee x_1] \supset [x_1 \vee x_1]]].$$

We have already indicated the analogy between statement letters of our formal system and atomic sentences of English. It is quite clear that we can define the *principal connective* of a wff and apply the truth-table method to wffs in a purely mathematical way to determine which of our wffs are *tautologies*. Notice that any instance of one of our axiom schemes will be a tautology. Furthermore, we can prove that *modus ponens* preserves the property of being a tautology. (The proof of this latter fact is essentially the same as that given for statements of English in Theorem 1 of Section 1.) These two facts immediately give us the result that all theorems are tautologies since our theorems are, by definition, obtained from axioms by successive applications of *modus ponens*. The rigorous proof of this is by induction in the metalanguage.

To illustrate formal deduction in P, we prove that $\vdash [x_1 \supset x_1]$. Recall that $[X \supset Y]$ is $[[\sim X] \vee Y]$ by definition.

$$[[x_1 \vee x_1] \supset x_1] \supset [[[\sim x_1] \vee [x_1 \vee x_1]] \supset [[\sim x_1] \vee x_1]]$$
$$[[x_1 \vee x_1] \supset x_1]$$
$$[[[\sim x_1] \vee [x_1 \vee x_1]] \supset [[\sim x_1] \vee x_1]]$$
$$[[\sim x_1] \vee [x_1 \vee x_1]]$$
$$[[\sim x_1] \vee x_1]$$

The first line of this deduction is an axiom of the last type listed in our axiom schemes. The second line is an axiom of the first type. The third line is obtained from the first two by *modus ponens*. The fourth line is an axiom of the second type (recall our defining abbreviations) and the fifth line is obtained from the third and fourth by *modus ponens*.

If we intended to engage in considerable formal deduction in **P**, then we would use many techniques to shorten proofs and render them more readable. This will be done for formal proofs of first-order theories after they are introduced. The reason why we do not do so for **P** will be clarified shortly.

We have the notion of deduction within the object language. This is the purely formal, mathematical process of deduction given in Definition 3 of Section 1.2. But we also have a notion of deduction within the metalanguage, which is the usual informal mathematical notion of deduction. Since deduction within the object language is a precisely defined mathematical operation, we can study its properties just as we can study the properties of any mathematical system. We can prove theorems about the operations within the formal system. These theorems about the formal system are called *metatheorems*. They are carried through in our intuitive logic, which underlies our thinking about mathematical structure.

The fact that all the theorems of our system **P** are tautologies, the proof of which we sketched several paragraphs ago, is a metatheorem about the system **P**. It is a theorem in the metalanguage about the set $K(P)$ of theorems of **P**.

It is possible that a given object language, about which we are proving metatheorems, has an interpretation as a part of English and that many of the statements we make about the object language can, in this sense, be made "within" the formal language itself; that is, within that part of English which is the interpretation of the formal language. For example, it is possible to explain, within English, the grammar of English itself. In such a case, it could happen that some theorems and metatheorems coincide. In other words, some metatheorems about the object language might be among the statements of the metalanguage that are also part of the interpretation of the object language. This would all depend on how we defined the interpretation of our object language in the first place.

It will naturally occur to the reader that it would be possible to envisage formalizing mathematically the metalanguage itself and studying its own internal structure. This is certainly true, but such a formalization could be done only within a meta-metalanguage, for we must *use* some language to communicate. Of course, it is conceivable that object language and metalanguage may be the same. Such is the case found in our example of explaining the grammar of English within English. However, Tarski has shown, by reasoning too involved for inclusion here, that in most cases we can avoid certain contradictions, which arise from the circularity of speaking about a language within the language itself, only when the metalanguage si strictly stronger than the object language (see Tarski [1], pp. 152 ff.). We delay further discussion of these delicate questions to Chapter 6 where we shall have the tools necessary to engage in a more precise analysis.

In closing this discussion, let us return once again to a consideration of our language **P**. Our observation that all the theorems of **P** are tautologies leads naturally to the question of whether or not the converse is true: are all tautologies theorems of **P**? The answer is "yes", but we will not give the details of the proof, which can be found in any standard work on mathematical logic such as Church [3] or Mendelson [1].

This second metatheorem, that all tautologies are theorems of **P**, is known as the *completeness theorem* for **P**. Logicians have extensively studied many different, partial (incomplete) versions of the statement calculus. Church [3] contains a long discussion of these questions.

Notice that, since the theorems of **P** are precisely the tautologies, we could have designated the tautologies as axioms (we still would have an effective test for the set of axioms) and let the set R of rules of inference be empty. This form is often given to **P** and we will use it extensively in the present study.

Now it is clear why we are not very eager to engage in protracted formal deductions in **P**. It is because the theorems of **P** are precisely the tautologies and it is easier to prove that a given wff is a tautology than to find a formal deduction for it. **P** is an example of a decidable formal system. Deduction is, in principle, unnecessary in any such decidable system (though the test for theorem determination may be much more complicated than the truth-table method).

1.4. First-order theories

Our goal in this section will be to describe a whole class of formal languages known as *first-order theories* or *first-order systems*. The importance of these languages is that they have a well-defined structure which is relatively simple while being adequate for a surprisingly large number of purposes.

Since we are describing not one particular language but a class of languages, we cannot specify the exact alphabet for each such language. Rather we shall describe a list of signs from which the particular alphabet of any given first-order theory will have to be chosen.

By an *individual variable*, we mean the small italic "x" with a positive integral numeral subscript. Examples are "x_1", "x_2", and so on.

By an *individual constant letter*, or simply *constant letter*, we mean the small italic "a" followed by any positive integral numeral subscript. Examples are "a_1", "a_2", and so on.

By a *function letter* we mean the small italic "f" together with positive integral numeral sub- and superscripts. Examples are "f_1^1", "f_3^2", etc. The reason for the double indexing will be made clear subsequently.

By a *predicate letter* we mean the capital italic letter "A" together with positive integral numeral sub- and superscripts. Examples are "A_2^1", "A_7^4", etc.

To these four *syntactical categories* of signs we add the following specific signs: "(" called *left parenthesis*; ")" called *right parenthesis*; "$\sim$" called *negation sign*; "$\vee$" called *disjunction sign*; and "," called *comma*. These signs plus the individual variables are called *logical signs*.

We now require that the alphabet of any given first-order theory include at least the following: all logical signs and at least one predicate letter. In addition, a particular first-order language may contain any number of constant letters, function letters, or additional predicate letters. A first-order theory is not defined until one has specified precisely which signs belong to its alphabet. (The possibility exists of a first-order theory having signs other than the above, but we defer this question until later.)

Some of our signs, such as the parentheses and comma, are signs of ordinary English and will thus be used in our metalanguage, as well as in our formal language. This will cause no confusion since the formal usage of these signs will be precisely defined.

It will often be convenient to speak of the variables and constant letters as being ordered in some well-defined manner. The ordering by increasing order of subscript is called *alphabetic order* and it will be the one most frequently used.

By a *term* of a first-order theory, we mean (i) an individual variable, (ii) a constant letter of the theory, or (iii) a function letter f_m^n of the theory followed by one left parenthesis, then n terms, separated by commas, and then a right parenthesis; i.e. an expression of the form

$$f_m^n(t_1, t_2, \ldots, t_n)$$

where the t_i are all terms.[†] (iv) These are the only terms by this definition. (Actually, we

---

[†] Here the signs "f_m^n", "t_1", and the like are metavariables that we use in order to talk about signs and expressions of our formal system. We use letters such as "A", "B", "x", "x_i", etc., which are not signs of our system, as metavariables to refer to arbitrary wffs, arbitrary variables, and the like. In some cases the use of these letters in the metalanguage will be more like the dummy letters used in connection with our system **P**, though we will not worry about distinguishing between these two uses. As usual, juxtaposition of variables and special signs in the metalanguage represents juxtaposition in the object language.

Many books use different type styles for metavariables. We shall not, since we have unequivocally designated shall later introduce other types of terms involving an extension of the present definition.)

Notice that the superscript number n of the function letter in part (iii) of the above definition is the same as the number of terms t_i involved in forming the new term. We say that the superscript of any function letter is the *argument number* of the function letter. Similarly, we call the superscript of a predicate letter its *argument number*. The argument number of a function letter tells us how many terms are necessary to combine with the function letter to form a new term. The particular terms t_i used in forming the new term are called the *arguments* of the new term.

The subscript numerals of the function letters simply distinguish between different function letters. Also, the subscript numerals of the individual variables and constant letters serve, in each case, to differentiate them. In the latter two cases, we need no argument number since the use we make of these two syntactical categories does not involve their having other things as arguments.

We define a wff of a first-order theory as follows: (i) Any predicate letter A_m^n of the system followed by n terms as arguments, i.e. an expression of the form $A_m^n(t_1, t_2, \ldots, t_n)$ where the t_i are all terms, is a wff. (ii) If X is a wff, then the expression $(y)X$ is a wff (where y is any individual variable) and the expression $(\sim X)$ is a wff. (iii) If X and Y are any two wffs, then the expression $(X \lor Y)$ is a wff. (iv) These are the only wffs. Wffs of type (i), from which our other wffs are built up, are called *prime* formulas.

Having now defined clearly the wffs of a first-order language, we must give some attention to the question of interpretation. We shall eventually consider a precise, mathematically defined notion of interpretation. For the moment, however, we proceed on the intuitive level.

The intuitive interpretation of a first-order system is that the predicate letters express properties or relations, depending on the argument number. For example, the predicate letter A_1^1 might be thought of as expressing the property "to be red". Then, where a_i is some constant letter thought of as naming some object, the wff $A_1^1(a_i)$ would mean, intuitively, "the object designated by a_i is red". The negation and disjunction signs have the same meaning in this system as they do in the statement calculus. Thus, "the object designated by a_i is not red" would be rendered as $(\sim A_1^1(a_i))$.

In any first-order system, we introduce the usual definitions for the signs "$\supset$", "$\land$", and "$\equiv$", following Definitions 1, 2, and 3 of Section 3, where the wffs of **P** are replaced, in each case, by the wffs of the first-order theory in question. Henceforth, whenever we speak of a first-order system, we suppose these definitions to have been made.

Using our connectives, we can now express new properties by using our original ones together with the names of other objects. "If the object a_1 is red, then the object a_2 is not red" can be expressed by the wff $A_1^1(a_1) \supset (\sim A_1^1(a_2))$. The number of possible combinations is clearly not finitely limited.

---

those signs which can be in the alphabet of a first-order system and these will never be used as metavariables.

Notice that the alphabet of any first-order system is infinite. Our individual variables *are* such signs as "x_1", "x_2", and the like, rather than being abbreviations for other expressions as was the case with our statement letters of **P**. Since we require that any first-order theory have all the individual variables among the signs of its alphabet, it follows that the theory has an infinite alphabet.

Technically, the sub- and superscripts on the variables, constant letters, function letters, and predicate letters are all numerals, names of integers. We now agree to use the words "subscript" and "superscript" to designate also the number that is named by a particular numeral subscript or superscript respectively. Thus, the number two is the subscript of the variable x_2, the number three is the superscript of the function letter f_2^3, and so on.

Notice that we here use parentheses for grouping in a way similar to our previous use of brackets. But our parentheses also have other uses such as applying a predicate letter to its arguments. To achieve an economy of notation, we have assimilated all these functions to parentheses and dispensed with brackets entirely.

A two-argument predicate letter represents a binary relation. For example, A_1^2 might express the relation "less than". Then, $A_1^2(a_1, a_2)$ would mean intuitively "a_1 is less than a_2". Relations of higher degree have more argument places. The subscript of the predicate letters simply distinguishes among different predicate letters.

From this brief discussion, we have seen one way of obtaining a statement from a predicate letter, namely by using individual constant letters as arguments. Another basic way is *quantification*, which we now explain.

The expression "(x_i)", where x_i is some individual variable, is thought of as expressing the words "for all x_i". The variables x_i may be thought of as playing the role of the pronouns of ordinary speech, just as the constant letters play the role of proper names. For example, if the predicate letter A_1^1 is thought of as expressing the property "to be red", the expression "$A_1^1(x_2)$" means "it is red" with ambiguous antecedent for "it". By affixing now our expression "(x_2)" we obtain "$(x_2) A_1^1(x_2)$", which means "Whatever thing you may choose, that thing (it) is red", or more succinctly, "Everything is red".

Notice that the expression "$A_1^1(x_2)$" does not represent a sentence, since the variable "x_2" is not thought of as being a proper name for some object as was the case with our constant letters. Thus, properly speaking, the phrase "it is red" or "x is red" has no subject since the pronoun "it" (or "x") cannot be regarded as a subject if it has no antecedent at least understood from the given context. We use the term *open sentence* to refer to an expression which is obtained from a sentence by replacing one or more substantives (nouns) by pronouns with ambiguous antecedents. A formal expression such as "$A_1^1(x_2)$" involving one or more predicate letters with individual variables as arguments is thought of as representing in our formalism the intuitive, informal linguistic notion of an open sentence.

Our individual variables are thus thought of as variables ranging over some given domain D called the *universe* or *universe of discourse*. The variables, unlike the constant letters, do not stand for a particular object, but they are thought of as naming ambiguously any arbitrary member of the given domain D. For our variables to have such an interpretation, we must specify the domain D, just as we must specify the meaning we assign to the predicate letters such as A_1^1 and the objects which are named by the constant letters.

There are thus two basic ways to obtain a sentence from an open sentence such as "x is red", which is represented in our formalism, let us say, by $A_1^1(x_2)$. One way is the obvious device of substituting a name for the variable, thus replacing the variable by a constant. If we replace "x" by "the Washington Monument", we obtain the false sentence "The Washington Monument is red". Again, if we have some constant letter, say "a_1", which is thought of as naming the Washington Monument, then $A_1^1(a_1)$ will mean "The Washington Monument is red" in our formalism.

The second method of obtaining a sentence from an open sentence is by *quantification of variables*, which is the application of our prefix "for all", using some variable x. We say that we quantify in the *name* of the given variable. Thus "For all x, x is red" is represented as previously given by "$(x_2) A_1^1(x_2)$". Hence, quantification is a new logical operation just like negation, disjunction, and the like. The latter operate on sentences to give new sentences, whereas quantification operates on open sentences to give sentences or open sentences (this latter case may occur when there is more than one variable in the expression to which quantification is applied).

Substitution of constants for variables is likewise a logical operation used for obtaining sentences (or open sentences when some variables are not quantified or replaced by constants) from open sentences.

The logic of first-order theories, which will be embodied in logical axioms and rules that we have yet to describe, can be thought of as a generalization of the logic of the sentence connectives, the sentential calculus. The sentential calculus deals with valid or universally true ways of operating on sentences with sentential connectives. The *predicate calculus*, as we shall define it, can be thought of as the analysis of the valid ways of combining the sentential connectives plus the additional operations of quantification and substitution. In the predicate calculus, moreover, we are operating with a larger class of expressions, representing open sentences as well as sentences.

Returning to our examples again, we observe that we can iterate and compound our various operations just as with the sentential calculus. We can say "$(x_2)(\sim A_1^1(x_2))$", "For all x_2, x_2 does not have the property expressed by 'A_1^1' ", or "$(\sim (x_2) A_1^1(x_2))$", "Not everything has the property A_1^1". Using other predicate letters we can make more complicated statements such as "$(x_1)(x_2)(x_3)((A_1^2(x_1, x_2) \wedge A_2^2(x_2, x_3)) \supset A_1^2(x_1, x_3))$". "For all x_1, x_2, x_3 if x_1 bears the relation A_1^2 to x_2, and if x_2 bears the relation A_2^2 to x_3, then x_1 bears the relation A_1^2 to x_3."

Particularly interesting is the combination "$(\sim (x_i)(\sim A))$" where x_i is any individual variable and A is some wff. This says, "It is not true that, for all x_i, A is not true." Otherwise said, "There exists some x_i such that A is true." By means of quantification and negation, we can express the notion of existence. This fact was first recognized by Frege [1].

Definition 1. Where A is any wff of a first-order system and x_i is any individual variable, we abbreviate $(\sim (x_i)(\sim A))$ by $(Ex_i)A$. We call "(x_i)" *universal* quantification and "(Ex_i)" *existential* quantification. We read the existential quantifier as, "There exists x_i such that". "Quantification", without modification, will henceforth mean either existential or universal quantification. Quantification is said to be *in the name* of the variable x_i appearing in the parentheses of the quantifier (Qx_i), be it existential or universal.

We suppose this definition made for all first-order theories.

From now on, and for the rest of this book, we shall omit parentheses at will, but consistent with the following convention. The negation sign and the quantifier (Qx), existential or universal, apply to the shortest wff that follows them. Next in line of increasing "strength" is the conjunction sign, then the disjunction sign, the conditional sign, and finally the biconditional. Let us take some examples: $(x_1) A_1^1(x_1) \vee A_2^1(x_1)$ is properly read as $((x_1) A_1^1(x_1) \vee A_2^1(x_1))$ and not $(x_1)(A_1^1(x_1) \vee A_2^1(x_1))$ since (x_1) is weaker than $\vee$. $A_1^1(x_1) \wedge A_2^1(x_1) \vee A_3^1(x_1)$ is read $((A_1^1(x_1) \wedge A_2^1(x_1)) \vee A_3^1(x_1))$, since $\wedge$ is weaker than $\vee$. $A_1^1(x_1) \vee A_3^1(x_1) \supset A_2^1(x_1) \equiv A_4^1(x_1)$ is unambiguously read as $(((A_1^1(x_1) \vee A_3^1(x_1)) \supset A_2^1(x_1)) \equiv A_4^1(x_1))$. $\sim \sim \sim A_1^1(x_1) \vee A_2^1(x_1)$ is read as

$$((\sim(\sim(\sim A_1^1(x_1)))) \vee A_2^1(x_1)).$$

Any wff B which has an occurrence within another wff A is called a *subformula* of A. By the way we have defined the wffs of first-order theories, any occurrence of a universal or existential quantifier (Qx) in a wff A applies to some particular subformula B of A. B will be the (unique) wff immediately following the given occurrence of (Qx). The occurrence of B immediately following the given occurrence of (Qx) is called the *scope* of that occurrence of (Qx).

For example, in the wff $(x_1)((x_1) A_1^1(x_1) \vee (\sim (x_1) A_1^1(x_1)))$ the scope of the first occurrence of (x_1) is the entire formula which follows it. The scope of the second occurrence of (x_1) is the first occurrence of $A_1^1(x_1)$ and the scope of the third occurrence of (x_1) is the second occurrence of $A_1^1(x_1)$.

We now define an occurrence of a variable x_i in a wff A to be *bound* if it occurs in the scope of some occurrence of a quantifier in its name or if it occurs within the parentheses of a quantifier (Qx_i). Otherwise, the occurrence is *free*. Clearly a given variable may have both bound and free occurrences in the same given formula. Nevertheless, if x_i has at least one free occurrence in A, we say that x_i is *free in A* or that it is a *free variable of A*. Also, if x_i has at least one bound occurrence in A, it is *bound in A* and is a *bound variable of A*. Thus, bondage and freedom are opposites for occurrences but not for variables.

Now from our description of the intuitive interpretation of the formalism of a first-order theory, and from the definition of wffs, it is clear that a wff of a first-order theory may fail to represent a sentence. As we have seen, it will represent an open sentence if it contains variables which are not quantified; in other words, if it contains free variables. We define a wff to be *closed* if it contains no free variables. The set of closed wffs of a given first-order theory F will be called the *sentences* or *propositions of F*. The closed wffs represent sentences of the vernacular under our intuitive interpretation. A closed wff of a first-order theory is thus the formal analogue of a sentence in the vernacular just as a wff which is not closed is a formal analogue of an open sentence.

We extend the notion of bondage and freedom to cover more general terms and subformulas: An occurrence of a term t or of a subformula B in a formula A is *bound* if at least one of its free variables falls within the scope of a quantifier in its name. Otherwise, the occurrence is free. A term or formula is bound in another formula if it has at least one bound occurrence in that formula, and free if it has at least one free occurrence.

Exercise. State which occurrences of which terms and subformulas are free or bound in each of the following:

$$(x_2) (x_1) A_1^2(x_1, x_2) \supset \sim A_1^2(x_2, f_1^1(x_1));$$
$$(x_3) (A_1^2(x_1, x_2) \supset A_2^2(x_1, f_1^2(x_2, x_3))) \vee A_1^1(x_3);$$
$$(x_4) (A_1^1(x_1) \supset A_2^1(x_2)).$$

From the way we have defined our wffs, we can apply quantifiers indiscriminately to any wff and obtain a wff. What intuitive meaning do we give to such wffs as $(x_1) A_1^1(x_2)$ where the variable x_2 is still free since we have applied a quantifier in the name of another variable? Intuitively we regard the quantifier as vacuous in this case. The displayed wff above means the same as $A_1^1(x_2)$.

Another case of vacuous quantification is double quantification such as in the wff $(x_1)(x_1) A_1^1(x_1)$. Intuitively, this wff means the same as the one in which the initial quantifier is dropped.

We will so formulate our rules and axioms of logic that these intended equivalences in meaning turn out to be provable.

It remains to give the intuitive interpretation of the function letters f_m^n. Once our domain D

of the individual variables x_i is chosen, the function letters are thought of as representing functions from n-tuples of elements of D into D. By D^n we mean the n-fold cartesian product of D with itself, the set of n-tuples of elements of D. Where the argument number of a function letter is n, the interpretation of a function letter f_m^n is a function from D^n into D.

For example, suppose our domain D is the natural numbers. Then we can think of f_1^2 as the operation of addition, which associates with any two elements x_1 and x_2 of D (remember the variables range over D) the sum $f_1^2(x_1, x_2)$. For a less mathematical example, we can let D be the set of all people living or dead and f_1^1 the function "father of". This would associate with every person x_1 his father $f_1^1(x_1)$.

We have given an intuitive explanation of the interpretation of the wffs and terms of a first-order theory. We now proceed to give a precise mathematical definition, which should be regarded primarily as a precise statement of our intuitive notion of an interpretation.

Definition 2. An *interpretation* $\langle D, g \rangle$ of a given first-order theory F consists of a given, non-empty set D together with a mapping g from the set consisting of the function letters, individual constants, and predicate letters of F into the set $D \cup \mathcal{P}(\mathcal{M}(D))$, which is the set D together with all the subsets of the set $\mathcal{M}(D)$ of all finite sequences of elements of D. The mapping g assigns to each predicate letter A_m^n of F some subset $g(A_m^n)$ of D^n; to each function letter f_j^i of F some subset $g(f_j^i)$ of D^{i+1}, $g(f_j^i)$ a functional relation, i.e. a mapping from D^i into D; and to each constant letter a_i some element $g(a_i)$ of the set D. We call $g(a_i)$ the object *named* by a_i.

Intuitively, D is our universe of discourse over which the individual variables range. The mapping g assigns relations (considered as sets of n-tuples) to predicate letters, operations (functions) to function letters, and elements of D to constants. The set D together with the selected elements of D, functions on D^n, and relations over D assigned by the mapping g to some constant letter, function letter, or predicate letter of F (respectively) is sometimes called a *structure for F*.

Now, for any interpretation $\langle D, g \rangle$ of a formal system F, the sentences or closed wffs of the system should be either true or false in the interpretation since it was part of our original definition of sentences that they were either true or false. We now proceed to define truth and falsity, for a formal system relative to a given interpretation, in a purely mathematical way.

Suppose a first-order system F given, and let some interpretation $\langle D, g \rangle$ be given. Let us define for each *infinite* sequence $s = s_1, s_2, \ldots, s_i, \ldots$ of elements of D a function g_s from the set of terms of F into the domain D. If t is an individual constant a_i, then $g_s(t)$ is the element $g(a_i) \in D$ that is named by a_i. If t is the individual variable x_i, then $g_s(x_i) = s_i$ (i.e. the ith member of the sequence). Finally, if t is of the form $f_j^n(t_1, \ldots, t_n)$, then $g_s(t) = g(f_j^n)(g_s(t_1), \ldots, g_s(t_n))$; that is, we apply the associated operation of f_j^n to the objects in D which correspond to the terms t_i that make up the term t. We have used the notation g_s to emphasize the dependence on the chosen sequence s.

Next we define what it means for a sequence s to *satisfy* a given wff A. If A is of the form $A_k^n(t_1, \ldots, t_n)$ and if the n-tuple

$$\langle g_s(t_1), g_s(t_2), \ldots, g_s(t_n) \rangle$$

is in the set $g(A_k^n)$ then we say that the sequence s *satisfies* the wff A. Otherwise, the sequence does not satisfy that wff. If A is of the form $(\sim B)$, then s *satisfies* A if and only if it does not satisfy B. If A is of the form $(B \lor C)$, then s satisfies A if and only if s satisfies B or C or both. If A is of the form $(x_i)B$, then s satisfies A if and only if every infinite sequence of elements of D which differs from s in at most the ith component satisfies B.

We now say that a wff A is *true* under a given interpretation if and only if *every* sequence s satisfies A. We say that A is *false* if and only if *no* sequence satisfies A.

According to our definition of truth, it is possible for a wff of a formal system to be neither true nor false under a given interpretation. It may be satisfied by some sequences but not by all. However, it is not possible for a closed wff to be neither true nor false. For any closed wff X, either every sequence satisfies X or no sequence satisfies X.

The rigorous proof of this last statement is by induction, and we give only a sketch. For this purpose, let us define the *closure* or the *universal closure* of a given wff. Let $t_1, t_2, \ldots, t_n$ be the free variables of a given wff X in increasing order of subscript (i.e. if x_r is t_j and x_m is t_i, then $j \leqslant i$ if and only if $r \leqslant m$). The universal closure of X is the wff

$$(t_n)(t_{n-1}) \ldots (t_1)X,$$

obtained by prefixing universal quantifiers in the name of each free variable of X in the indicated order. Actually, the question of order does not really matter, but we choose some order so that the closure will be uniquely defined for a given wff. We also define the *existential closure* of X to be the wff obtained by prefixing existential quantifiers in the name of all free variables in the indicated order; that is, the wff

$$(Et_n)(Et_{n-1}) \ldots (Et_1)X.$$

If X is a closed wff, then X is its own universal closure and its own existential closure.

The first observation concerning our definition of truth and the concepts we have just defined is the following: Any wff X whatever is true if and only if its universal closure is true. By definition a sequence s satisfies $(x_i)X$ if and only if *every* sequence which differs from s in at most the ith place satisfies X. But if every sequence satisfies X to begin with (i.e. if X is true), then it immediately follows that every sequence will satisfy $(x_i)X$. Also, if every sequence satisfies $(x_i)X$ then certainly every sequence satisfies X (remember, any sequence s differs from itself in at most the ith component). Iterating this argument to any number of applications of universal quantification to X, we obtain that the universal closure of X is true if and only if X is true.

Thus, one finds that the truth of any wff X of any system F under a given interpretation $\langle D, g \rangle$ is equivalent to the truth of some closed wff, namely the universal closure of X.

Another important observation concerning the satisfaction relation between sequences and formulas is the following: Let A be any formula all of whose free variables are included in the list $x_{i_1}, \ldots, x_{i_n}$. Then any two sequences s and s' for which $s_{i_j} = s'_{i_j}$, $1 \leqslant j \leqslant n$, either satisfy or fail to satisfy A together. In other words, the values of a sequence at indices corresponding to variables which are not free in A do not affect the satisfaction of A by s. Thus, even though sequences are infinite in length, we are never really concerned with more than a finite number of values at any given time since any formula has only a finite number of free variables.

Using these observations, we now want to sketch the inductive proof that any closed wff X in any system F is either true or false under any given interpretation $\langle D, g \rangle$ of F. Notice that this is equivalent to proving that whenever one sequence s satisfies X, then every sequence does. The induction is on the number of sentence connectives and quantifiers of X.

3*

If X has none of these, then it must be a prime formula, and since it is closed it must contain only variable-free terms as arguments. X is thus of the form $A_j^n(t_1, t_2, \ldots, t_n)$ where the t_i contain no variables. In this case, $g_s(t_i)$ is the same for every sequence s. Hence, either all sequences will satisfy X or none will, and the theorem holds in this case.

Assuming, inductively, that the theorem holds for all wffs with fewer than n sentence connectives and quantifiers, we suppose X has n connectives and quantifiers. If X is prime, the proof is the same as above. Otherwise, X is of the form $(\sim B)$, $(B \lor C)$, or $(x_i)B$, and the theorem holds for B and C since they each have less than n quantifiers and sentence connectives. Let us consider each case. If X is of the form $(\sim B)$, then B is closed since $(\sim B)$ has the same free variables as B. Thus, applying the theorem to B, either every sequence satisfies B and thus does not satisfy $(\sim B)$ (by the definition of satisfaction), or else no sequence satisfies B in which case every sequence satisfies $(\sim B)$, again by the definition of satisfaction. If X is of the form $(B \lor C)$, then suppose there is no sequence satisfying either B or C. Then none satisfies $(B \lor C)$ and it is false. Otherwise, there is at least one sequence satisfying one of B or C, say C. Now, C must be closed since $(B \lor C)$ is. Applying the induction hypothesis to C, we conclude that every sequence satisfies C. Thus, by the definition of satisfaction, every sequence satisfies $(B \lor C)$.

Finally, we consider the case where X is of the form $(x_i)B$. Either x_i is free in B or not. If not, then B is closed (since $(x_i)B$ is) and the result is immediate since the quantifier now changes nothing. Otherwise, $(x_i)B$ is the universal closure of B and x_i is the only free variable in B. Thus, for any sequence s, the value s_i is the only relevant one for determining the satisfaction of B by s. Suppose, now, that at least one sequence s satisfies $(x_i)B$. Then every sequence s' which differs from s in at most the ith place satisfies B (the definition of satisfaction). In particular, s satisfies B. But every sequence which differs from s in any place other than the ith one also satisfies B by our second observation above (i.e. because x_i is the only free variable of B). Thus, every sequence satisfies B and B is true. Hence, by our first observation, its universal closure $(x_i)B$ is true and is thus satisfied by every sequence. This completes the proof.

The fact that closed wffs are either true or false under any interpretation justifies applying the term "sentence" or "proposition" to them.

Exercise. Prove that a sequence s satisfies a wff $(Ex_i)B$ for a given interpretation if and only if there is at least one sequence differing from s in at most the ith place and satisfying B.

The possibility of a rigorous definition of truth was first conceived and executed by Tarski (see Tarski [1], p. 152). Now that we have such a notion at hand, we can define what we mean by a logically valid wff of a first-order theory. We say that a wff X of a first-order theory F is *logically valid* or *universally valid* if X is true for every interpretation of F.

The notion of logical validity for a first-order theory is analogous to the notion of tautology for the system **P** of the sentential calculus. In fact, it is a generalization of the notion of tautology. We ask the reader to show this.

Exercise 1. Show that any wff of a first-order system F which is tautological in form will be logically valid as a wff of F according to the given definition.

Exercise 2. Find some examples of wffs of first-order systems which are valid, but which are not tautologies.

It is important to understand that our definition of truth is rigorous. However, the reader should also see that the definition only makes precise the notion he would normally have of truth. As such, this intuitive notion will often suffice in understanding a given discussion of logical questions. Thus, to take an example, if we consider the wff $(x_1)(Ex_2) A_1^2(x_1, x_2)$ in some formal theory and if we let the domain D be the natural numbers and the relation $g(A_1^2)$ be the relation "less than" (the set of all ordered pairs of natural numbers x, y such that x is less than y), then the wff says that for every natural number x_1 there is some natural number x_2 greater than x_1. This statement is true in the natural numbers.

Moreover, it is clear that there is a qualitative difference between the definition of logical validity and the definition of tautology. For tautology, we have a purely mechanical test, the truth table, which allows us to decide whether a given wff is a tautology or not. But the definitions of interpretation, truth, and validity all depend on general notions of set theory. There is no simple way of deciding whether a given wff is valid or not. In some cases, as with tautologies, we can decide. Church has proved, however, that there is no general method of decision by which one can decide for all wffs of any system F whether or not they are logically valid (see Church [1]).

The concepts of truth, interpretation, and validity as we have defined them are as legitimate as any abstractly defined mathematical concept. But for practical purposes it would be very difficult to have to return again and again to these abstract definitions in order to prove facts about validity. We thus conceive of the following plan: We specify certain axioms and rules of inference and require that they hold for all first-order theories. A given first-order theory may have other axioms, but the axioms we specify are *logical axioms* required to hold in any system. These are called the *axioms and rules of the predicate calculus*. The logical axioms will be a decidable set and the rules will be simple, formal, decidable rules like the rule of *modus ponens* of **P** (in fact, *modus ponens* will be one of our rules). Furthermore, it will turn out that all logically valid wffs and only logically valid wffs can be formally proved from the axioms and rules of the predicate calculus. Thus, we can replace the notion of validity by the notion of formal deducibility or provability in the predicate calculus.

The virtue of this proposed plan is that, while the notion of validity depends for its definition on general set theory, the notion of formal deducibility does not. From the way we have defined formal deducibility in a formal system, it is clear that the only mathematical notions involved are those which are essentially number-theoretic, having to do with the length of proofs, and so on. Metatheorems about formal deducibility will usually involve no more tools than elementary number theory and the principle of mathematical induction. But metatheorems about validity easily involve highly nonconstructive principles of general set theory such as the axiom of choice.

This last point is very important for the purposes of our study in this book. We will be treating, in future chapters, different formal languages in which mathematics and general set theory can be expressed. These languages will, for the most part, be first-order theories. If our only approach to a logical discussion of these languages was in terms of interpretations and validity, there would be little virtue in the formal axiomatic approach. Instead, we shall proceed by formal deduction within these languages, using our decidable axioms and rules. Thus, theorems within these formal languages will be those wffs for which we actually exhibit a purely formal, mechanical deduction; no concepts of general set theory shall be necessary to justify the notion of proof.

Of course nothing prevents us from studying these set-theoretical languages from the point of view of general set theory itself. Chapter 6 of this book contains a detailed discussion of this question, and the interested reader can proceed directly to this chapter if he so wishes.

We now address ourselves to the question of formulating the formal axioms and rules of the predicate calculus. The reader should note that our definitions are purely formal and do not involve appeal to general set-theoretic concepts.

We have used letters such as "A", "B", "x", and the like, which are not part of the alphabet or expressions of first-order systems, as variables in the metalanguage to represent such constructs as arbitrary wffs and arbitrary variables. We now use such forms as "$A(x)$" to represent a wff which may contain the variable x free. Similarly, "$A(x, y)$" represents a wff which may contain x and y free (of course x and y may be the same variable here). If $A(x)$ is a wff which may contain x free, then $A(t)$ will represent the result of replacing the term t for x in all of the free occurrences (if any) of x in $A(x)$.

Given a wff $A(x)$, we say that another term t is *free for* x in $A(x)$ if every new occurrence of t in $A(t)$ is free. For example, in the wff $(x_2) A_1^2(x_1, x_2)$, the term x_2 is not free for x_1, since it will become bound if we substitute x_2 for x_1 in this wff. Similarly, the term $f_1^2(x_3, x_2)$ is not free for x_1, since one of its variables, namely x_2, will become bound if the term is substituted for x_1 in the wff. Any variable other than x_2 and any term not containing x_2 free is free for x_1. Obviously, any variable is free for itself in any formula.

Given a wff $A(x)$ and a free occurrence of the variable x in $A(x)$, we say that a term t is *free for the occurrence of* x in question if no free occurrence of a variable in t becomes bound by a quantifier of $A(x)$ when t is substituted for the given occurrence of x. A term t is free for x in the sense of the preceding paragraph if it is free for every free occurrence of x in $A(x)$. It is possible for a term t to be free for some free occurrences of x and not for others.

Notice that substitution is not generally a symmetrical operation. If we obtain $A(y)$ from $A(x)$ by substitution where the variable y is free for x, it does not follow that we can obtain $A(y)$ from $A(x)$. For example, $A_1^2(x_1, x_1)$ is obtained from $A_1^2(x_1, x_2)$ by substitution of x_1 for x_2 but we cannot reverse the procedure. Whenever two wffs are obtainable each from the other by opposite substitutions, we say they are *similar*. More precisely, if y is a variable free for x in $A(x)$ and if y has no free occurrences in $A(x)$, them $A(y)$ is similar to $A(x)$. Thus, $A_1^2(x_1, x_2)$ is similar to $A_1^2(x_1, x_3)$.

We now require that the axiom set of any first-order system F satisfy the following: (1) All wffs of F which are tautological in form are axioms. (2) If $A(x)$ is some wff and x a variable, then every wff of the form $(x) A(x) \supset A(t)$ is an axiom where t is any term free for x (remember that $A(t)$ is obtained from $A(x)$ by replacing t for x at all of the latter's free occurrences, if any, in $A(x)$). (3) Every wff of the form

$$(x)(B \supset A(x)) \supset (B \supset (x) A(x))$$

where x is any variable that is *not* free in B, and $A(x)$ is any wff, is an axiom. (4) These are the only logical axioms.

The following are the rules of inference of any first-order system. (1) *Modus ponens*; that is, from A and $A \supset B$ we can infer B where A and B are any wff. (2) For any wff A, we can infer the wff $(x)A$. This is the rule of universal generalization (abbreviated "UG").

It is presumed that a first-order system may have other *proper* axioms (meaning proper to the particular theory in question), but the rules of inference are the same. Any first-order system having only the logical axioms is called a *predicate calculus*. By *the* predicate calculus we mean the theory of formal deduction using our logical axioms and rules. By a *theorem of the predicate calculus* we mean any theorem of any predicate calculus. By the *pure predicate calculus* we mean the first-order predicate calculus with no function letters, no constant letters, and all the predicate letters.

In any formal system, a formal deduction is a finite sequence of wffs such that every element of the sequence is either an axiom or follows from prior members of the list of axioms by a rule of inference. Thus, the theorems of any predicate calculus are the wffs that can be obtained by formal deduction from our logical axioms by means of our rules of inference.

From now on, to the end of this chapter, the terms "system" and "formal system" will be restricted to mean "first-order system" unless otherwise specified.

It may seem surprising that such a simple set of axioms suffices to yield, as theorems, precisely the universally valid wffs, but this is so. We now wish to examine this question more closely.

Given a first-order system, we now have two parallel notions. We have the notion of a provable wff or theorem, and the notion of a true wff under a given interpretation. For any such first-order theory F, those properties of F which are defined by means of interpretations of F are called *semantical* properties of F. Those properties which are defined in terms of the deductive structure of F are called *syntactical* properties or proof-theoretic properties of F. What we are then interested in is the precise relationship between syntax and semantics.

It is easy to see that the theorems of any predicate calculus must be logically valid. First, we establish that any tautology is universally valid (cf. preceding exercise). Next, consider a wff of the form $(x_i) A(x_i) \supset A(t)$ where t is free for x_i in $A(x_i)$. This is an axiom under scheme (2) of our logical axioms. This will be universally valid only if it is true under every interpretation; that is, if every sequence in every interpretation satisfies it. This will be the case only if, for every interpretation, every sequence satisfying $(x_i) A(x_i)$ satisfies $A(t)$.

Thus, let some interpretation $\langle D, g \rangle$ be chosen, and consider any sequence s satisfying $(x_i) A(x_i)$. By our definition of satisfaction, this means that every sequence differing from s in at most the ith place satisfies $A(x_i)$. Thus, whatever the object $g_s(t)$ may be, the sequence s must *a fortiori* satisfy $A(t)$ (a rigorous proof of this is by induction on the structure of $A(t)$). Hence, any formula of the form $(x_i) A(x_i) \supset A(t)$ is true in any interpretation and thus universally valid.

Exercise. Sketch a proof of the fact that wffs of the form (3) of our logical axioms are universally valid.

Next, we observe that our rules of inference preserve universal validity. This is obviously true for *modus ponens* in view of the truth table for the conditional. We have already remarked that any wff is true if and only if its universal closure is. Thus, the rule of generalization also preserves universal validity.

Since a theorem of a predicate calculus is obtained from the axioms by our rules of inference, it follows that the theorems must be universally valid (again, we skirt an induction on the length of the proof of the theorem).

This situation is clearly analogous to our system **P** in which the rules of inference (namely *modus ponens*) preserved the property of being tautological, and all of our axioms were tautologies. It then followed that all theorems of **P** were tautologies. The converse, that all tautologies of **P** were theorems of **P**, was stated but not proved. Likewise, we state, but do not prove:

THEOREM 1. *For any first-order system F, the universally valid formulas of F are precisely those theorems of F deducible from the logical axioms of F by our rules of inference. Every universally valid wff of F is thus a theorem of F.*

The proof of this nontrivial theorem was first given by Gödel [1]. This article is reprinted in van Heijenoort [1], p. 582.

The only real justification for our logical axioms and rules of inference is Theorem 1. The extreme importance of the theorem lies in the fact that we can describe the universally valid wff syntactically as well as semantically. That is, we have a decidable set of purely formal axioms and formal rules which generate the universally valid formulas of any given first-order system F.

The importance of the axioms being a decidable set has already been emphasized, and should not be overlooked. It is easy to designate a set of axioms for the universally valid formulas of a system F in a nonformal, undecidable way; just let the valid formulas be axioms, for example, and have no rules of inference. This may seem analogous to designating the tautologies as axioms, but again we emphasize that the tautologies are a decidable set of wffs. We can determine whether a given wff is a tautology by the truth-table method just as surely as we can determine whether or not a given wff really is or is not of the form $(x_i) A(x_i) \supset A(t)$.

In view of Theorem 1, one might hope to prove that the valid formulas are a decidable set after all by proving that the set of theorems of the predicate calculus is decidable (for these two sets are the same). As we have already mentioned, however, Church has proved the theorems of the predicate calculus to be undecidable (see Church [1]). There is, in short, no mechanical test which will allow us generally to determine whether a given wff is universally valid (or, which is the same thing by Theorem 1, a theorem of the predicate calculus). For this reason, technique and skill in logical deduction are of some importance.

Before continuing our general discussion of first-order theories, let us exhibit a formal deduction in the predicate calculus. We show, for example, $\vdash A_1^1(x_1) \supset (Ex_1) A_1^1(x_1)$ in any theory F having A_1^1 as a predicate letter:

$$(x_1)(\sim A_1^1(x_1)) \supset (\sim A_1^1(x_1))$$
$$((x_1)(\sim A_1^1(x_1)) \supset (\sim A_1^1(x_1))) \supset ((\sim(\sim A_1^1(x_1))) \supset (\sim(x_1)(\sim A_1^1(x_1))))$$
$$(\sim(\sim A_1^1(x_1))) \supset (\sim(x_1)(\sim A_1^1(x_1)))$$
$$A_1^1(x_1) \supset (\sim(\sim A_1^1(x_1)))$$
$$\left\{ \begin{array}{l} (A_1^1(x_1) \supset (\sim(\sim A_1^1(x_1)))) \supset (((\sim(\sim A_1^1(x_1))) \supset (\sim(x_1)(\sim A_1^1(x_1)))) \\ \qquad \supset (A_1^1(x_1) \supset (\sim(x_1)(\sim A_1^1(x_1))))) \end{array} \right.$$
$$((\sim(\sim A_1^1(x_1))) \supset (\sim(x_1)(\sim A_1^1(x_1)))) \supset (A_1^1(x_1) \supset (\sim(x_1)(\sim A_1^1(x_1))))$$
$$A_1^1(x_1) \supset (\sim(x_1)(\sim A_1^1(x_1)))$$

The above sequence of seven wffs is a formal proof whose last line is the desired wff. We leave as an exercise to the reader the task of verifying that this sequence of wffs really meets the criteria of a formal proof using only our logical axioms and rules of inference.

Recall from Section 1.2 that a given line in a formal proof is *justified* by the particular criteria for members of a formal proof that it satisfies. Thus, the *justification* for a line of a proof is (i) that it is an instance of one of our axiom schemes or (ii) that it is inferred from some prior members by one of our rules of inference. Although it is not strictly necessary, it is helpful in verifying formal proofs if one states the justification for each line. It is also useful to number the lines and to abbreviate wffs to avoid rewriting long wffs. We illustrate this by showing that, in any first-order theory F, $\vdash (x_i)(A \lor B(x_i)) \supset (A \lor (x_i) B(x_i))$, where the variable x_i is not free in the wff A and $B(x_i)$ is any wff of F. In stating our justifications, we abbreviate "tautology" by "Taut", "universal generalization" by "UG", and "*modus ponens*" by "MP".

1. $(x_i)(A \lor B(x_i)) \supset (A \lor B(x_i))$ Axiom 2, t is x_i
2. $(A \lor B(x_i)) \supset (\sim A \supset B(x_i))$ Taut
†3. $[1] \supset ([2] \supset ((x_i)(A \lor B(x_i)) \supset (\sim A \supset B(x_i))))$ Taut
‡4. $R[3]$ 1, 3, MP
5. $(x_i)(A \lor B(x_i)) \supset (\sim A \supset B(x_i))$ 2, 4, MP
6. $(x_i)((x_i)(A \lor B(x_i)) \supset (\sim A \supset B(x_i)))$ 5, UG
7. $[6] \supset ((x_i)(A \lor B(x_i)) \supset (x_i)(\sim A \supset B(x_i)))$ Axiom 3, x_i is not free in $(x_i)(A \lor B(x_i))$
8. $R[7]$ 6, 7, MP
9. $(x_i)(\sim A \supset B(x_i)) \supset (\sim A \supset (x_i) B(x_i))$ Axiom 3, x_i not free in $\sim A$ since it is not free in A
10. $[8] \supset ([9] \supset ((x_i)(A \lor B(x_i)) \supset (\sim A \supset (x_i) B(x_i))))$ Taut
11. $R[10]$ 8, 10, MP
12. $(x_i)(A \lor B(x_i)) \supset (\sim A \supset (x_i) B(x_i))$ 9, 11, MP
13. $(\sim A \supset (x_i) B(x_i)) \supset (A \lor (x_i) B(x_i))$ Taut
14. $[12] \supset ([13] \supset ((x_i)(A \lor B(x_i)) \supset (A \lor (x_i) B(x_i))))$ Taut
15. $R[14]$ 12, 14, MP
16. $(x_i)(A \lor B(x_i)) \supset (A \lor (x_i) B(x_i))$ 13, 15, MP

This last theorem is really a metatheorem, for we have not specified the wffs A and $B(x_i)$, but only required that they satisfy certain conditions. Of course, for any particular wffs satisfying the conditions, and for any system F, the proof would be line for line the proof we have given.

Our first theorem also gives rise to a metatheorem that

$$\vdash A(x_i) \supset (Ex_i) A(x_i)$$

in any formal system F where $A(x_i)$ is any wff of the theory. The proof in this general case will be obtained by replacing the wff $A(x_i)$, whatever it may be, for the particular wff $A_1^1(x_1)$ in the preceding proof. In fact, we obtain an even stronger metatheorem as the following exercise shows.

Exercise. Let F be any first-order system and $A(x)$ a wff of F. If the term t of F is free for x in $A(x)$ and if $A(t)$ results from $A(x)$ by substituting the term t for all free occurrences of the variable x in $A(x)$, then $\vdash A(t) \supset (Ex) A(x)$.

† The numbers in brackets form the name of the wff occurring at the line of the proof having the bracketed number. One is to imagine that we have written out the wff that would occur if we replaced the number by the indicated line of the proof.
‡ This means "the wff occurring on the right side of the principal connective of line 3". This device, like the bracketed numbers, is to shorten the writing of complicated wffs.

Prove, using our logical axioms and rules, that $\vdash_F (x) A(x) \equiv (y) A(y)$ and $\vdash_F (Ex) A(x) \equiv (Ey) A(y)$ in any system F, where $A(x)$ and $A(y)$ are similar.

Prove, using our logical axioms and rules, that $\vdash_F (x)(y) A \equiv (y)(x) A$ in any system F.

Definition 3. Given two wffs A and B of a first-order theory F, we say that A *implies* B *in* F if and only if $\vdash_F A \supset B$. If F is a predicate calculus, we say that A *logically implies* B or simply A *implies* B. By Theorem 1, A implies B means that the conditional $A \supset B$ is logically valid.

Logical implication is a generalization of tautological implication defined in Section 1.1, Definition 4. Since tautologies are theorems of any predicate calculus, a tautological implication is also a logical implication. Obviously, the converse is not true in general, and logical implication is a broader relation than tautological implication.

Definition 4. Given two wffs A and B of a first-order theory F, we say that A *is equivalent to* B *in* F if $\vdash_F A \equiv B$. If F is a predicate calculus, we say that A is *logically equivalent to* B or simply A is *equivalent to* B. By Theorem 1, this means that $A \equiv B$ is logically valid.

This is again a parallel generalization of the relation of tautological equivalence given in Definition 5 of Section 1.1.

We now return to our general discussion of first-order theories.

1.5. Models of first-order theories

Definition 1. Let F be some first-order system. By a *model* for F we mean any interpretation of F in which the proper axioms of F are all true. (Remember that the proper axioms of F are those axioms of F, if any, other than our logical axioms. The logical axioms of F are automatically true in any model for F, since they are universally valid and thus true under every interpretation.)

It does not follow that every first-order system has a model. The question of whether or not a system F does have a model is closely related to the question of consistency.

Definition 2. A first-order system F is called *inconsistent* or *contradictory* if there is some wff A of F such that the wff $(A \wedge (\sim A))$ is a theorem of F. A wff of the form $(A \wedge (\sim A))$ is called a *contradiction*. A system is *consistent* if it is not inconsistent.

Since the set of theorems of a first-order system may not be decidable, a system may well be inconsistent without our knowing it. The following shows why inconsistent systems may cause trouble.

THEOREM 1. *A system F is inconsistent if and only if every wff of F is a theorem.*

Proof. If F is inconsistent, let $(A \wedge (\sim A))$ be a provable contradiction. Let X be any wff. Now, $\vdash (A \wedge (\sim A)) \supset X$, since this is a tautology. Hence, since $\vdash (A \wedge (\sim A))$, we obtain $\vdash X$ by modus ponens. But X was any wff, and so every wff is provable.

Conversely, if every wff of F is provable, then let X be any wff. $(X \wedge (\sim X))$ is also a wff and therefore provable. Thus, a contradiction is provable in F.

In an inconsistent system, everything is provable.

Now let us think again how we defined the notion of a model. A model is a certain kind of interpretation, one which makes every proper axiom of the given system true. As we have already observed in connection with validity, our two rules of inference, *modus ponens* and generalization, both preserve truth. That is, they yield true wffs when applied to true wffs (where truth is defined relative to any given interpretation). Thus, all the theorems of any system will be true in any model of the system. This yields the following theorem:

THEOREM 2. *No inconsistent system F has a model.*

Proof. Let F be inconsistent. Then some contradiction $(A \wedge (\sim A))$ is a theorem of F. Now, the wff $\sim (A \wedge (\sim A))$ is a tautology and thus universally valid. It is true in every interpretation. But, in any interpretation, $\sim (A \wedge (\sim A))$ is true if and only if $(A \wedge (\sim A))$ is false. Thus, $(A \wedge (\sim A))$ is false in every interpretation. If F had a model $\langle D, g \rangle$, then every theorem of F, and thus $(A \wedge (\sim A))$, would be true in $\langle D, g \rangle$. But $(A \wedge (\sim A))$ is false in every interpretation and thus false in $\langle D, g \rangle$. Thus, $(A \wedge (\sim A))$ would be both true and false in $\langle D, g \rangle$. But no wff X can be both true and false under an interpretation for it is impossible for every sequence to satisfy X and no sequence to satisfy X (remember that D must be nonempty and so there are sequences). Thus, the assumption that F has a model is contradictory and F has no model.

COROLLARY. *If F has a model, it is consistent.*

Proof. This is the contrapositive of Theorem 2.

The wffs that are false in every interpretation are called *logically false* wffs. Obviously, the negation of every logically false wff is logically true and the negation of every logically true wff is logically false.

Now, for every closed wff X of any first-order system, X is either true or false (and not both) under any interpretation. Furthermore, $\sim X$ is false if and only if X is true. It is natural for us to think of the theorems of a first-order system as being the set of truths under some interpretation. But this will be possible only if the system is *complete*; i.e. if, for every closed wff, either $\vdash X$ or $\vdash \sim X$. Of course, any inconsistent system is complete. We are interested, however, in consistent, complete systems. We now prove that any predicate calculus is a consistent but incomplete (i.e. not complete) theory.

THEOREM 3. *Any predicate calculus is consistent.*

Proof. Given any wff of a first-order system, we define its *associated statement form*, abbreviated asf. We obtain the asf of a given wff X in a purely formal manner by (1) suppressing all terms and quantifiers of the wff together with accompanying commas and parentheses; (2) replacing each predicate letter by a statement letter of the system **P**, replacing everywhere the same predicate letter by the same statement letter, and using different statement letters for different predicate letters; (3) replacing the remaining parentheses by brackets, left brackets for left parentheses and right brackets for right parentheses. (We order the replacement of predicate letters by statement letters by ordering the predicate letters of X; first according to argument number, and then according to subscript number within each class of those having the same argument number. We then begin by replacing the first predicate letter by x^*, the second by x^{**}, and so on, according to our other restrictions.) For example, the associated statement form of $(x_1)(A_1^2(f_1^1(x_1), x_2) \supset A_2^1(x_4))$ is $[x^{**} \supset x^*]$. What we obtain is a wff of **P** in every case.

Now, observe that the asf of any of our logical axioms is a tautology. For the tautologies this is obviously true. For axioms of the second type we just get a form $[X \supset X]$, and for axioms of the third type the form $[[X \supset Y] \supset [X \supset Y]]$. In each case we get a tautology.

Furthermore, it is clear that if the asf of A is a tautology and if the asf of $(A \supset B)$ is a tautology, then the asf of B must be a tautology. Thus, the rule of *modus ponens* preserves the tautological property of the asf.

Similarly, the rule of generalization preserves the tautological property of the asf. In fact, the asf of A is the same as the asf of $(x)A$. It thus follows that all theorems of any predicate calculus will be such that their asf is a tautology. But the asf of any contradiction $(A \wedge (\sim A))$ is not a tautology and so no contradiction can ever be a theorem of a predicate calculus. Hence, every predicate calculus is consistent.

THEOREM 4. *No predicate calculus is a complete theory.*

Proof. Let A_m^n be some predicate letter of a predicate calculus F (there must be at least one). Consider the wff

$$(x_1)(x_2) \ldots (x_n) A_m^n(x_1, x_2, \ldots, x_n).$$

This wff, call it X, is closed. Its asf is simply x^*. This is not a tautology and so X is not a theorem of the predicate calculus F. The asf of $(\sim X)$ is $[\sim x^*]$ and this is not a tautology either. Thus, neither X nor $\sim X$ are theorems of the predicate calculus F. But F was any predicate calculus and so no predicate calculus is a complete system.

The method of proof of Theorem 3 and Theorem 4 hinges on the fact that having a tautology for an asf is a necessary condition for a wff to be a theorem of a predicate calculus (and thus a valid wff). This condition is not sufficient, however, or we would have a decision method for the predicate calculus. That is, there are wffs whose asf is a tautology, yet are not valid wffs. Of course, necessity does give us a negative test which is of some value.

We recall that a formal system is said to be axiomatic (or axiomatized) if its set of axioms is a decidable set. Since the logical axioms of any first-order system form a decidable set, a first-order system F is axiomatic if its set of proper axioms is decidable. A first-order system F is axiomatizable if there is another first-order system F' with the same wffs and the same theorems of F and whose proper axioms form a decidable set. In short, a first-order system is axiomatizable if there exists an axiomatization of it which yields the same set of theorems.

Exercise. Give an example of a wff which is not valid but whose asf is a tautology.

Notice that any consistent, complete, axiomatizable system is decidable. We merely order all formal proofs in some convenient way and grind them out one by one. Any wff X is provable if and only if its universal closure is provable. Given any wff X, we take its universal closure $\bar{X}$ and $\sim \bar{X}$. Since these are closed wffs, we must eventually turn up a proof either of

$\overline{X}$ or $\sim \overline{X}$. If $\vdash \overline{X}$, then $\vdash X$ by application of *modus ponens* and our axiom of type 2 for a first-order system. If $\vdash \sim \overline{X}$, then $\overline{X}$ cannot be a theorem and hence neither can X. Hence, we can decide whether a given wff is a theorem or not.

Most interesting first-order theories of any great degree of expressiveness are neither complete nor decidable. Again, Chapter 6 contains a more detailed discussion.

The Corollary to Theorem 2 stated that every first-order theory which has a model is consistent. Obviously, an interesting question is whether the converse holds here. Does every consistent first-order theory have a model? The answer is "yes" and the theorem is a very profound one indeed. It says, essentially, that we cannot talk consistently without talking about something. We state the following theorem:

THEOREM 5. *Every consistent first-order theory has a model.*

For proofs of this, consult Robinson [1] or Mendelson [1].

This theorem is called the *completeness theorem of first-order logic*. What is asserted to be complete is not any given first-order system, but rather our logical axioms and rules which together constitute the underlying logic of all first-order systems. Theorem 5 says that, given any first-order theory F which has no model, then we can establish this fact using only our logical axioms and rules by formally deducing a contradiction in F. In other words, our logic is completely capable of detecting any theory which does not have a model. (Of course, we may not be clever enough to find the proof of contradiction, but that is another matter entirely.)

It is interesting that Theorem 1 of Section 1.4, which states that every valid formula of a first-order theory is a theorem of it, can be deduced from Theorem 5. To see this, we prove as lemmas several theorems which do not depend on Theorem 1 of Section 1.4.

Recalling the notion of dependence of formulas in proofs examined in Section 1.2, we now establish a fundamental result of proof theory called the *deduction theorem*.

THEOREM 6. *If $A \vdash_F B$, and if no application of the rule of generalization applied to a wff which depends on A, and in which the quantified variable was free in A, has occurred in the proof, then $\vdash_F A \supset B$.*

Proof. Again we apply induction to the length of the proof in question. If the proof is of length 1, then the proof consists of the one member B. If B is A, then $(A \supset A)$ is a tautology and thus a theorem of F. This gives $\vdash_F A \supset B$. If B is an axiom of F, we deduce $\vdash_F A \supset B$ from B itself and the tautology $(B \supset (A \supset B))$.

We assume that the proposition holds for deductions of length less than n and thus consider a deduction of length n. If B is A or an axiom, we have the same argument as just given. Suppose that B is inferred from prior wffs C and $(C \supset B)$ by *modus ponens*. C and $(C \supset B)$ are the result of deductions from A of length less than n, since they precede B, and so an application of the induction hypothesis yields $\vdash_F A \supset C$ and $\vdash_F A \supset (C \supset B)$. Now, $(A \supset C) \supset ((A \supset (C \supset B)) \supset (A \supset B))$ is a tautology as can be checked by the truth-table method. Applying *modus ponens* twice, we obtain $\vdash_F A \supset B$. Finally, suppose B is obtained from a prior wff C by the rule of generalization. Then B is $(x)C$ for some variable x. By the conditions assumed in the hypotheses of our theorem, either C does not depend on A or x is not free in A. If C does not depend on A, then $\vdash_F C$ by Corollary 2 of Theorem 1, Section 1.2. Since B is obtained from C by generalization, we have immediately $\vdash_F B$ and thus $\vdash_F A \supset B$, again using the tautology

$$(B \supset (A \supset B)).$$

If C does depend on A, then x is not free in A. Also, C is the result of a deduction from A of length less than n and so the induction hypothesis yields $\vdash_F A \supset C$. Applying generalization, we obtain $\vdash_F (x)(A \supset C)$, where x is not free in A. Now, $((x)(A \supset C) \supset (A \supset (x)C))$ is a logical axiom, since x is not free in A, and so *modus ponens* yields $\vdash_F A \supset (x)C$; that is, $\vdash_F A \supset B$, since B is $(x)C$. Our proposition holds for n, and the corollary is established.

COROLLARY. *If $A \vdash_F B$ and if A is closed, then $\vdash_F A \supset B$.*

Proof. The hypotheses of Theorem 6 are immediately satisfied.

Theorem 6 is called the deduction theorem because it establishes a fundamental connection between our metamathematical relation of deducibility "$\ldots \vdash ---$", the formal symbol "$\supset$" and the metamathematical relation "$\vdash \ldots \supset ---$". Notice that the two relations "$\ldots \vdash ---$" and "$\vdash \ldots \supset ---$" are not the same, since Theorem 6 contains certain restrictive conditions. These restrictive conditions are necessary, for we have $A \vdash_F (x)A$ in any first-order system F. However, it is generally not true that $A \supset (x)A$ is provable if the variable x is free in A.

By *modus ponens*, $\vdash_F A \supset B$ implies $A \vdash_F B$ in any first-order system F, and so the relation "$\vdash_F \ldots \supset ---$" is strictly stronger than the relation "$\ldots \vdash_F ---$" in most first-order systems F. (In contradictory systems, of course, everything is provable.)

We use the notation $A \vDash_F B$ to stand for "$A \vdash_F B$ and the hypotheses of the deduction theorem are satisfied". We thus have $A \vDash_F B$ if and only if $\vdash_F A \supset B$ in any first-order system F. In fact, $X, A \vDash_F B$ it and only it $X \vDash_F (A \supset B)$ by the proof of theorem 6.

THEOREM 7. *Let F be any first-order system and X a closed wff which is not a theorem of F. Then if the wff $(\sim X)$ is added to the axioms of F, the system F' thus obtained is consistent.*

Proof. Suppose F' is inconsistent. Then every wff, in particular X, is provable in F'. To say that X is provable in F' means precisely that $(\sim X) \vdash_F X$ holds. But $(\sim X)$ is closed and so, by the Corollary to Theorem 6 we obtain $\vdash_F ((\sim X) \supset X)$. Applying *modus ponens* to this and the tautology $((\sim X) \supset X) \supset X$, we obtain $\vdash_F X$ which contradicts our hypotheses. Hence, F' must be consistent.

Notice that the system F of Theorem 7 is consistent, since there is a wff, namely X, which is not a theorem of F.

We can now see easily how the completeness theorem implies that all valid wffs must be theorems of any system. Let F be any first-order system and let X be any valid wff of F which is not provable. The universal closure $\overline{X}$ of X is not provable either. But $\overline{X}$ is closed, and so by Theorem 7, we can add $(\sim \overline{X})$ as an axiom and obtain thereby a new system F', which is consistent. Since F' is consistent, it has a model by Theorem 5. A model of F' must make all the axioms of F', in particular $(\sim \overline{X})$, true. But if $(\sim \overline{X})$ is true, $\overline{X}$ must be false. Thus,

X is not true, because any wff X is true if and only if its universal closure $\bar{X}$ is true. But X is valid and thus true under every interpretation. Hence, X is both true and not true, a contradiction establishing that X must have been provable in the first instance. Since X is any valid formula, it follows that all valid formulas are theorems of F and Theorem 1 of Section 1.4 is proved.

Theorem 1 of Section 1.4 is sometimes referred to as the "weak completeness" of our logic. The notion of completeness there is that our logical axioms and rules are sufficient to enable us to prove all logically valid formulas as theorems in any first-order theory.

Theorem 7 also has the following interesting corollary:

COROLLARY. *Let F be a first-order theory and let X be any wff of F true in every model of F. Then $\vdash_F X$.*

Proof. Assume that X is true in every model of F, but that it is not provable in F. F is consistent, since not every wff is provable. Moreover, the universal closure $\bar{X}$ of X is not provable in F, since $\vdash \bar{X} \supset X$ in any system. Thus, we can add $(\sim \bar{X})$ as an axiom and obtain a consistent system F'. But F' has a model $\langle D, g \rangle$, since it is consistent and $(\sim \bar{X})$ must be true in this model. Moreover, every axiom of F is an axiom of F' and so $\langle D, g \rangle$ is also a model for F. But X, and thus $\bar{X}$, is true in every model of F and hence in $\langle D, g \rangle$. Thus, $(\sim \bar{X})$ is false in $\langle D, g \rangle$, contradicting our first conclusion. Hence, our assumption of the unprovability of X is false and $\vdash_F X$.

Exercise. Prove that, in any first-order system F, $A \vdash_F B$ if and only if B is true in every model of F in which A is true.

Theorem 7 brings up the important notion of one system being an *extension* of another.

Definition 3. A first-order system F' is an *extension* of a first-order system F if the alphabet and theorems of F are each subsets of the alphabet and theorems of F'. We write $F \subset F'$ to mean that F' extends F. If $F \subset F'$ while both have the same alphabet, then we say F' is a *simple* extension of F. If $F \subset F'$ and there are wffs or theorems of F' that are not wffs or theorems of F, then the extension F' is said to be *proper*.

The method of Theorem 7 can be used to prove the following useful theorem:

THEOREM 8. *Every consistent first-order theory has a consistent, complete, simple extension.*

Proof. Since the wffs of any first-order theory are denumerable, we begin with some fixed enumeration of all closed wffs of F. Let $B_1, B_2, \ldots, B_n \ldots$ be the enumeration in question. Now B_1 is either provable or not. If it is, then we proceed to B_2. If it is not, then we add $(\sim B_1)$ as an axiom and obtain by Theorem 7 a consistent simple extension of F (which may or may not be proper). We then proceed to B_2. In either case, we let F_1 be the system that results after our consideration of B_1. It is a consistent simple extension (perhaps proper) of F. We do the same with B_2, and so on. We let F_n be the system that results after considering B_n, and let P_n be the set of axioms of F_n. F_n is a consistent extension of F_{n-1} for all $n \geq 2$, and F_1 is a consistent extension of F. Let F_∞ be the system which has the same symbols and wffs as F and whose axioms are the set

$$P_\infty = \bigcup_{n=1}^{\infty} P_n.$$

Now, F_n must be consistent, for otherwise we can deduce a contradiction in F_∞. But a formal deduction is of finite length and thus involves only a finite number of axioms of F_∞. These are all contained in F_i for some i, and so a contradiction must be forthcoming in F_i. But all of the systems F_n are consistent and so we have a contradiction establishing the consistency of F_∞.

Finally, F_∞ must be complete, for it is a simple extension of F and thus has the same wffs and closed wffs as F. For every closed wff B_n of F_∞, either $\vdash_{F_n} B_n$ or $(\sim B_n)$ is added as an axiom to F_n. Thus, either B_n or $(\sim B_n)$ is provable in F_∞, which is an extension of all the F_n. This completes the proof.

Definition 4. Let S be the set of wffs of any first-order theory F, and let X be any subset of S. We say that the set X has a model if the first-order theory F^*, having S as its set of wffs and X as its proper axioms, has a model. In other words, X has a model if there is some interpretation (not necessarily a model) of F in which all the wffs of X are true.

Definition 5. A set X of wffs of a first-order theory F is said to be *inconsistent* if the theory F^*, having the same wffs as F and X as its set of proper axioms, is inconsistent. X is inconsistent if and only if it has no model.

The fact, useful in the proof of Theorem 8, that proofs are of finite length means that any inconsistent set X must be inconsistent on some finite subset. For if a contradiction is deducible from the hypotheses X, it must be deducible from some finite subset, since proofs are of finite length. Using Theorem 5, we thus obtain the *compactness theorem*:

THEOREM 9. *If a set X of wffs is such that every finite subset of it has a model, then X has a model.*

Proof. If every finite subset of X has a model, then every finite subset of X is consistent. The set X is thus consistent, since it is not inconsistent on any finite subset. But every consistent set X has a model, and our theorem is established.

The compactness theorem has many useful applications to algebra and analysis. The interested reader should consult A. Robinson [1] and [2].

Finally, we state (without proof) a modern form of the famous Löwenhein–Skolem theorem:

THEOREM 10. *If a system F has a model, then it has a finite or denumerable model; that is, a model $\langle D, g \rangle$ in which the set D is finite or denumerable. Furthermore, if D' is any set with cardinality greater than or equal to D and if F has a model with domain D, then F has a model with domain D'.*

COROLLARY 1. *Every consistent first-order theory has a denumerable model.*

Proof. If the system is consistent, it has a model and thus, by Theorem 10, a finite or denumerable model. If the model is in some finite domain D, then any denumerable domain D' has a cardinality greater than or equal to D and the system thus has a model with domain D'.

COROLLARY 2. *Every consistent, first-order theory has models of every infinite cardinality.*

Proof. If the theory is consistent, it has a denumerable model, and every infinite cardinality is greater than or equal to the cardinality of a denumerable domain.

The reader should be apprised of the fact that extending models of a given system to domains of a higher cardinality is essentially trivial. It amounts to showing that, once we have a given model, we can throw in any number of extra objects without disturbing the original model. The more significant part of Theorem 10 is thus the part asserting that any consistent system has a countable model. It is in this form that the Löwenhein–Skolem theorem is most often stated. For a proof of this part of Theorem 10, the reader should consult Church [3].

Theorem 10 is basically true because the set of wffs of any first-order theory is denumerable. Theorem 10 has some surprising consequences, for in later chapters we shall deal with first-order theories which are set theories, and in which we can prove the existence of uncountable sets. Yet, these set theories have denumerable models as all first-order theories do. Chapter 6 contains a detailed discussion of this point.

Exercise 1. Prove that, for any first-order theory F, and any set X of hypotheses, $X \vDash_F y$ if and only if, for every model $\langle D, g \rangle$ of F, every sequence s which satisfies every wff in X also satisfies y.

Exercise 2. In any first-order theory F, a proper axiom p_1 of F is said to be *independent* if it is not provable in the theory F^* obtained from F by deleting p_1 as an axiom. Prove: p_1 is independent in F if and only if there exists a model of F^* in which p_1 is false. Such a model is called an *independence model* for p_1.

Definition 6. Any set X of wffs of a first-order language L is said to be *independent* if each wff y in X is independent (in the sense of Exercise 2 above) in the theory having the same non-logical symbols as L and having X for its set of proper axioms.

This notion will be useful to us in the future.

Independence is a useful property of an axiom set as a point of simplicity and economy, but it is not so crucial from a logical standpoint as are other properties such as consistency. In fact, there are often times when a nonindependent axiomatization is more elegant and more readily understandable.

1.6. Rules of logic; natural deduction

The form we have given to the axioms and rules of the predicate calculus is especially adequate for proving metatheorems about first-order theories. Some of the metatheorems we have stated depend on nothing more than simple principles, such as mathematical induction, for their proof. Others involve highly nonconstructive principles of set wfeory. As we have already mentioned, it would be inappropriate to use these highly nonconstructive metatheorems to prove theorems in first-order systems which are themselves involved with expressing set-theoretical principles. The prime object of study in this book is precisely such "foundational" systems, and so we are interested, for our purposes, in developing the technique of formal reasoning within first-order logic. Of course we still may refer to our nonconstructive metatheorems in talking about a particular system, especially in trying to get some idea of what a model of it looks like. But such metamathematical discussion must be clearly distinguished from the purely formal and constructive approach of proving theorems within the system.

As it turns out, our form of the axioms and rules of the predicate calculus is not particularly useful for the technique of formal deduction. It is too far removed from intuitive reasoning for this. Witness, for example, the samples of formal deduction given in Section 1.4. We can see the beginnings of a more natural kind of deduction with the introduction of the deduction theorem, Theorem 6 of Section 1.5.

Let us reflect on how one intuitively proves a proposition of the form "If X, then Y" where X and Y are statements. Traditionally, one begins by assuming X is true; that is, taking X as an hypothesis and showing that the truth of Y follows. One establishes $\vdash X \supset Y$ by establishing $X \vdash Y$. The deduction theorem tells us under what conditions this method is valid. Notice that our proof in Section 1.4 of $\vdash (x_i)(A \vee B(x_i)) \supset (A \vee (x_i)B(x_i))$, x_i not free in A, did not proceed by this method, since we had not yet proved the deduction theorem.

Another natural method of intuitive logic not directly provided for by our axioms and rules of the predicate calculus is the handling of existential quantification. How might we establish a proposition of the form $(Ex) B(x) \supset A$? We might first assume $(Ex) B(x)$ as an hypothesis. Then we might say, "since there is some x such that B is true, call it a; that is, let it be designated by some arbitrary new dummy constant a". Assuming, then, that $B(a)$ holds (where $B(a)$ result from $B(x)$ by replacing a for x in all its free occurrences), we deduce A, where A does not contain the dummy constant a. We then conclude that $\vdash (Ex) B(x) \supset A$.

It is possible to prove that just such a procedure as this one is permissible with the rules we have already presented. We need first to state a few definitions:

Definition 1. Let F be any first-order system. By a *dummy constant letter* for F, we mean any constant letter which is not a constant letter of F.

A first-order theory F may well have a countably infinite number of constant letters, thus using all of the constant letters a_1, a_2, etc. However, we can always suppose that F has an unlimited supply of dummy constant letters. We can, for example, choose all the odd-numbered constant letters a_1, a_3, ... to be in the theory, and thus leave the even-numbered ones available as dummy constant letters.

Definition 2. Given a first-order system F, let F' be an extension of F that is the same as F except for containing some (any nonzero finite number of) constant letters not in F. By a *dummy well-formed formula* of F, abbreviated dwff, we shall mean a wff of any such F' which is not a wff of F.

A dwff of F is exactly like a wff except for containing at least one dummy constant letter as a term. From now on, the word "formula" will be used to mean either a wff or a dwff.

The purpose of introducing dummy constants and dwffs into our logic is to allow for the more direct and flexible handling of the existential quantifier as indicated in our brief discussion above. The operation of removing the existential quantifier by substituting a new dummy constant in place of the existentially quantified variable will be called "rule c" or the "choice rule". Deductions using this rule will be called c-deductions.

Definition 3. By a *c-deduction from the hypotheses* X we mean a finite list of wffs or dwffs of F such that, for each member Y of the list, one of the following conditions holds: (1) Y is in X, (2) Y is an axiom of F or Y is a dwff which is a logical axiom (of the extension F') all of whose dummy constants have already appeared in the proof, (3) there is a prior member of the list of the form $(Ex) A(x)$ and Y is of the form $A(b)$ where b is a dummy constant which does not appear in any dwff of the list prior to Y (in other words, b is *new* in the proof), (4) Y is inferred from prior members of the list by MP or by UG except that UG is never applied to a variable x which is free in a formula of the form $(Ez) A(z)$ to which operation (3) has been previously applied and on which the given formula c-depends (see following definition). We write $X \vdash_c A$ to stand for "A is the last line of a c-deduction from the hypotheses X".

Definition 4. An occurrence of a dwff or wff *c-depends* on an occurrence of a wff or dwff if it depends on that occurrence of the formula in the sense of Definition 4, Section 1.2 in any deduction involving possible applications of rule c. Formulas *c-depend* on other formulas if at least one occurrence of one c-depends on an occurrence of the other.

From now, but only to the end of Theorem 1 below, we restrict the notion of dependence to apply only to uses of the rules MP and UG. This is only to emphasize the explicit uses made of rule c and to examine precisely the relationship between deductions which involve rule c and those which do not. Also, the deduction theorem (Theorem 6 of Section 1.5) has so far been proved to hold only where the notion of dependence involves MP and UG.[†]

In this terminology, the restrictions on the application of UG in Definition 3 insure, in particular, that UG is never applied to a variable x free in any formula $B(b)$ which has been immediately inferred by rule c from a prior formula, and on which the given formula depends.

THEOREM 1. *In any first-order system F, if, for some set of hypotheses X, $X \vdash_c A$ and none of the X nor A are dwffs, then $X \vdash A$ where UG is applied to some formula c-dependent on and variable free in some hypothesis in X only if there was such an application of UG in the original c-deduction.*

Proof. We make use of the following lemma which we prove using only our logical rules and axioms and the deduction theorem.

LEMMA. $\vdash (z) (B(z) \supset A) \supset ((Ez) B(z) \supset A)$ *where z is not free in A, in any system F.*

Proof. 1. $(z) (B(z) \supset A))$ Hyp
    2. $B(z) \supset A$ 1, Log Ax 2, MP
    3. $\sim A \supset \sim B(z)$ 2, Taut, MP
    4. $(z) (\sim A \supset \sim B(z))$ 3, **UG** ([3] depends on [1] but z is not free in [1])
    5. $\sim A \supset (z) (\sim B(z))$ 4, Log Ax 3 (z not free in $\sim A$), MP
    6. $(Ez) B(z) \supset A$ 5, Taut, MP

We have now established [1] $\vdash$ [6] and so $\vdash_F$ [1] $\supset$ [6] by the deduction theorem.

[†] A deduction theorem involving rule c is forthcoming in Theorem 9 of this section.

Returning now to the proof of the theorem, we let $(Ey_1) B_1(y_1)$, ..., $(Ey_k) B_k(y_k)$ be the list of the wffs or dwffs to which the choice rule has been applied in the proof in order of application of the rule. Let b_1, ..., b_k be the new dummy constants thereby introduced. Obviously, $X, B_1(b_1), \ldots, B_k(b_k) \vdash A$ since the choice rule will have served only to give us the formulas $B_i(b_i)$. In fact, since we have not applied UG to any variable free in any $B_i(b_i)$ and to a formula which depends on any $B_i(b_i)$, the conditions of the deduction theorem are met and we have $X, B_1(b_1), \ldots, B_{k-1}(b_{k-1}) \vdash B_k(b_k) \supset A$ by the deduction theorem.

Let us now replace the dummy constant b_k everywhere it appears in the proof by an entirely new variable z (this is possible since the number of variables in the proof is necessarily finite). Once we have completed this formal replacement operation, we will still have a valid proof since neither the form of axioms nor the validity of the application of the rules MP and UG will have been changed. We thus have $X, B_1(b_1), \ldots, B_{k-1}(b_{k-1}) \vdash (B_k(z) \supset A)$ where z is not free in A. (Since A was, by hypothesis, a wff, no dummy constants appeared in A, which thus remains unchanged by our replacement operation. The wffs in X and the formulas $B_i(b_i)$, $1 \leqslant i \leqslant k-1$ are also unchanged since b_k does not appear in them.) Applying UG we obtain $X, B_1(b_1), \ldots, B_{k-1}(b_{k-1}) \vdash (z)(B_k(z) \supset A)$.) Notice that z was entirely new to the original proof, and so z does not occur in any of the $B_i(b_i)$, $1 \leqslant i \leqslant k-1$, nor in any of the hypotheses X actually appearing in the deduction.

Appealing now to the lemma, we have $X, B_1(b_1), \ldots, B_{k-1}(b_{k-1}) \vdash ((Ez) B_k(z) \supset A)$. But, we also have $X, B_1(b_1), \ldots, B_{k-1}(b_{k-1}) \vdash (Ey_k) B(y_k)$ since this latter formula was the last one to which the choice rule was applied in the original deduction. But, $B(y_k)$ is similar to $B_k(z)$. Hence, $\vdash (Ey_k) B_k(y_k) \equiv (Ez) B_k(z)$, (see the exercise on page 32), and applying modus ponens twice, we obtain $X, B_1(b_1), \ldots, B_{k-1}(b_{k-1}) \vdash A$.

By successively eliminating, in the same way, the other hypotheses $B_i(b_i)$, we arrive at the desired conclusion. Moreover in the final deduction of A from the hypotheses X, there will be an application of UG to some formula c-dependent on and variable free in some hypothesis in X only if there was such an application of UG in the original c-deduction (the application of UG to the variable z does not violate this since z was a completely new variable and thus one which did not appear in any of the hypotheses X actually used in the deduction).

The reason we need constant letters outside a system F for dummy constant letters is that the proper axioms of F may well assume special properties about the constant letters of F. If this is so, and certainly there is no reason to have constant letters in a system unless some assumptions are made about them, then the constant letters of F are not really ambiguous names at all, but names of specific objects. To use such constant letters in removing the existential quantifier would be similar to reasoning that, because we have proved that there are irrational numbers, then some particular constant, such as zero, is irrational. It would be an instructive (and not difficult) exercise for the reader to see exactly where the proof of Theorem 1 breaks down if we admit constants other than dummy constants in applications of rule c.

What Theorem 1 tells us is that we can use rule c as freely as we want in deductions, as long as we observe the restrictions on UG. If we proceed in this manner, every proof using rule c can be transformed into a proof without it, thus a proof using our original logical axioms and rules.

Exercise. Use rule c to prove in the predicate calculus that $\vdash (B \supset (Ex)A) \equiv (Ex)(B \supset A)$ where x is not free in B.

We now want to go even further in the direction of a more natural deduction by presenting a proof system for first-order systems which uses only rules and no logical axioms at all (except for tautologies). We will subsequently show that the deductive power of our new proof procedure is the same as with the logical axioms and rules.

Definition 5. Given a first-order system F, by a *proof in F* we now mean a finite sequence $B_1, \ldots, B_k$ of wffs or dwffs of F together with a *justification* for each *line* of the proof. By a line of the proof we mean an ordered pair $\langle n, B_n \rangle$ where B_n is the nth member of the sequence. A justification is a statement in English which accompanies a given line of the proof. The line $\langle n, B_n \rangle$ of a proof is called the *nth line* of the proof, and B_n is called the *formula of* the nth line. For each line $\langle n, B_n \rangle$ of a proof in F, one and only one of the following must hold: (1) B_n is a wff which is a proper axiom of F and the justification for the nth line consists of designating which of the proper axioms of F B_n is; (2) B_n is a tautological formula and it is a wff or else a dwff whose dummy constant letters each appear in some previous line of the proof. The justification for the nth line is that B_n is a tautology (we write "Taut"); (3) B_n is a wff or else a dwff whose dummy constant letters each appear in some previous line of the proof, and the justification for the nth line is that $\langle n, B_n \rangle$ is an hypothesis (we write "H"); (4) B_n is immediately inferred from the formulas of explicitly designated prior lines in the proof by one of the rules of inference given below, and the justification for the nth line consists in designating the prior lines and the rule of inference in question. We say also that the line $\langle n, B_n \rangle$ is *immediately inferred* from the explicitly designated prior lines in question.[†]

Notice that any finite sequence of wffs is capable of being considered a proof in a trivial way. Just let the justification for each line be that it is an hypothesis. Of course it will not be true that any sequence of wffs can be a proof if we insist on a particular type of justification. We will use this fact to generate exactly the same theorems with our new rules as with our old rules and axioms.

We now turn to the statement of our rules of inference in order to complete Definition 5. In the following, the metavariables represent formulas (wff or dwff) unless a specific restriction is indicated. A scheme of the form

$$\frac{X, Y, \ldots}{Z}$$

means that we can immediately infer the formula Z from the formulas X, Y, etc.

$$\text{MP:} \frac{(A \supset B), A}{B} ; \qquad\qquad\qquad \text{eE:} \frac{(Ex)\, A(x)}{A(b)}$$

where b is some dummy constant letter not appearing in $(Ex)\, A(x)$ nor in any wff or dwff

---

[†] We will sometimes abuse our language by identifying a line of a proof with the formula of that line. In particular, we will sometimes refer to a formula B_n as an hypothesis when it is really the line $\langle n, B_n \rangle$ which is the hypothesis. No confusion will result if we keep in mind that B_n may be the formula of two different lines of a proof and these lines may have different justifications.

previously in the proof, and $A(b)$ represents the result of substituting b for x at all free occurrences of the variable x in $A(x)$.

$$e\forall : \quad \frac{(x)\,A(x)}{A(t)}$$

where t is any term of F free for x or else a dummy constant letter previously introduced into the proof by an application of eE, and $A(t)$ represents the result of substituting t for the variable x in all the free occurrences of x in $A(x)$.

$$iE: \quad \frac{A(t)}{(Ex)\,A(x)}$$

where t is any term of F which is free for x in $A(x)$ or any dummy constant letter, and $A(t)$ is the result of substituting t for the variable x in all the free occurrences of x in $A(x)$.

$$i\forall : \quad \frac{A(x)}{(x)\,A(x)}$$

where the variable x is not free in any hypothesis on which $A(x)$ depends, and where x is not free in any wff or dwff $(Ey)\,B(y)$ to which rule eE has been previously applied in the proof unless $A(x)$ is a wff and depends only on hypotheses that are wffs. (In this latter case neither $A(x)$ nor any hypotheses on which it depends can contain dummy constant letters.)

$$eH: \quad \frac{A,\; B}{(A \supset B)}$$

where A is any hypothesis on which B depends, occurring before B in the proof.

In each of the above rules in which "previously" is used, it is understood to mean "previous to the line $\langle n, B_n \rangle$ which is being inferred from other (necessarily prior) lines".

Definition 6. Let $\langle n, B_n \rangle$ occur as an hypothesis in a given proof in a first-order system F. We say that the line $\langle i, B_i \rangle$ of the proof, $n \leqslant i$, *depends on the hypothesis* $\langle n, B_n \rangle$ if and only if (1) $n = i$, or (2) $\langle i, B_i \rangle$ is immediately inferred from prior lines of the proof at least one of which depends on $\langle n, B_n \rangle$, except that $\langle i, B_i \rangle$ does not depend on $\langle n, B_n \rangle$ if B_i is of the form $(B_n \supset B_k)$, $n < k$, and the justification for $\langle i, B_i \rangle$ is that it is inferred from $\langle n, B_n \rangle$ and $\langle k, B_k \rangle$ by eH.

Notice that Definition 6 defines dependence in such a way that a line of a proof may depend only on a line which is an hypothesis.

Definition 7. A *theorem* of a first-order system F is a wff of F which can be obtained as the formula of the last line $\langle n, B_n \rangle$ of a proof in F such that $\langle n, B_n \rangle$ depends on no hypotheses whatever.

Notice that the definition for a theorem of a first-order system excludes dwffs as theorems. Even tautologies that are dwffs of F are not theorems of F, though wff tautologies certainly are, as always. Still, our rules permit us to introduce a dwff tautology X into a proof without introducing further dependence on hypotheses (provided that the dummy constants of X have been previously introduced into the proof).

In citing the rules of inference as justification for a given line of a proof, we give the numbers of the prior lines of the proof from which the given line follows and the name of the rule in question. Notice also that a line X of a proof does not necessarily depend on an hypothesis B that occurs before X in the proof. It depends on B only if B has somehow contributed to obtaining X, as is clear from Definition 6 of this section. Moreover, any hypothesis B depends on itself. At each line X of a deduction, we indicate all the hypotheses on which X depends by displaying in parentheses the number of the line of each such hypothesis to the left of the given line X.

The names we have given the rules are meant to suggest the formal operation involved. "i" stands for "introduction" and "e" for "elimination". "E" stands for "existential quantifier" and "$\forall$" for "universal quantifier".

We call our new set of rules *natural deduction rules*. For the remainder of this work, most instances of formal deduction will take place in our natural deduction rule system. However, we may still appeal to the original rules and axioms in proving metatheorems about systems. We want now to see that our natural deduction rules are both *adequate*, meaning that everything provable by our old rules and axioms is provable with the natural deduction rules, and *sound*, meaning that everything provable by our natural deduction rules is provable with our original set of rules and axioms. We begin by a descriptive examination of our natural deduction system, comparing it to our original system.

The first of our natural deduction rules is just *modus ponens*. The second is the elimination of the existential quantifier in favor of a new dummy constant letter. Theorem 1 has already provided the justification for this procedure on the basis of our original rules and axioms. Notice that it is only by rule eE that a given dummy constant letter can be first introduced into a proof. The rule $e\forall$ is obvious and is justified by *modus ponens* and our previous logical axioms. The rule iE is also one we have previously seen to be valid from an example of formal deduction in Section 1.4. That is, we easily prove $\vdash A(t) \supset (Ex) A(x)$ according to our old rules and axioms where $A(x)$ and $A(t)$ are related as in the statement of rule iE. The rule eH of hypothesis elimination is just the deduction theorem. The rule $i\forall$ of universal quantifier introduction is somewhat complicated by the various restrictions imposed on the natural deduction rules. The first restriction, that the variable in whose name universal quantification is introduced must not be free in any hypothesis on which the formula in question depends, is necessary to insure that our rule eH (the deduction theorem) is valid. The other restrictions have to do with the rule eE rather than eH. Let us take a closer look at the eE restrictions in $i\forall$.

First, we notice that we have a certain flexibility in the eE restriction, because we have a disjunction of two possible restrictions. One or the other of these two must be satisfied, but it is not necessary that both be satisfied. The variable x in whose name universal quantification is applied must not be free in any prior wff $(Ey) B(y)$ to which eE has been previously applied in the proof, or else the formula $A(x)$ to which quantification is applied must be a wff and depend only on wff hypotheses. If we did not have such restrictions for eE in $i\forall$, we could reason falsely in the following manner:

(1) 1. $(x_1)(Ex_2) A_1^2(x_1, x_2)$ H

(1) 2. $(Ex_2) A_1^2(x_1, x_2)$ 1, $e\forall$

(1) 3. $A_1^2(x_1, a_1)$ 2, eE

(1) 4. $(x_1) A_1^2(x_1, a_1)$ 3, $i\forall$ (falsely!)

(1) 5. $(Ex_2)(x_1) A_1^2(x_1, x_2)$ 4, iE

    6. $(x_1)(Ex_2) A_1^2(x_1, x_2) \supset (Ex_2)(x_1) A_1^2(x_1, x_2)$ 1, 5, eH

The only false step here is in line 4 where we introduce the universal quantifier. The variable x_1 is free in the wff of line 2 to which the rule eE was applied, and so one of our restrictions is violated. It would be permissible to violate this restriction if it were not also true that line 3, to which the rule $i\forall$ is applied in this case, contains a dummy constant letter and is thus a dwff. Thus, neither of our alternate conditions is satisfied. Below is an example of a proof in which $i\forall$ is correctly applied.

(1) 1. $(x_1)(Ex_2)(x_3) A_1^3(x_1, x_2, x_3)$ H

(1) 2. $(Ex_2)(x_3) A_1^3(x_1, x_2, x_3)$ 1, $e\forall$

(1) 3. $(x_3) A_1^3(x_1, a_1, x_3)$ 2, eE

(1) 4. $A_1^3(x_1, a_1, x_3)$ 3, $e\forall$

(1) 5. $(Ex_2) A_1^3(x_1, x_2, x_3)$ 4, iE

(1) 6. $(x_3)(Ex_2) A_1^3(x_1, x_2, x_3)$ 5, $i\forall$

(1) 7. $(x_1)(x_3)(Ex_2) A_1^3(x_1, x_2, x_3)$ 6, $i\forall$

8. $(x_1)(Ex_2)(x_3) A_1^3(x_1, x_2, x_3) \supset (x_1)(x_3)(Ex_2) A_1^3(x_1, x_2, x_3)$ 1, 7, eH

Here the rules are correctly applied. The application of $i\forall$ in line 6 satisfies both of our restrictions concerning eE (it is only necessary that one of the two be satisfied), and the variable x_3 is not free in line 1. In the application of $i\forall$ in line 7, the variable x_1 is free in a previous line (line 2) and to which the rule eE is applied (in line 3). However, when we apply $i\forall$ in line 7, all constants introduced by eE have been eliminated and the hypotheses (namely line 1) on which line 6 depends are also all wffs (i.e. they contain no dummy constants). Thus, our second restriction concerning eE is satisfied and $i\forall$ can be applied in the name of x_1 (again, upon required checking, we see that x_1 is not free in line 1).

Notice that our conclusion in the first (incorrect) proof is not logically valid. Think of A_1^2 as being the "less than" relation on real numbers. Then for every real number, there is a greater real number, but it is not true that there is a real number greater than every real number. On the other hand, the conclusion of our correct deduction depends on no hypotheses and was deduced without the aid of any proper axioms. It is thus a theorem of the predicate calculus and is universally valid. We will, of course, need to justify that our new rules really do give the same results as our old ones.

It would be possible to formulate our rules in such a manner as to forego the use of dummy constants in connection with the rule eE and use free variables instead. However, the rules then become much less visual and practical, because our various other restrictions, particularly those in rule $i\forall$, require that we remain aware of which variables have been introduced by an application of eE and which have not. But with dummy constant letters, which are visually different from free variables, checking our rules is much easier. The dummy constants serve as "markers" when we come to apply $i\forall$. If the formula to which we wish to apply $i\forall$ contains dummy constants, then we must check that the variable in whose name we wish to generalize does not occur free in any previous formula to which eE has been applied. If the formula in question contains no dummy constants, we have only to check that it depends on no hypotheses which do contain dummy constants or that the variable in question is not free in any formula $(Ey) B(y)$ to which eE has been previously applied. Of course, we always have to check that the variable in whose name we generalize is not free in any hypothesis on which the wff in question depends. The reader will find that checking these things becomes rather natural after practice though descriptive statements of the procedure appear verbose.

Another point with respect to our restrictions for eE in $i\forall$ is that our system is really more flexible than is absolutely necessary. We can obtain an adequate set of rules if we replace $i\forall$ by the weaker rule

$$i\forall^*: \frac{A(x)}{(x)\,A(x)}$$

where the variable x is not free is any hypothesis on which $A(x)$ depends, and where $A(x)$ is a wff and depends only on hypotheses which are wffs. The weaker rule $i\forall^*$ is the same as $i\forall$ except that we have suppressed one of our alternatives for eE.

The fact that our set of natural deduction rules with the weaker rule $i\forall^*$ really is adequate will be presently justified. Our reason for making the observation here is that it will help us to see that our natural deduction rules allow us to introduce previously proved theorems at any point in a deduction without introducing further dependence on hypotheses. Under our old rules and axioms, such citing of previously proved theorems as justification for a line of a formal deduction has depended on the fact that the juxtaposition of two formal deductions is a formal deduction (see Exercise 4, Chapter 1, p. 12). That is, according to our old rules, if X is some sequence of wffs which constitutes a proof in any given system F, and if Y is another such sequence, then XY is also a formal proof in F. This justifies the introduction of a theorem at any point in a proof, since the formal proof of a theorem can be interposed at any line of a proof. Now under our natural deduction rules, a theorem depends on no hypotheses, and so if the proof of a theorem A can be legitimately interposed at any point of a deduction, it introduces no further *dependence* on hypotheses (the hypotheses themselves may be introduced, but dependence on them will have been eliminated by the time A is proved). However, for our natural deduction rules we have some complications not present in our old system of rules and axioms, and this fact requires that we examine carefully under what conditions it is legitimate to juxtapose two deductions. Let us take an example.

Suppose that we have a proof X in which the dummy constant b is introduced by an application of eE. Suppose now that we have a proof Y of the wff A and that Y also involves introduction of the dummy constant b. The juxtaposition XY is not a proof, for now the introduction of b in the proof of A violates one of our rules, namely eE; this occurrence of b in the sequence Y is no longer new to the proof XY, since it has been introduced in X, which precedes every formula of Y in the sequence XY.

Of course, it is immediately clear that this is inconsequential. If Y is a proof of the wff A, then there exists a proof Y' of A where the dummy constants introduced by applications of eE can be judiciously chosen to be new to any given proof X. Thus, the juxtaposition XY' will not violate our rule eE, and the citing of the theorem of A can be justified. We know that there always will be a proof of A which avoids the difficulty of the new constants in any given case.

Clearly there are no difficulties with the rules $e\forall$ and iE that would prevent us from juxtaposing two proofs involving any finite number of applications of these rules. However, observe that with $i\forall$ we have difficulties similar to those of eE. Suppose, for example, that we have a proof Y of the wff A as a theorem and there is an application in Y of $i\forall$ applied to a variable z and a dwff B. Since B is a dwff, this application of $i\forall$ is legitimate in the proof Y only if z is not free in any wff or dwff to which eE has been previously applied in Y. Suppose, however, that z is free in some wff or dwff $(Ew)C(w)$ to which eE has been applied in the proof X. Then if we juxtapose X and Y to form the sequence XY, our restriction on the application of $i\forall$ to z and B in Y is now violated though it was not before; for now the wff $(Ew)C(w)$ of X contains z free and occurs before B in the new sequence XY, and there has been an application of eE to $(Ew)C(w)$ prior to B in the sequence XY. In the present case, it is not so obvious that there is another proof Y' of A which avoids this tedious difficulty. Of course, one feels that there ought to be some way of avoiding it, since the previous use of eE in the sequence X is obviously not related to the later one in Y.

Let us now observe that we have no difficulties of the above kind if the rule $i\forall$ is replaced by the weaker rule $i\forall^*$, for then the universal quantifier is applied only to wffs that depend on wff hypotheses. Clearly any such application of $i\forall^*$ in any proof sequence Y is independent of any proof sequence X with which Y may be eventually juxtaposed.

We have already remarked that the introduction of hypotheses presents no problems for the citing of prior theorems in a deduction, since theorems depend on no hypotheses. We have thus justified the citing of prior theorems in the system of rules that are the same as our natural deduction rules, except that $i\forall$ is replaced by $i\forall^*$.

We now prove the adequacy of our natural deduction rules where we use only our weaker rule $i\forall^*$. This also justifies the citing of previous theorems in our natural deduction system.

THEOREM 2. *In any first-order system F, $\vdash (x)A(x) \supset A(t)$, where t is any term free for x in the wff $A(x)$, and $A(t)$ results from $A(x)$ by substituting t for x in all of the latter's free occurrences in $A(x)$.*

Proof. (1) 1. $(x)A(x)$ H
(1) 2. $A(t)$ 1, $e\forall$
3. $(x)A(x) \supset A(t)$ 1, 2, eH

THEOREM 3. *In any first-order system whatever,*

$$\vdash (x)(A \supset B(x)) \supset (A \supset (x)B(x))$$

where x is not free in A, and A and $B(x)$ are wffs.

Proof. (1)   1. $(x)(A \supset B(x))$ H
(1)   2. $A \supset B(x)$ 1, $e\forall$
(3)   3. A H
(1, 3) 4. $B(x)$ 2, 3, MP
(1, 3) 5. $(x)B(x)$ 4, $i\forall$ (x not free in 1 or 3)
(1)   6. $A \supset (x)B(x)$ 3, 5, eH
7. $(x)(A \supset B(x)) \supset (A \supset (x)B(x))$ 1, 6, eH

We have proved Theorem 2 and Theorem 3 for any system F and so wffs of the indicated form are theorems of any system. These are, of course, logical axioms of our previous formulation of rules for the predicate calculus. In proving Theorems 2 and 3, we have used only our weak rule $i\forall^*$, since we have not even used universal generalization. Besides the two types of logical axioms just proved, we had tautologies, which we also have in our natural deduction rules. Also, we have, in both cases, a rule of *modus ponens* (the rule MP) and generalization (the rule $i\forall^*$).[†] Our natural deduction rules thus yield all of our previous logical rules and axioms, and so every theorem provable according to our prior rules is provable according to our natural deduction rules.

Of course there is one further complication that must now be considered. In our natural deduction system, the use of the existential quantifier is defined explicitly, and we have no right to consider it as definable in terms of negation and the universal quantifier unless we can prove this fact from our rules. That is to say, we now consider that "*E*" figures in our alphabet for first-order systems, and our definition of wffs must be extended to include expressions obtained from wffs A by formally applying (Ex) to get $(Ex)A$ where x is any variable. Occurrences of variables in the scope of (Ex) are bound, etc. What we now need to prove is that $\vdash (Ex)\,A(x) \equiv (\sim (x)(\sim A(x)))$ for any wff $A(x)$ in any system F whatever. Also, we need to prove a general theorem of the substitutivity of logical equivalence in order to show that we really can always replace the existential quantifier by its equivalent in terms of negation and the universal quantifier. Moreover, the reader should observe that all this will be proved where every application of universal generalization satisfies our weak rule $i\forall^*$, as is the case for Theorem 2 and Theorem 3 of this section. Once this program is complete, the adequacy of our natural deduction rules is clearly established, since every proof according to our old rules and axioms is shown to be directly translatable into a proof using our natural deduction rules; in fact a proof in which only the weaker rule $i\forall^*$ is used.

We first prove the following theorem:

THEOREM 4. *In any first-order system* F, $\vdash(\sim (x)(\sim A(x))) \supset (Ex)\,A(x)$ *where* $A(x)$ *is any wff.*

Proof. (1) 1. $A(x)$ H
(1) 2. $(Ex)\,A(x)$ 1, iE
3. $A(x) \supset (Ex)\,A(x)$ 1, 2, eH
4. $(A(x) \supset (Ex)\,A(x)) \supset (\sim (Ex)\,A(x) \supset \sim A(x))$ Taut
5. $\sim (Ex)\,A(x) \supset \sim A(x)$ 3, 4, MP
(6) 6. $\sim (Ex)\,A(x)$ H
(6) 7. $\sim A(x)$ 5, 6, MP
(6) 8. $(x)(\sim A(x))$ 7, $i\forall$ (x not free in [6])
9. $\sim (Ex)\,A(x) \supset (x)(\sim A(x))$ 6, 8, eH
10. $(\sim (Ex)\,A(x) \supset (x)(\sim A(x))) \supset (\sim (x)(\sim A(x)))$
$\supset (\sim \sim(Ex)\,A(x)))$ Taut

[†] Notice that all our restrictions for $i\forall^*$ will always be satisfied whenever we apply $i\forall^*$ to a wff X which is a theorem. Since a theorem depends on no hypotheses, the eH restriction on $i\forall^*$ will be satisfied. Since a theorem is not a dwff, and since all hypotheses on which it depends (there are none) are wffs, the eE restriction for $i\forall^*$ is also satisfied. In any proof of a theorem using our old rules and axioms, UG will be applied only to theorems and axioms, and it thus follows that $i\forall^*$ will be as strong as UG for the purpose of translating proofs of theorems from our old axioms and rules into the natural deduction rules using only $i\forall^*$.

11. $\sim (x)(\sim A(x)) \supset (\sim \sim (Ex) A(x))$ 9, 10, MP
12. $\sim \sim (Ex) A(x) \supset (Ex) A(x)$ Taut
13. $[11] \supset ([12] \supset \text{Concl})$ Taut
14. $[12] \supset \text{Concl}$ 11, 13, MP
15. Concl 12, 14, MP

Here we see the use of bracketed numbers again to avoid rewriting lines of the proof which have already occurred. "Concl" means "conclusion"; that is, the statement to be proved. As we progress in our techniques of formal deduction, we shall begin to omit certain steps and give as a justification for the lines that appear the collective justification for the omitted and presented steps.

In the previous proof, for example, we might have jumped from line 9 directly to line 11 by giving as a collective justification for line 11 (which would then be line 10): 9, Taut, MP. We call such formal proofs, in which some lines are omitted, *quasiformal*. If he wishes, the reader may take it as a standing exercise in this book to supply the missing lines to quasiformal proofs.

Of course, informal proofs given in most mathematical literature are not even quasiformal. They are informal arguments which tend to convince the reader that a formal proof does exist and which permit the knowledgeable reader to supply the missing steps. In our treatment of deduction in this book, there will be a decreasing component of formalism. In Chapter 3, we give fairly complete formal or quasiformal proofs. This is also the case in the beginning of Chapter 5. Then we gradually relax and revert to the more usual informal "discussion" type of proof familiar in mathematical literature. This approach should enable the reader to appreciate more fully the notion of a proof. He should be more adept at translating from formal to informal and back again.

We now complete our treatment of the existential quantifier by proving the following theorem:

THEOREM 5. *In any first-order system F,* $\vdash (\sim (x)(\sim A(x))) \equiv (Ex) A(x)$, *where* $A(x)$ *is any wff.*

Proof. (1) 1. $(Ex) A(x)$ H
(1) 2. $A(b)$ 1, eE, b a new dummy constant
(3) 3. $(x)(\sim A(x))$ H
(3) 4. $\sim A(b)$ 3, e∀, b is free for x
5. $(x)(\sim A(x)) \supset \sim A(b)$ 3, 4, eH
6. $A(b) \supset \sim (x)(\sim A(x))$ 5, Taut, MP
(1) 7. $\sim (x)(\sim A(x))$ 2, 6, MP
8. $(Ex) A(x) \supset \sim (x)(\sim A(x))$ 1, 7, eH
9. $(\sim (x)(\sim A(x))) \equiv (Ex) A(x)$ 8, Th. 4,   Df. ≡, Taut, MP

Here we have given a quasiformal proof, omitting a few steps, of the converse of Theorem 4. In citing theorems in proofs we use the abbreviation "Th.", as well as the abbreviation "*Df.*" for citing definitions. Both Theorem 4 and Theorem 5 use only $i\forall^*$. In particular, this justifies our citing of Theorem 4 in the proof of Theorem 5.

Theorem 5 establishes the equivalence of $(Ex) A(x)$ with

$$(\sim (x)(\sim A(x))).$$

It thus shows that we can recover our definition of (Ex) in terms of negation and universal quantification from our natural deduction rules. Of course, when we defined the existential quantifier in terms of negation and the universal quantifier, it meant that we could always replace (Ex) by $\sim (x) \sim$. In order to establish the same thing here, we need the substitutivity of logical equivalence.

In Theorem 3 of Section 1.1, we proved the substitutivity of tautological equivalence. What we must now prove is that this principle of substitutivity holds for the predicate calculus. When we have done this, the full definability of (Ex) as $\sim (x) \sim$ will have been established. We need some preliminary lemmas.

THEOREM 6. *In any first-order theory,* $\vdash (x)(A \equiv B) \supset ((x)A \equiv (x)B)$, *where A and B are any wff's.*

Proof. (1)   1. $(x)(A \equiv B)$ H
        (1)   2. $A \equiv B$ 1, $e\forall$
        (3)   3. $(x)A$ H
        (3)   4. A 3, $e\forall$
        (1, 3)   5. B 2, 4, Taut, MP
        (1, 3)   6. $(x)B$ 5, $i\forall$
        (1)   7. $(x)A \supset (x)B$ 3, 6, eH
        (8)   8. $(x)B$ H
        (8)   9. B 8, $e\forall$
        (1, 8)   10. A 2, 9, Taut, MP
        (1, 8)   11. $(x)A$ 10, $i\forall$
        (1)   12. $(x)B \supset (x)A$ 8, 11, eH
        (1)   13. $[7] \wedge [12]$ 7, 12, Taut, MP
        (1)   14. $(x) A \equiv (x)B$ 13, $Df \equiv$
                 15. Concl 1, 14 eH

From now on, we shall not use the vernacular to mention specific conditions relating to the rule $i\forall$ as we did in line 8 of the proof of Theorem 4. It is up to the reader to see that each application of our rules is justified. Thus, in lines 6 and 12 of the proof of Theorem 6 of this section, the variable x is not free in the hypotheses on which the wff involved in the application of $i\forall$ depends, but we do not state this explicitly. It is to be considered part of the notation "$i\forall$" that the application of the rule must satisfy all restrictive conditions, and we mentioned these in previous proofs only for emphasis. Again, we note that only $i\forall^*$ is used in proving Theorem 6.

THEOREM 7. *In any first-order system,*

$$\vdash (x)(A(x) \equiv B(x)) \supset ((Ex) A(x) \equiv (Ex) B(x))$$

where A(x) and B(x) are any wff's.

Proof. (1)   1. $(x)(A(x) \equiv B(x))$ H
     (2)   2. $(Ex)A(x)$ H
     (2)   3. $A(b)$ 2, eE
     (1)   4. $A(b) \equiv B(b)$ 1, e∀
   (1, 2) 5. $B(b)$ 3, 4, Taut, MP
   (1, 2) 6. $(Ex)B(x)$ 5, iE
     (1)   7. $(Ex)A(x) \supset (Ex)B(x)$ 2, 6, eH
     (8)   8. $(Ex)B(x)$ H
     (8)   9. $B(c)$ 8, eE
     (1)   10. $A(c) \equiv B(c)$ 1, e∀
   (1, 8) 11. $A(c)$ 9, 10, Taut, MP
   (1, 8) 12. $(Ex)A(x)$ 11, iE
     (1)   13. $(Ex)B(x) \supset (Ex)A(x)$ 8, 12, eH
     (1)   14. $[7] \wedge [13]$ 7, 13, Taut, MP
     (1)   15. $(Ex)A(x) \equiv (Ex)B(x)$ 14, Df. ≡
           16. Concl 1, 15, eH

Again, the dummy constants b and c in the proof of Theorem 7 are required to satisfy the conditions of being new constant letters and the other relevant restrictions. We suppose these requirements summed up in the notation "*eE*".

We are now in a position to prove the following theorem:

THEOREM 8. *In any first-order system, if* $\vdash A \equiv B$, *where A and B are any wffs, then* $\vdash X \equiv X'$ *where X' is obtained from the wff X by replacing B for A at zero, one, or more occurrences of A in X.*

Proof. The proof is by induction on the number of sentence connectives and quantifiers in X. If X has no sentence connectives or quantifiers, then X is a prime formula. There are no well-formed (proper) parts to a prime formula, and so X is A. Thus, either X' is B or X' is A. In either case, the desired result follows.

We now suppose the theorem true for all wffs X with fewer than n quantifiers and sentence connectives. We must prove the assertion where X has n quantifiers and sentence connectives. If X is a prime formula or if X is A, then the argument is the same as in the foregoing case. Thus, we suppose that A is a proper part of X and that X is not a prime formula. In this case, X is of the form (1) $(x)C$, or (2) $(Ex)C$, or (3) $(\sim C)$, or (4) $(C \vee D)$ where C and D are wffs which necessarily have fewer than n quantifiers or connectives. In order to complete the demonstration, we consider each of these cases.

If X is of the form $(x)C$, then A is a part (not necessarily proper) of C, since A is a proper part of X. Let C' be the result of replacing A by B in C (in zero, one, or more occurrences). Then X' is $(x)C'$. Now suppose that $\vdash A \equiv B$. Then, by induction hypothesis $\vdash C \equiv C'$, since C has fewer than n occurrences of quantifiers and sentence connectives. Now, since $C \equiv C'$ is a theorem, it depends on no hypotheses. Thus, we can apply *i∀* (in fact *i∀**) to it and obtain $\vdash (x)(C \equiv C')$. Now, applying MP to this and Theorem 6, we obtain $\vdash (x)C \equiv (x)C'$; that is, $\vdash X \equiv X'$.

The proof for the case that X is of the form $(Ex)C$ is exactly the same except that here we use Theorem 7.

If X is of the form $(\sim C)$, then X' will be the wff $(\sim C')$ and, by inductive hypothesis, $\vdash C \equiv C'$ if $\vdash A \equiv B$. But

$$\vdash (C \equiv C') \supset ((\sim C) \equiv (\sim C')),$$

since this last wff is a tautology. Applying MP we obtain the desired results.

In the last case, we have the tautology

$$\vdash (C \equiv C') \supset ((D \equiv D') \supset (C \vee D \equiv C' \vee D'))$$

which, with an argument analogous to our previous example (both C and D must have fewer than n connectives and quantifiers), yields the desired result.

Thus, our assertion is established for the case of n quantifiers and connectives and the theorem follows by mathematical induction.

Exercise. Let X be a wff of some first-order theory and let A be a wff which is a subformula of X. We say that A *has simple occurrence* in X if no free variable of A is bound in X. Show that in any first-order theory $\vdash (A \equiv B) \supset (X \equiv X')$, where X' is obtained from X by replacing B for A at zero, one, or more occurrences of A in X, A has simple occurrence in X, and B has simple occurrence in X'.

Again, we remark that all of the above theorems have been established by using only the weak rule $i\forall^*$.

Since Theorem 3 establishes $\vdash (Ex) A(x) \equiv (\sim (x)(A \sim (x)))$ for all wffs $A(x)$ in any system F, it follows from Theorem 8 that we can always replace "(Ex)" by "$\sim (x) \sim$" just as when (Ex) was defined notation. The situation is analogous to our method of defining some of the sentential connectives in terms of others. We could just as easily have all five of our sentential connectives as basic signs of our alphabet, since our definitions of "$\supset$", "$\wedge$", and "$\equiv$" in terms of "$\vee$" and "$\sim$" are all tautological equivalences, and thus equivalences of the predicate calculus.

The adequacy of our rules is now fully established. We will no longer be concerned with whether or not a deduction satisfies our weaker rule $i\forall^*$. Although we know that the exclusive use of $i\forall^*$ will yield an adequate system of rules, we prefer the flexibility of our full rule $i\forall$.

Exercise. Use the exercise on page 32 and Theorems 6 and 7 above to prove that the universal closures of two similar wffs are logically equivalent, as well as the respective existential closures.

Notice that Theorem 8, in conjunction with the exercise on similarity on page 32, tells us the following: If $A(x)$ and $A(y)$ are similar, then $(Qx) A(x)$ can be replaced by $(Qy) A(y)$ in any formula X to obtain an equivalent formula X', where (Qx) and (Qy) represent either universal or existential quantification. We call this replacement rule the "change of name of a bound variable". It tells us that two formulas which differ only by the name of quantified variables are equivalent.

We now turn to the problem of the soundness of our natural deduction rules. What we need to show is that anything provable by our natural deduction system is also provable by our original rules and axioms. We will establish this by showing how to translate any natural deduction proof into a proof involving our original axioms and rules.

It might seem at first glance that Theorem 1 is already a justification of the soundness of natural deduction since it shows how to translate a c-deduction into a deduction without use of rule c, and the use of rule c as a primitive rule is clearly the main innovation involved with our natural deduction rule system. However, what must be justified is the way rule c is used in conjunction with the introduction and elimination of hypotheses involved in our natural deduction system. In short, we need a deduction theorem for deductions involving the use of rule c.

Definition 8. Let $B_1, \ldots, B_n$ and A be wffs or dwffs of a given first-order theory F. We write $B_1, \ldots, B_n \vDash_c A$ to mean that A is the last line of a c-deduction from the hypotheses B_i in which no application of UG to a variable free in one of the B_i, and to a formula which c-depends on that B_i, has occurred.

It follows immediately from Definition 8 and Theorem 1 of this section that, in any system F, if $B_1, \ldots, B_n \vDash_c A$ where the B_i and A are all wffs, then $B_1, \ldots, B_n \vDash A$.

We are now in a position to state and prove our deduction theorem involving rule c:

THEOREM 9. *If, for some first-order theory F, $B_1, \ldots, B_n$, $B \vDash_c X$ and $A_1(c_1), \ldots, A_k(c_k)$ are the dwffs in order of first occurrence in the proof that result from an application of rule c, then $\vec{B}_i \vDash_c (B \supset X)$ and all of the (results of) applications of rule c which occur in this new c-deduction are among $(B \supset A_1(c_1)), \ldots, (B \supset A_k(c_k))$.[†] Moreover, the formula $(B \supset X)$ c-depends on any of the formulas $(B \supset A_j(c_j))$ or B_i in the new c-deduction only if X c-depended on the formula $A_j(c_j)$ or B_i in the original deduction.*

Proof. The proof is by induction on the length of the original deduction. Clearly the theorem holds for deductions of length 1. We thus suppose it holds for deductions of length less than m and consider a deduction of length m.

Again, the result is immediate if X is an hypothesis or an axiom.

If X results by application of MP to formulas $C \supset X$ and C, then $\vec{B}_i \vDash_c B \supset (C \supset X)$ and $\vec{B}_i \vDash_c (B \supset C)$ by induction hypothesis. Thus, $\vec{B}_i \vDash_c B \supset X$ by tautology and *modus ponens*.

If X is inferred from $C(y)$ by UG applied to y, then, again by induction hypothesis, $\vec{B}_i \vDash_c (B \supset C(y))$. If y is not free in B, then y is not free in any of the formulas $(B \supset A_j(c_j))$ on which $(B \supset C(y))$ c-depends since X cannot c-depend on any $A_j(c_j)$ in which y is free. Thus, $\vec{B}_i \vDash_c (y)(B \supset C(y))$ and thus $\vec{B}_i \vDash_c B \supset (y)C(y)$ by a logical axiom and *modus ponens*.

If y is free in B, then $C(y)$ does not c-depend on B. Thus, $\vec{B}_i \vDash_c C(y)$ with the same applications of rule c and where y is not free in any $A_j(c_j)$ on which $C(y)$ depends. Let z be an entirely new variable. Then we also have $\vec{B}_i \vDash_c C(z)$ with a deduction of equal or lesser length (and in particular, of length less than m). The B_i and the $A_j(c_j)$ are unchanged since y was not free in any B_i or $A_j(c_j)$ on which $C(y)$ c-depended[‡]. Thus, we have $\vec{B}_i \vDash_c C(z)$ with a deduction of length less than m and, trivially, $\vec{B}_i, B \vDash_c C(z)$ with a deduction of length less than m, whence $\vec{B}_i \vDash_c (B \supset C(z))$ by induction hypothesis. z is not free in B. Moreover (and here is the only point of this contortion), the only possible applications of rule c are of the indicated kind. Thus,

---

† $\vec{B}_i$ means $B_1, \ldots, B_n$.

‡ Let us recall that, by Theorem 1 of Section 1.2, we can always suppose that the first deduction only involves formulas on which $C(y)$ c-depended.

$\vec{B}_i \vDash_c (z)(B \supset C(z))$, whence $\vec{B}_i \vDash_c (B \supset (z)C(z))$ by a logical axiom and *modus ponens*, and finally $\vec{B}_i \vDash_c (B \supset (y)C(y))$ by similarity.

Finally, if X is inferred by rule c, then X is $A_k(c_k)$ and, by induction hypothesis, $\vec{B}_i \vDash_c (B \supset (Ey)A_k(y))$. If y is free in B, let z be an entirely new variable. By similarity, we have the equivalence of $(Ez)A_k(z)$ and $(Ey)A_k(y)$ and so $\vec{B}_i \vDash_c (B \supset (Ez)A_k(z))$ where z is not free in B. Thus (see exercise following Theorem 1), $\vec{B}_i \vDash_c (Ez)(B \supset A_k(z))$, z not free in B. Now, by an application of rule c, we obtain $\vec{B}_i \vDash_c (B \supset A_k(c_k))$ (since c_k was new in the original deduction, it is new here). This completes the proof of the theorem.

Definition 9. By an *n-deduction* in any first-order system F, we mean a finite list of wffs or dwffs which is like a *c*-deduction from hypotheses except that every instance of UG applied to a variable y and a formula C must satisfy precisely the following condition: y is not free in any hypothesis on which C depends and either (1) y is not free in any wff or dwff to which rule c has been previously applied in the proof or (2) A is a wff and all of the hypotheses B_i on which A *c*-depends are wffs. We write $\vec{B}_i \vDash_n A$ to mean that A is the last line of an *n*-deduction from the hypotheses B_i.

The UG condition contained in Definition 9 is precisely that of our natural deduction system. In fact, clearly $\vec{B}_i \vDash_n A$ if and only if A is deducible from the hypotheses B_i in our natural deduction system. We will now establish that $\vec{B}_i \vDash_n A$ implies $\vec{B}_i \vDash_c A$.

Notice that an *n*-deduction in which every application of UG satisfies condition (1) of Definition 9 is a *c*-deduction. In fact, condition (1) is slightly more restrictive than the UG restrictions contained in Definition 3. The reason is that we want a natural deduction system in which the notion of dependence applies only to dependence on hypotheses and not to dependence on formulas in general. But the more liberal condition on UG application contained in Definition 3 involves dependence on formulas other than hypotheses.

THEOREM 10. *Where F is any first-order system, and where B_i and A are wffs or dwffs, if $\vec{B}_i \vDash_n A$, then $\vec{B}_i \vDash_c A$, where A has precisely the same c-dependencies on hypotheses as in the original n-deduction.*

Proof. The proof is by induction on the number of applications of UG which do not satisfy condition (1) of Definition 9. If there are no such, then the deduction is already a *c*-deduction and the theorem holds.

Suppose now, as induction hypothesis, that the theorem holds for less than m applications of UG not satisfying condition (1), and let us consider the case of m such applications. Let $A_1(y_1), \ldots, A_m(y_m)$ be the m formulas in order in the deduction to which UG has been applied and for which condition (1) is not satisfied. Then, in each case, condition (2) is satisfied and all of the $A_j(y_j)$ are necessarily wffs. Obviously $\vec{B}_i, (y_m)A_m(y_m) \vDash_n A$ with less than m applications of UG not satisfying condition (1) and so $\vec{B}_i, (y_m)A_m(y_m) \vDash_c A$ by induction hypothesis.

By Theorem 9, we thus have $\vec{B}_i \vDash_c ((y_m)A_m(y_m) \supset A)$. But we also have $\vec{B}_i \vDash_n A_m(y_m)$ with less than m applications of UG not satisfying condition (1), and so $\vec{B}_i \vDash_c A_m(y_m)$, again by induction hypothesis. But $A_m(y_m)$ is a wff and *c*-depends only on wff hypotheses. Moreover, y_m is not free in any hypothesis on which $A_m(y_m)$ *c*-depends. Hence, by Theorem 1 and Definition 8, we have $\vec{B}_i \vDash A_m(y_m)$ and thus $\vec{B}_i \vDash (y_m)A_m(y_m)$. Applying *modus ponens* we thus obtain $\vec{B}_i \vDash_c A$, and the theorem is proved.

COROLLARY. *In any first-order system F, if $\vec{B}_i \vDash_n A$ where A and the B_i are all wffs, then $\vec{B}_i \vDash A$. In particular, if $\vDash_n A$, then $\vdash A$.*

Proof. $\vec{B}_i \vDash_n A$ implies $\vec{B}_i \vDash_c A$ by Theorem 10, which implies $\vec{B}_i \vDash A$ by Theorem 1 and Definition 8.

The Corollary to Theorem 10 establishes the soundness of our natural deduction system by showing that anything provable in it is provable by our original axioms and rules. Moreover, Theorems 1, 9, and 10 together with the relevant definitions give a concrete method of translating natural deduction proofs into deductions based on our original rules and axioms. In general, this translation process is long and distasteful which is the reason why we have established the equivalence of our two systems. Henceforth, we can use the original system for proving metatheorems about systems and the natural deduction rules for deduction within systems without having to worry about translating back and forth.

Exercise. Using our natural deduction rules, establish the following as theorems of the predicate calculus where $A(x)$ and $B(x)$ are any wffs:

$$\vdash (x)\,(A(x) \wedge B(x)) \equiv ((x)\,A(x) \wedge (x)\,B(x));$$
$$\vdash (Ex)\,(A(x)) \vee (B(x)) \equiv ((Ex)\,A(x) \vee (Ex)\,B(x));$$
$$\vdash (x)\,(A(x) \supset B(x)) \supset ((x)\,A(x) \supset (x)\,B(x)).$$

Notice that we have not given any special rule of logic concerned with the principle of "proof by contradiction", or *reductio ad absurdum*. The reason is that this "method" of proof is an easy consequence of our rules. In a proof by contradiction, we seek to establish $\vdash A$, for some wff A, by establishing that $\vdash \sim A \supset (B \wedge \sim B)$ for some wff B. That is, we show that the assumption that A is false allows us to deduce a contradiction. But once $\vdash \sim A \supset (B \wedge \sim B)$ is established, we can deduce $\vdash A$ as follows: First, we infer, by tautology and *modus ponens*, that

$$\vdash \sim (B \wedge \sim B) \supset \sim \sim A.$$

Then, applying further tautologies, we obtain $\vdash (B \vee \sim B) \supset A$. But $B \vee \sim B$ is a tautology and thus a theorem of any system. Applying *modus ponens* we obtain $\vdash A$ as desired.

When giving formal or quasiformal proofs, we shall go through these steps and actually obtain the desired result without appealing to any rule of proof by contradiction. Later on in our study, when we countenance proofs of a less formal nature, we shall merely observe that a certain hypothesis has led to contradiction, and infer immediately that its negation is therefore provable.

This completes our discussion of rules and techniques of formal deduction. We shall not give protracted examples of formal deduction at this time, since the content of future chapters will furnish us with many working examples.

Exercise. Consider any first-order theory F. Prove that the relation of equivalence in F, "$\vdash_F \ldots \equiv - - -$", induces a Boolean algebra on the equivalence class of wffs in which the theorems are the equivalence class determining the maximal element of the algebra and the negations of the theorems determine the minimal element. This Boolean algebra is called the Lindenbaum algebra of F. We also get a Boolean algebra if we consider only the closed wffs of F, and use again the relation of logical equivalence. Many of our theorems of logic in Section 1.5 follow from Boolean algebraic results. For example, Theorem 8 of Section 1.5, which states that every consistent first-order theory has a consistent, complete extension, follows from the Boolean algebraic (or ring-theoretic) result that every proper ideal is contained in a maximal ideal. This fact can also be used to prove the completeness theorem, Theorem 5 of Section 1.5. In fact, a certain generalized form of the completeness theorem is equivalent to the Boolean maximal ideal theorem.

1.7. First-order theories with equality; variable-binding term operators

We have used the notation $A(x)$ to represent a wff in which the variable x may occur free. Similarly, $A(t_1, \ldots, t_n)$ will represent a wff in which the variables t_i may occur free.

Definition 1. A *first-order theory with equality* is a first-order theory F containing some binary predicate letter A_i^2 which satisfies the following two conditions: (1) $(x)\, A_i^2(x, x)$ is an axiom where x is any variable; (2) $(x)(y)(A_i^2(x, y) \supset (X \equiv X'))$ is an axiom where x and y are variables, X' is obtained from X by replacing y for x in zero, one, or more free occurrences of x in X, and y is free for x in the occurrences of x it replaces. We usually denote such a predicate letter by "$=$" and write "$(x = y)$" or "$x = y$" instead of $A_i^2(x, y)$.

Principle (2) is the substitutivity of equality, analogous to the substitutivity of equivalence which we have already discussed. Principle (1) is the reflexivity of equality. Since any first-order theory with equality is presumed to have (1) and (2) as axioms, the term "proper axioms" for such a theory will always refer to axioms other than (1) and (2).

Exercise. Using the two principles (1) and (2), prove the symmetry and transitivity of equality; that is, $\vdash_F x = y \supset y = x$ and

$$\vdash_F (x = y \land y = z) \supset x = z$$

where x, y, and z are any terms, and F is a first-order theory with equality.

Definition 2. F is a first-order theory in which *equality is definable* if there is some wff $A(x, y)$ of F with exactly two free variables x and y each free for the other in $A(x, y)$, and such that: (1) $\vdash_F (x)\, A(x, x)$, where $A(x, x)$ is obtained from $A(x, y)$ by substituting x for y in all the latter's free occurrences in $A(x, y)$;

$$(2) \quad \vdash_F (x)(y)(A(x, y) \supset (X \equiv X')),$$

where X and X' are related as in Definition 1 of this section.

Most of the systems we will study in later chapters are systems with equality, or systems in which equality is definable.

As follows from Definition 1 and the exercise immediately following it, the interpretation of the predicate letter "$=$" must be an equivalence relation on D for any model $\langle D, g \rangle$. If the interpretation is, in fact, the identity relation on D, then we say that $\langle D, g \rangle$ is a *normal model* for F.

5*

It is important to note that any model $\langle D, g \rangle$ of a first-order theory F with equality gives rise to a normal model of F in a natural way. For $\langle D, g \rangle$ to be a model of F, the interpretation of "$=$" must be an equivalence relation E on the domain D, and we must have substitutivity of equality by our property (2) of Definition 1. In short, the axioms for equality must be true of the relation E in the structure $\langle D, g \rangle$, and therefore the equivalence relation E must be a congruence relation for the operations and relations of the structure $\langle D, g \rangle$. This means that, for elements $d_1, \ldots, d_n$ and $b_1, \ldots, b_n$ of the domain D, $d_i E b_i$ implies that

$$g(f_m^n)(d_1, \ldots, d_n) \, E g(f_m^n)(b_1, \ldots, b_n)$$

and also that $g(A_m^n)(d_1, \ldots, d_n)$ holds if and only if $g(A_m^n)(b_1, \ldots, b_n)$ holds for any n-ary function letter f_m^n and any n-ary predicate letter A_n^m respectively of F. We can thus take the set D/E of equivalence classes as objects of a new domain and use the induced operations and relations to form a new model $\langle D/E, g^* \rangle$ of the same system F. This new model will obviously be one in which the interpretation of "$=$" is identity, and so it is a normal model. We speak of this new model as the *contraction* of $\langle D, g \rangle$ to a normal model. The equivalence classes of any equivalence relation over a set D partition D into disjoint classes, and so the cardinality of the contracted model is always less than or equal to the cardinality of the original model.

Obviously we have the same contraction to normal models for theories in which equality is definable, since the same properties of substitutivity, and the like, hold for a defined equality (see Definition 2 of this section) as they do for a predicate letter of equality.

The normal contraction $\langle D/E, g^* \rangle$ is *elementarily equivalent* to $\langle D, g \rangle$ in the precise sense that exactly the same formulas are true in the two models.

The fact that every model of a theory with equality has a normal contraction amounts to a completeness theorem for logic with equality. If we require the standard semantics of the equality sign to be the interpretation of $=$ as the diagonal relation $\{\langle d, d \rangle \mid d \in D\}$, then we have shown that every model of a theory involving $=$ in which the conditions (1) and (2) of Definition 1 are satisfied is elementarily equivalent to a model where $=$ receives its standard interpretation. Moreover, wffs of the forms of conditions (1) and (2) are always true in any normal model of a theory involving $=$. Hence, any theory with equality having no normal model has no model whatever and is thus inconsistent.

Notice that, whereas we have models of an arbitrarily high cardinality for any consistent first-order system (Theorem 10, Section 1.5), we do not have normal models of an arbitrarily high cardinality. Consider, for example, a first-order system U with no function letters, with "$=$" as the sole predicate letter, and whose axioms are the axioms of equality plus the one proper axiom $(Ex_1)(x_2)(x_2 = x_1)$. This system (which is clearly consistent) will have a normal model only in a one-element domain. But if the interpretation of "$=$" in some interpretation $\langle D, g \rangle$ is only an equivalence relation E, then we have one big equivalence class, the domain D itself, which can obviously be arbitrarily large without affecting the model. Contracting to a normal model will still yield only one element, namely the whole class D.

Thus, a consistent system may have only finite normal models. However, Tarski has shown that if a first-order system has an infinite normal model, then it has a normal model in every infinite domain. It follows from this result that any consistent first-order system, which has no finite normal models, has normal models of every infinite cardinality.

In first-order theories in which equality is definable, we can define a special type of existential quantifier "$(E! x)A$", which means intuitively "there is one and only one x such that A"

or simply "there is a unique x such that A". The definition is as follows: Let $A(x)$ be any wff and let y be the first variable in alphabetic order different from x and new to $A(x)$. Let $A(y)$ be the same as $A(x)$ except for containing free occurrences of y where $A(x)$ contains free occurrences (if any) of x. Then $(E! \, x) \, A(x)$ for $(Ex) \, (y) \, (x = y \equiv A(y))$. It is clear why we need the properties of our relation "$=$" in order to accomplish this definition.

For first-order theories in which equality is definable, function letters can always be eliminated in favor of predicate letters. Instead of

$$f_m^n(y_1, \ldots, y_n) = y_{n+1},$$

we can write $A_k^{n+1}(y_1, \ldots, y_n, y_{n+1})$ for some appropriate predicate letter for which the following is a theorem:

$$\vdash (y_1) \ldots (y_n) \, (E! \, y_{n+1}) \, A_k^{n+1}(y_1, \ldots, y_n, y_{n+1}).$$

Exercise. Let $A(x)$ be any wff of a first-order theory in which equality is definable. Show that, in F, $\vdash (E! \, x) \, A(x) \equiv (Ex) \, A(x) \wedge (x) \, (y) \, (A(x) \wedge A(y) \supset x = y)$ and $\vdash (E! \, x) \, A(x) \equiv (Ex) \, (A(x) \wedge (y) \, (A(y) \supset y = x))$, where y is free for x in $A(x)$.

We want also, in this section, to consider certain terms of a more general type than those thus far considered for first-order theories. We shall make use of such terms in systems introduced in later chapters.

Our function letters are symbols which combine with other terms to yield a term. The application of a function letter to a term does not bind any variables. No notion of quantification is involved. But there is a very natural way in which we can consider terms formed not from other terms, but from wffs. Usually (and always in the cases we will consider) such an operation involves binding a variable.

Let $A(x)$ be a wff of a first-order system in which x is a free variable. Suppose $A(x)$ means "x is red" under our current interpretation. We might then want to speak about "a red thing". "A red thing" must be a term, not a wff; but how can we formulate such a notion? Quantification is an operation that yields another wff when applied to a wff. We need some way to pass from the wff $A(x)$ to a term t.

The only answer is to introduce some new type of operator in our language. Suppose we introduced an operator "τx" which always yields a term when applied to a wff and which binds the variable x in its scope. Application of "τx" to a wff $A(x)$ would be generally interpreted as picking out some particular object a from the class of objects of D that satisfies the property expressed by $A(x)$ under the interpretation $\langle D, g \rangle$. Thus, "$\tau x A(x)$" would mean "a red thing" under our interpretation of $A(x)$ as "x is red". If $A(x)$ means "x is prime", then "$\tau x A(x)$" would mean "a prime number".

This is known as a *selection operator.* Another operator is the *description operator* "ιx", which associates with the wff $A(x)$ the unique thing a such that $A(a)$, if such a unique thing exists, or some conventional object otherwise. If, for example, $A(x)$ means "x is even and prime", then "$\iota x A(x)$" would be "the even prime number".

The most familiar example of a variable-binding term operator to mathematicians is the *abstraction operator* of set theory. For a property $A(x)$, the term "$\{x \mid A(x)\}$" represents a set, namely the set of all objects x such that $A(x)$ is true. Thus, if $A(x)$ means "x is a prime number", then "$\{x \mid A(x)\}$" stands for the set of prime numbers.

Of course, these are all examples and illustrations of the notion of a variable-binding term operator. Let us now give a formal definition.

Definition 3. By a *variable-binding term operator*, abbreviated "*vbto*", of a first-order theory F, we mean a symbol v which is explicitly added to the alphabet of F and which combines with a wff A and a variable x to yield a term t (the particular formal manner of combination is unimportant, but must be specified clearly when defining the wffs of F). The variable x is said to be *bound* in the term t, and it is considered bound in any wff X that contains the term t. The rules concerning substitution, bound and free variables, freedom for, and so on apply to wffs containing t, and to t, respectively.

Though we have thought it better to avoid fixing the grammar of *vbto*s once and for all, we will generally write "$vxA(x)$" to represent the term formed by applying the unspecified *vbto* v to the variable x and wff $A(x)$.

We need now to extend our definition of an interpretation $\langle D, g \rangle$ of a formal system in order to incorporate *vbto*s. We need to define, for a given infinite sequence s of elements of D, what the object $g_s(t)$ is, where g_s is our function from terms of F to objects of D defined relative to the sequence s, and t is a term defined by means of a *vbto* v. To do this we need to define the interpretation of a *vbto* under the mapping g.

Definition 4. Let a first-order theory F, a *vbto* v of F, and an interpretation $\langle D, g \rangle$ of F be given. The mapping g assigns to v a function g_v from $\mathcal{P}(D)$ to D. g_v thus assigns an object in D to each subset of D.[†]

We now extend the definition of our functions g_s in order to define $g_s(t)$ where t is a term defined by a *vbto* v, and s is any infinite sequence of elements of D as before. This extension is accomplished by the following definition:

Definition 5. Let A be some wff of a first-order theory F and let t be the term formed by a *vbto* v from A, where v binds the variable x_i. Let an infinite sequence s of elements of D be given. Then, $g_s(t) = g_v(Y)$ where Y is the set of all elements $d \in D$ such that s' satisfies A where $s'_i = d$ and $s'_j = s_j$ for all $j \neq i$. We suppose this condition added to the recursive definition of g_s previously given.

Using the extended definition of g_s, we now know what it means for a sequence s to satisfy a formula involving *vbto*s. We thus know what it means to say that such formulas are true. We have therefore totally determined the semantics of *vbto*s.

Let us illustrate this semantics with an example. Let v be the abstraction operator of set theory and consider the term $t = \{x_1 \mid x_1 = x_2\}$ in an appropriate language. For any sequence s having $s_2 = d_2$, $g_s(t) = g_v(Y)$ where Y is the set of all d in D such that the couple $\langle d, d_2 \rangle$ is in $g(=)$. For a normal model, Y will be the set $\{d_2\}$. Otherwise, it will be the set of all those elements of D which bear the relation $g(=)$ to d_2. If, now, we consider the term $r = \{x_1 \mid x_1 = 2\}$, then Y will be the same for every sequence, namely the set of all elements of D which bear the relationship $g(=)$ to 2. For a normal model, this will be $\{2\}$. Finally, $g_v(Y)$ is, in each case, some element d' in D.

[†] In order to accommodate this definition, the codomain of the mapping g must now be extended to include the set $\mathcal{P}(D) \times D$.

In thinking about the semantics of *vbtos*, as well as in other contexts, the following definitions are useful.

Definition 6. Let A be some wff of a first-order theory with exactly the variables $x_{i_1}, \ldots, x_{i_n}$ free in alphabetical order (i.e. if $j < k$, then $i_j < i_k$). Further, let some interpretation $\langle D, g \rangle$ be given. Then the *truth set* of A (or the *relation expressed* by A) relative to $\langle D, g \rangle$ is the subset of D^n consisting of those n-tuples $\langle d_1, \ldots, d_n \rangle$ in D^n such that any sequence s with $s_{i_j} = d_j$ satisfies A. We extend this definition to closed wffs by declaring the truth set of a true proposition to be D and of a false proposition to be the null set $\varLambda$.

Definition 7. A subset Y of D^n which is the truth set of some wff of F is said to be *definable* in F over $\langle D, g \rangle$. It is *para-definable* in F over $\langle D, g \rangle$ if there exists some wff A of F for which the following condition holds: where $x_{i_1}, \ldots, x_{i_k}$ are the free variables of A, there exist elements $d_{j_1}, \ldots, d_{j_m}$ of D, $n = k - m$, such that Y is the set of all n-tuples $\langle d_1, \ldots, d_n \rangle$ in D^n such that any sequence s satisfies A if s takes the fixed values $d_{j_1}, \ldots, d_{j_m}$ for the indices $i_{j_1}, \ldots, i_{j_m}$ and the values $d_1, \ldots, d_n$ in alphabetic order for those indices among the $i_1, \ldots, i_k$ which do not occur among the $i_{j_1}, \ldots, i_{j_m}$.[†]

Every definable set is obviously para-definable.

It is clear from Definition 5 that the value $g_s(t)$ of variable-bound terms t depends only on the value of g_v for para-definable subsets of D. Thus, the value of g_v for subsets of D other than para-definable ones is essentially arbitrary.

1.8. Completeness with *vbtos*

Since we have extended our semantic system to define interpretations for *vbtos*, the question naturally arises as to whether or not we can also extend our deductive system of axioms and rules in order to maintain logical completeness for languages with *vbtos*. It is easily seen that we do not have logical completeness without some such extension. For example, $vx_1 A_1^1(x_1) = = vx_2 A_1^1(x_2)$ is universally valid in any language with equality having *vbto* v. But it cannot be proved from our present logical axioms and rules.

The positive solution to this problem has been obtained by Corcoran, Hatcher, and Herring [1] for languages with equality, and independently by Newton da Costa for more general languages and in particular for languages without equality (cf. da Costa [2] and Druck and da Costa [1]).

For first-order languages in general (with or without equality), to obtain completeness of the logic it is necessary and sufficient to add the following two axiom schemes as logical axioms where v is any *vbto* of the language:

V.1. $(x)(A(x) \equiv B(x)) \supset X \equiv X'$ where X' is like X except that we have replaced $vxA(x)$ by $vxB(x)$ at zero, one, or several free occurrences of $vxA(x)$ in X and $vxB(x)$ is free for $vxA(x)$ for those occurrences it has replaced.

[†] Some authors extend the use of the word "definable" to cover what we have here called para-definable (see Chang and Keisler [1], p. 211). Indeed, this was the usage in Corcoran, Hatcher, and Herring [1], but I have since found reason to change on this point.

V.2. $X \equiv X'$ where X' is like X except that we have replaced zero, one, or several free occurrences of $vxA(x)$ in X by free occurrences of $vyA(y)$ where $A(x)$ and $A(y)$ are similar.

Thus, any first-order theory with *vbtos* in which these two schemes are theorems is both strongly complete, meaning that every consistent set of formulas has a model, and weakly complete, meaning that every formula which is universally valid according to our extended semantics is a theorem.

For many of the systems containing *vbtos* considered in this book, the schemes V.1 and V.2 are theorems which follow from the proper axioms of the system. In one or two cases, we will find it necessary to add versions of these schemes explicitly as axioms.

Exercise 1. Prove that, in any system in which equality is definable, the conjunction of V.1 and V.2 is equivalent to the conjunction of:

$$\text{V.1'. } (x)\,(A(x) \equiv B(x)) \supset vxA(x) = vxB(x) \quad \text{and} \quad \text{V.2'. } vxA(x) = vyA(y)$$

where, again, $A(x)$ and $A(y)$ are similar.

Exercise 2. Prove that the conjunction of V.1′ and V.2′ is equivalent to the single axiom scheme: $(x)(y)(x = y \supset X \equiv Y) \supset (vxX = vyY)$, where v is a *vbto*, x and y are distinct variables and X (respectively Y) may contain x but not y free (respectively y but not x free). This is the Truth Set Principle of Corcoran, Hatcher, and Herring [1], p. 179 (cf. also Corcoran and Herring [1]).

It is also possible to have *vbtos* which bind any fixed finite number of variables in a definite order. The simplest syntactic form is to display the variables before the formula as in quantification. This is accomplished by the following definition:

Definition 1. Let v be an n-ary *vbto*, and A a formula of a system F. Let $x_{i_1}, \ldots, x_{i_n}$ be any n distinct variables of the language. Then $vx_{i_1} \ldots x_{i_n} A$ is a variable-bound term in which the variables $x_{i_1}, \ldots, x_{i_n}$ are bound.

The interpretation g_v of an n-ary *vbto* v is a mapping from $\mathcal{P}(D^n)$ to D. We can now define the interpretation modulo a sequence s for such n-ary variable-bound terms.

Definition 2. Given a term $t = vx_{i_1}, \ldots, x_{i_n} A$ and a sequence s, then $g_s(t)$ is defined to be $g_v(Y)$ where $Y \subset D^n$ is the set of all n-tuples $\langle d_1, \ldots, d_n \rangle$ such that s' satisfies A where s' is like s except for having $d_1, \ldots, d_n$ respectively for $s_{i_1}, \ldots, s_{i_n}$.

The same methods used to show the completeness of unary *vbtos* with the two added schemes V.1 and V.2 can be used to show that completeness for n-ary *vbtos* can be obtained by addition of the obvious n-ary generalizations of V.1 and V.2.

Exercise. Formulate explicitly the n-ary generalizations of V.1′ and V.2′ for languages with equality, and prove the equivalence of these two schemes with the obvious n-ary generalization of the Truth Set Principle.

The introduction of n-ary *vbtos* naturally incites the imagination to wonder how far we can go in generalizing the logical operators of our language. The answer follows from the realization that all of our logical operators, including the sentential connectives and the universal quantifier, are special cases of the following general type of operator: an operator v which binds m variables and combines with n terms and p formulas to form either a term or a formula. For example, disjunction $\vee$ would be a formula-maker of type $\langle 0, 0, 2 \rangle$ (it binds no variable, uses no terms, and uses two formulas). The universal or existential quantifier would be a formula-maker of type $\langle 1, 0, 1 \rangle$ and an n-ary *vbto* would be a term-maker of type $\langle n, 0, 1 \rangle$.

A systematic exposition and detailed description of the syntax and semantics of languages having logical operators of all types is to be found in Gagnon [1]. In particular, Gagnon proves the soundness and completeness of such languages (with equality) with the addition of two schemes which are generalizations of V.1′ and V.2′.

This theory of arbitrary variable-binding term and formula-makers might be called the theory of *extensional* logical operators, because the semantic system used depends only on the extensions (sets denoted by) the logical operators. Non-extensional operators would involve modal, probability, or other such kinds of intensional operators.

A recent paper of Newton da Costa (see da Costa [3]) presents an overview of first-order model theory with *vbto*s.

In closing our discussion of *vbto*s, it should be stressed that this precise definition of truth and validity for wffs involving *vbto*s is a way of rendering exact our intuitive use of these term operators. When we use these in formal systems, we proceed by purely formal deduction and so the formal manipulation with these terms does not involve such appeals to general set theory as we have made in defining their interpretation. We have defined in a purely formal manner the notion of bound and free variables in *vbto*s and the corresponding formulation of the notions of "free for", and the like is also formal. (We will not bother to restate each of these notions, since the extended formulation is immediate.)

Thus, the formal use of variable-binding term operators in first-order theories is defined on a basis just as solid and just as devoid of general set theory as any other formal notion, such as quantification. In fact, we introduce the use of *vbto*s only to reduce the preoccupations with purely formal and linguistic matters in our future treatment of first-order theories. In most cases, we can theoretically dispense with their use. This point will become clearer when we consider specific sysems such as system **F** of Chapter 3.

1.9. An example of a first-order theory

Let us conclude this chapter by giving an example of a first-order theory. The theory **S** we describe is known as *first-order number theory* or *first-order arithmetic*. We want to construct a formal system **S** the intuitive interpretation of which will be the system N of natural numbers (including zero).[†] It is not immediately evident how one should go about this. We know that N has operations of addition and multiplication and that these satisfy the associative, commutative, and related properties. It is easy to imagine how to formulate such *algebraic* properties. For example, the commutative property of addition could be expressed by

$$(x_1)(x_2)((x_1 + x_2) = (x_2 + x_1))$$

with the terms and wffs appropriately defined. The difficulty derives from the fact that the set of

---

[†] We shall always use the term "natural numbers" for the nonnegative integers.

natural numbers is infinite and we thus need some way of expressing the well-ordering principle in our first-order language.

The solution to this problem lies in consideration of the famous Peano postulates for the natural numbers.[†] Informally stated, these Peano postulates are as follows: (1) 0 is a natural number; (2) for every natural number n, there is exactly one natural number n' called the successor of n; (3) 0 is not the successor of any natural number n; (4) the successors of two different natural numbers are different (in other words, the successor function is 1–1); (5) every set containing 0 and the successor of any natural number it contains contains all the natural numbers. In other words, the natural numbers are the smallest set satisfying the first four properties. These properties are obviously true of the natural numbers as we intuitively conceive of them. The sense in which they completely characterize the natural numbers will be the object of further clarification in Chapter 3 and Chapter 6. We now define **S** by using certain formal analogues of the Peano postulates.

S is a first-order theory with one binary predicate letter "=", two binary function letters "+" and "·", one singulary function letter "'", and one constant letter "0".[‡] Actually, we should specify which of the predicate letters A_m^n the sign "=" is, which of the function letters f_m^n the sign "+" is, and so on. However, it does not really matter whether we do so or not, as long as our basic alphabet is explicitly described as we have done.

Moreover, instead of writing "$+(x, y)$" and "$·(x, y)$" we write "$(x+y)$" and "(xy)". Finally, we write "t'" instead of "$'(t)$" where t is a term. Let us define rigorously our wffs in order to avoid ambiguity.

By a term of **S** we mean any expression t which is one of the following forms and no other: (i) t is an individual variable; (ii) t is the constant letter 0; (iii) there are terms x and y such that t is of the form $(x+y)$ or (xy) or x'; (iv) these are the only terms of **S**. By a wff of **S** we mean an expression X such that either (i) X is of the form $(t = r)$ where t and r are terms or (ii) X is of the form $(x)A$ or $(Ex)A$ where x is a variable and A is a wff or (iii) X is of the form $(\sim A)$ or $(A \vee B)$ where A and B are wffs. (iv) These are the only wffs of **S**. The proper axioms of **S** are as follows.

S.1 $(x_1)(x_1 = x_1)$

S.2 $(x_1)(x_2)(x_1 = x_2 \supset x_2 = x_1)$

S.3 $(x_1)(x_2)(x_3)((x_1 = x_2 \wedge x_2 = x_3) \supset x_1 = x_3)$

S.4 $(x_1)(x_2)(x_1 = x_2 \supset x_1' = x_2')$

S.5 $(x_1)(x_2)(x_3)(x_4)((x_1 = x_2 \wedge x_3 = x_4) \supset (x_1+x_3) = (x_2+x_4) \wedge (x_1x_3) = (x_2x_4))$

S.6 $(x_1)(\sim (x_1' = 0))$

S.7 $(x_1)(x_2)(x_1' = x_2' \supset x_1 = x_2)$

S.8 $(x_1)((x_1+0) = x_1)$

S.9 $(x_1)(x_2)((x_1+x_2') = (x_1+x_2)')$

S.10 $(x_1)((x_10) = 0)$

S.11 $(x_1)(x_2)((x_1x_2') = (x_1x_2)+x_1)$

S.12 For any wff $A(x)$ of **S** the wff $(A(0) \wedge (x)(A(x) \supset A(x'))) \supset (x)A(x)$ is an axiom where x is any variable and $A(0)$ and $A(x')$ are obtained from $A(x)$ by substitution for all the free occurrences of x in $A(x)$.

[†] These postulates were fully and explicitly given by Dedekind in Dedekind [1].

[‡] In speaking of a one-place function or operation, we prefer the etymologically more pleasing "singulary" to the presently more common "unary". However, we will feel free to use both terms, and we do so with the understanding that they are completely synonymous.

This completes the list of proper axioms of **S**. We have an infinite set of axioms, since **S**.12 is an axiom scheme.

The axioms **S**.6, **S**.7, and **S**.12 are formal analogues of the Peano postulates 3, 4, and 5. The first two Peano postulates are already part of our language by the inclusion of the constant letter 0 and the singulary function letter ′. (It is common practice to omit initial universal quantifiers in stating axioms. Since the universal closure of any theorem is a theorem by $i\forall$, the closed form immediately follows. Such differences in presentation are immaterial.)

The first three axioms are the reflexive, symmetric, and transitive properties of equality. **S**.4 and **S**.5 express the substitutivity of equality with respect to our basic function letters. The general substitutivity of equality can be proved as a metatheorem by using induction (in the metalanguage) on the number of function letters in the terms considered. Thus, **S** is a system in which equality is definable.

Exercise. Prove the last assertion. That is, prove that

$$(x)(y)(x = y \supset (X \equiv Y))$$

is a theorem of **S** where Y is a wff obtained from the wff X by replacing y for x in zero, one, or more occurrences of x in X, and where y is free for x in those occurrences of x that it replaces. This, together with **S**.1, yields the result that equality is definable in **S**.

We could, of course, have chosen to state this metatheorem as an axiom scheme and thus make **S** a theory with equality. The axioms **S**.2 to **S**.5 could then have been omitted, since they are special cases of this metatheorem. Such differences in presentation are immaterial as is the question of whether or not to include initial universal quantifiers in stating axioms.

Axioms **S**.8 and **S**.9 are known as the *recursive definitions* of addition. Of course we are already given that addition is defined, since we have a binary function letter for it. Therefore, in the system **S**, these equations serve to determine certain necessary properties of addition. Similarly, **S**.10 and **S**.11 give necessary properties of multiplication. By means of these definitions, together with the axiom of induction **S**.12, the usual properties of addition and multiplication of natural numbers can be deduced.

This brings us to a discussion of **S**.12. Under the *standard model* of **S**, the one in which the domain is the set N of natural numbers, 0 names zero, the successor function represents the addition of the unit 1, addition represents addition, multiplication represents multiplication, and equality stands for identity. Any wff $A(x)$ with exactly one free variable will express some set of natural numbers, its truth set X. **S**.12 thus says that if this set X contains 0 and the successor of any natural number it contains, then everything (and thus, under the standard model, every natural number) is in the set. Consequently, **S**.12 would seem to be the formal analogue of the last Peano postulate. This is, however, not quite true. There are obviously only denumerably many wffs of **S** and hence only denumerably many wffs with one free variable. Thus, there are only denumerably many different truth sets that the wff $A(x)$ of **S**.12 can express. But the set $\mathcal{P}(N)$ of all subsets of N is nondenumerable. Hence, there are nondenumerably many subsets of N that are excluded by **S**.12. There is simply no way to "talk about" them in **S**. **S**.12 thus represents a weak form of Peano postulate (5). This seemingly innocent fact has some surprising consequences, which are examined in Chapter 6.

For an example deduction in **S**, let us prove the following theorem:

THEOREM 1. $\vdash (x_1)(x_1 = 0 + x_1)$.

Proof.   1. $0 + 0 = 0$ S.8, $e\forall$

2. $0 = 0 + 0$ S.2, $e\forall$, 1, MP

(3) 3. $x_1 = 0 + x_1$ H

(3) 4. $x_1' = (0 + x_1)'$ 3, S.4, $e\forall$, MP

5. $0 + x_1' = (0 + x_1)'$ S.9, $e\forall$

6. $(0 + x_1)' = 0 + x_1'$ 5, S.2, $e\forall$, MP

(3) 7. $x_1' = 0 + x_1'$ 4, 6, S.3, $e\forall$, Taut, MP

8. $[3 \supset 7]$ 3, 7, eH

9. $(x_1)[8]$ 8, $i\forall$

10. $[2 \wedge 9]$ 2, 9, Taut, MP

11. $(x_1)(x_1 = 0 + x_1)$ 10, S.12, MP

Using Theorem 1 as a lemma, we can prove easily that

$$\vdash 0 + x_1 = x_1 + 0.$$

This theorem will be the first step of an inductive proof of the commutative law of addition.

Exercise. Complete the program sketched by giving a formal proof of the commutative law of addition, proving whatever lemmas you judge to be necessary.

We now introduce some definitions for **S**. For example:

Definition 1.   1 for $0'$.

Definition 2.   2 for $1'$.

As we have previously explained, such definitions are to be regarded as metalinguistic abbreviations of formal statements. Thus, a wff involving "1" is to be imagined as written in the formal language using $0'$. In other words, definitions do not introduce new symbols into the alphabet of the system, but give us metalinguistic ways of abbreviating for ourselves the purely formal expressions.

We can also introduce such abbreviations for wffs. For example:

Definition 3. $(x \neq y)$ for $(\sim (x = y))$ where x and y are any terms.

Definition 4. $(x < y)$ for $(Ez)(z \neq 0 \wedge y = x + z)$, where z is the first new variable not occurring in the terms x and y. (By "first" new variable we mean the one with the lowest subscript.)

Such verbal statements of conditions on the variables and terms are necessary in order for us to know, in any given situation, how to replace a defined expression by a unique, purely formal one in the original notation. We have stipulated that the bound variable z be the first new variable which satisfies the other given conditions so that, in every case, a given expression will be the abbreviation of only one wff of the language. However, in view of our rule of change of name of bound variables, any new variable would really do, since any wff which differs from the one we are abbreviating only in having a different variable z will be equivalent to the original one. In view of this, we shall not bother in the future to pin down the order of introduction of bound variables in abbreviated wffs (there could be several such bound variables), but will state only the other relevant conditions. The reader may take it as an exercise to supply some agreeable order of introduction of such bound variables if he desires to do so.

The set of axioms of **S** is not independent. In particular, the axiom **S**.5 is nonindependent. We can prove **S**.5 from our other axioms by using in particular **S**.4, the recursive definitions of arithmetic and multiplication, and the principle **S**.12 of mathematical induction.

We have called the natural numbers the standard model for **S**. Of course, we have not rigorously proved that N really is a model according to our abstract, precise definition of this notion. Rather, we have taken it as intuitively obvious. In either case, it is certainly not wholly unreasonable to wonder if there might not be other models of **S** different from N. However, we must ask in what sense we mean "different from" N. We know that there are many different equivalent ways of describing N. They are equivalent in that there is an obvious likeness of structure common to all structures satisfying different descriptions. We need some way of making all of this precise.

Given a first-order theory F with equality and two normal models $\langle D, g \rangle$ and $\langle D^*, g^* \rangle$ of F, we say that the two models are *isomorphic* if there is a bijective (1–1 onto) mapping h from D to D' such that the following holds: If $A(y_1, \ldots, y_n)$ is a wff of F whose free variables are the variables y_i then the n-tuple $\langle d_1, \ldots, d_n \rangle$ of elements of D is in the relation X_n expressed by $A(y_1, \ldots, y_n)$ in $\langle D, g \rangle$ if and only if the n-tuple $\langle h(d_1), \ldots, h(d_n) \rangle$ is in the relation X_n^* expressed by $A(y_1, \ldots, y_n)$ in $\langle D^*, g^* \rangle$. Clearly, isomorphic models are elementarily equivalent. It is also clear that the converse does not hold; isomorphism is strictly stronger than elementary equivalence.

We can thus ask whether there are models of **S** which are not isomorphic to N. The answer, somewhat surprising at first, is that there are. In fact, there are at least $2^{\aleph_0}$ of them which are not elementarily equivalent to each other, as we show in Chapter 6.

A first-order system F is said to be *categorical* if any two normal models of F are isomorphic. Two isomorphic models of a system must be of the same cardinality, since there is a bijection between them. This is why we require normality in our definition of a categorical first-order system. Without this restriction, there would be no categorical systems, since any consistent system has non-normal models of arbitrarily high cardinality as we have already seen.

Even with our restriction to normal models, it is necessary that a system have only finite models in order to be categorical. This can be seen as true by recalling Tarski's result that any system with an infinite normal model has normal models in all infinite domains. We therefore define the notion of *categoricity in power*: Given a cardinal number α we say that a system F is α-categorical if any two normal models of cardinality α are isomorphic. There are systems categorical in some infinite cardinalities. The first-order theory of dense, linear order without endpoints is $\aleph_0$-categorical, all denumerable models being isomorphic to the ordering of the rational numbers. In fact, there are theories α-categorical for every infinite α (cf. the discussion of these questions in Mendelson [1], for example).

Exercise 1. Prove that the system **U** of Section 1.7, whose only normal models are in a one-element set, is categorical.

Exercise 2. Prove that **S** has no finite normal models.

Chapter 2

The Origin of Modern Foundational Studies

2.1. Mathematics as an independent science

It is well known that mathematicians hold and have held different views concerning the nature of mathematics. Some mathematicians tend to look toward the physical sciences, and to physics in particular, as the ultimate source of mathematical problems and ideas. Others regard mathematical intuition as something concerned primarily with the abstract structure of mathematical objects themselves (whatever these are considered to be) and independent of the other sciences. Some mathematicians have a strong sense of the "reality" of abstract mathematical objects, such as the set of natural numbers or the set of real numbers, whereas others do not have such feelings. Probably all these various feelings and viewpoints were present in mathematics from the beginning. For example, Greek mathematics saw the development of axiomatic geometry and the rudiments of formal logic, yet the Greeks did not even invent a number system, but worked rather with lengths of line segments and their ratios. They clearly recognized points and lines as abstract, nonphysical entities and yet they certainly knew that their geometry applied to practical problems of space measurement. The differential calculus was invented simultaneously by Newton, obviously motivated by a strong sense of physical reality, and Leibniz, who was much more oriented toward logic and formal mathematics.

In any case, there are several fundamental points on which most mathematicians would agree regardless of their personal philosophic convictions concerning the nature of mathematics. The first is that mathematics is abstract, and that it consists primarily of reasoning with and contemplating abstractions. The second is that the truth or falsity of a proposition in mathematics is determined by a process of deduction, of showing that the proposition can be proved on the basis of some given first principles or assumed truths (the reason for assuming one thing as true rather than another is not in question here). This process seems to differ from other sciences in at least one respect: every other science, even one as abstract as physics, ultimately depends upon a certain amount of manipulation of the physical world. One tests hypotheses and laws empirically by "seeing" if they work. But in mathematics, one can prove theorems, thus establishing mathematical truth, without any manipulation of the physical world at all. Even the trivial type of manipulation involved in writing down symbols and looking at them can be ignored by good mathematicians.

This difference between mathematics and other sciences is important for the question of the foundations of science. However abstract the thought of a theoretical physicist may be, he can appeal to the external physical world as the ultimate basis upon which his science rests. After all, light exists even if our discourse about it (or our model for it) is paradoxical, or even contradictory. But there is no corresponding physical reality to serve as an ultimate basis for mathematics. One can reason by analogy: if the physical sciences rest on the base of physical reality, upon what does mathematics rest?

This type of independence of mathematics from the other sciences has been dramatized from time to time by various developments in mathematics. One of the most famous (though its significance is perhaps exaggerated) was the discovery of the non-Euclidean geometries in the eighteenth and nineteenth centuries. Of course, if one views Euclid's geometry as an abstract axiomatic system, there is nothing surprising about the discovery that one of the axioms is independent of the others, and that consequently there are models for the remaining axioms for which the negation of the independent axiom holds. The shock produced at the time by the non-Euclidean geometries is *prima facie* evidence that Euclid's system was not regarded in this manner. Rather it was considered as a catalogue of truths about space, and so the possibility of another geometry seemed to imply the possibility of (or even the necessity for) another space, i.e. another physical reality. Attempts were made by Gauss and others to decide which of the geometries was the "right" one for space. But it was clearly impossible, in any case, for all the geometries to be true of the same physical reality in the same way. Thus, the discovery of the non-Euclidean geometries served to popularize the notion that a mathematical system could have an intrinsic integrity independent of any notion of physical reality whatsoever, namely by being self-consistent.

The widespread acceptance of such a view of the independence of mathematical systems from the physical world was undoubtedly linked to a newly emerging scientific philosophy in the nineteenth and twentieth centuries. The older absolutist point of view began to disappear from physics itself. The realization grew that "space" and "empirical reality" were themselves human constructs, and that man is much further from the raw stuff of external reality than he imagines. The twentieth century has witnessed the extension of these philosophic and scientific attitudes, and the result has been a much more prudent philosophy of science in all quarters.

One might say, then, that the first ingredient of modern foundational studies is the view of mathematics as a science independent of physical reality, a science whose objects of study are abstract systems capable of being self-consistent without necessarily being true of any particular reality (though we must remember the completeness theorem of first-order logic, which tells us that they will be true of some structure albeit a nonphysical one).

2.2. The arithmetization of analysis

A second ingredient in modern foundational studies is the development of analysis and set theory which took place in the nineteenth and twentieth centuries. This development had the effect of separating the purely arithmetical (or algebraic) aspects of number from the geometric aspects. Algebra had been developed not by the Greeks but by the Hindu–Arabic civilizations of the East. It was Descartes who "invented" analysis by fusing geometry and algebra into a single discipline, analytic geometry. This enabled mathematicians to "see" functions by looking at their graphs. The real number line was regarded as a continuum that was both geometric and algebraic, and one proved facts about functions by looking at and talking about their graphs. A function was never wholly separated from the curve that was its geometric counterpart. Well into the nineteenth (and, indeed, even into the twentieth) century we find mathematics authors using quasigeometric proofs of analytic theorems.

The arithmetization of analysis, brought about by Dedekind, Weierstrass, and others, succeeded in developing an algebraically self-contained notion of real number without any appeal to geometric intuition. The definition of the real numbers proceeded from the rational numbers by considering real numbers as certain infinite sets of rationals. The rationals were, in turn, definable as ratios of integers and the integers were easily constructed from the natural numbers. Beside avoiding appeal to geometric intuition in proofs of theorems in analysis, Dedekind's formulations showed that it might be possible to construct all mathematics from a few broad, basic principles. This feeling was especially prevalent when Peano and Dedekind succeeded in giving an abstract form to the theory of natural numbers. Also notable was the contribution of Cauchy, who gave precise definitions of the notions of limit and convergence. The notions had given trouble to mathematicians ever since the invention of the calculus. Cauchy showed how appeals to infinitesimals and to geometric intuition could be avoided in the treatment of the notion of convergence.

Subsequent to the arithmetization of analysis was the generalization of geometry to topology, which, itself, turned into a self-contained, pure discipline. Fréchet's fusion of the two in the notion of a metric space allowed for an exact separation of the geometric and analytical content in theorems of analysis.

It was soon clear, however, that the foundation to the edifice of analysis built by these new constructions involved much more, for instance, than the natural numbers It was rather the natural numbers plus a large amount of set-theoretic reasoning. One had substituted appeal to set-theoretic intuition for appeal to geometric intuition. Many mathematicians expressed apprehension concerning the overwhelming generality of set-theoretic reasoning, and in particular the way in which infinite sets were handled. This apprehension was greatly increased with the appearance, at the turn of the century, of a series of logical contradictions (known inexplicably as "paradoxes") in set-theoretic reasoning. These contradictions necessitated a much closer look at the basis upon which the edifice of analysis rested.

The intuitive set theory which generated the paradoxes had been initiated and developed by Cantor. His main contribution was the general theory of infinite sets, and Cantor's notions were essential to the Dedekind–Weierstrass construction of the real numbers. Cantor's theory of infinite sets served to separate mathematics even further from other sciences, for here were clearly abstractions to which there could be no corresponding physical reality. There is no indication whatever that there are infinite collections of physical objects. Even such sets as the set of all atoms in the universe are considered extremely large but finite. When the real numbers were thought of as a geometric continuum, one could still appeal to some sort of physical model as a basis for analysis. Time, for example, was often considered a physical model for the real line. Whatever value, positive or negative, the reader may wish to attach to such physical analogies, the point here is that with infinite set theory they are no longer possible.

The difficulty in visualizing and intuiting infinite sets gave rise to attempts to axiomatize the operations of set theory. In this way, one could at least be explicit about what operations were allowable even if one could not always form a clear mental picture about what was going on. Zermelo's 1908 paper was a milestone in this direction.

2.3. Constructivism

Not only because of the paradoxes of set theory, but also because of their disbelief in the existence of infinite sets, a number of mathematicians took (and continue to take) a constructive approach to analysis. Brouwer, Poincaré, Kronecker, and Weyl are a few of the mathematicians who took some kind of constructivist position. The theory of the natural numbers can be developed without considering them to be a completed set or collection to which other set-theoretic operations (such as power-set; that is, forming the set of all subsets) may be applied. One can conceive of a never-ending hierarchy of numbers $0, 1, \ldots, n, \ldots$ where each has a unique successor and where each is obtained from a unique predecessor, except for the initial starting point 0. The usual number-theoretic functions of addition, multiplication, exponentiation, and the like can be defined recursively and their properties studied without ever supposing that there is some set N that gathers together this never-ending hierarchy into one collection. The integers can be easily constructed from the natural numbers, and the rational numbers constructed as ratios of integers. All this can be done in a constructive manner, without any appeal to infinite sets. It is also clear that certain irrationals can be effectively constructed. This is the case whenever we can exhibit some constructively defined sequence of rationals that converges to an irrational number. Thus, $\sqrt{2}$, $\sqrt{3}$, e, and π would all be constructive irrationals. (This notion must not be confused with the *constructible* numbers, in the sense classically studied since ancient times, corresponding to those lengths constructible by ruler and compass. Constructibility in this sense is narrower than the constructivity we are describing here.) In short, a constructive theory of the real numbers is what we do whenever we are involved with computational mathematics as opposed to mathematics involving infinite sets.

Constructivists view discussion about infinite sets as a fiction. They feel that it may help some mathematicians by enabling them to think of the real line as completed, but will have a practical effect only in that it may help them discover truths about the constructive reals.

Recent work on constructive analysis shows that a certain amount of classical analysis can be done in a constructive way. There are constructive analogues to many classical theorems.

Constructivism represents a philosophical position concerning the foundations of mathematics, but it is also of interest to nonconstructivists to know just how much of modern mathematics can be performed constructively. In Chapter 6, we consider a more precise definition of the notion of a constructive or effective process.

Relevant to any discussion of constructivism and the foundations of mathematics is the program of Hilbert. Hilbert was not a constructivist, but he granted that there were difficulties involved in dealing with infinite sets. His idea was to formulate a system in which the notions of infinite set and of operations on infinite sets were expressible, and to prove by purely constructive means that the system was noncontradictory. In this way, the use of infinite mathematics would be justified in a constructive way, since we would have shown constructively that no contradiction could be forthcoming in infinite mathematics. Hilbert's program was cleverly conceived, but it foundered on unforeseen difficulties, which are the subject of Chapter 6 of this book.

2.4. Frege and the notion of a formal system

In Chapter 1, we discussed formal logic and first-order theories from a modern point of view, but we did not discuss their historical origins. Leibniz and others had often contemplated the idea of developing a special language for mathematics, a uniquely mathematical and logical language. Attempts in this direction were largely unsuccessful until the nineteenth century. (Of course algebra with its use of variables and special symbols was already a significant step in this direction. Even a simple algebraic equation can become hopelessly complicated if it is stated only in the vernacular language.) It was in the nineteenth century that Boole, DeMorgan, Schröder, Peirce, Peano and others began working on the problem with renewed interest, and with more success. By far the most successful of these innovators was G. Frege who was perhaps the first logician to develop a formal language and formal logic in the modern sense. Frege's *Begriffschrift*, published in 1879, was barely recognized at the time, and the full force of Frege's work has been appreciated only in recent years. Yet in this, and in Frege's later works, the framework of modern formal logic is clearly elaborated.

Attempting to fit the development of formal logic into the general scheme we have sketched in this chapter, one might view things as follows: The concentration of mathematicians involved in the arithmetization of analysis was on content. They were concerned almost exclusively with their understanding of mathematical objects. Less thought was given about the logic being used to prove things about those objects. No attempt was made, for example, to say what a proof *was*. Of course proofs were certainly given for theorems that were asserted, but there was no analysis of the notion of proof itself.

One consequence of this deficiency was that the notion of validity, deriving from the axiomatic approach to mathematics, could hardly receive adequate treatment, since no precise notion of "proof" or "logic" was at hand. The difficulties involved were increased by the obvious inadequacy for mathematics of scholastic logic, the only existing formal logic, excluding Boole's publications which began in 1847. However, Boole's systems were not really formal, and they were comparable in power to the scholastic logic, though differing in form from them.

Frege's contribution was two-fold. First, he originated and developed the logic of relations (which was independently discovered by DeMorgan, Schröder, Peirce, and, in various degrees of clarity, by other logicians of the time). This new logic proved to be an essential step forward in formal logic and constituted the basic difference between modern logic and the scholastic Aristotelian logic.

Secondly, Frege introduced the basic notion of a formal system or "calculus of signs". This fundamental advance consisted in considering a mathematically defined algebraic system of a certain kind as an abstract representation of the notion of a language. Such a formal language differs from an intuitively given natural language in that the grammar of the former is completely described and defined in precise terms. In this way, statements about the language are no longer subject to the ambiguity inherent in a grammatical discussion of a natural language such as English.

Frege's version of a foundation for mathematics may now be roughly described in this way: Because mathematics is independent of physical reality, its truths must also be independent of this reality. Such truths are true simply by virtue of the way in which we use words. These truths are universal and prior to all experience as they do not depend on any "contingent" realities. Let us call such truths (if, indeed, there are any) *analytic*. One then attempts to give a foundation to mathematics by showing that all truths of mathematics are analytic. This is to be done by deriving mathematics from a set of basic principles which are themselves universal logical principles (i.e. analytic truths). Moreover, the derivation of mathematics from these principles is to be expressed in a formal language in which the logical rules of derivation will be given explicitly. These rules are clear statements of universal principles of logical inference. A *proof*, in such a formal language, will be some finite sequence of symbolic expressions of the language such that each member of the sequence either represents one of our originally given universal principles or follows from prior members of the sequence by one of our explicitly stated rules of logical inference. A *theorem*, thus a truth of mathematics, will be represented by any expression obtainable as the last line of a formal proof. (There is a bit of an anachronism in this description because Frege actually used a complex two-dimensional notation for proofs, but we will not enter into these considerations here. The essentials of our sketch of Frege's program do not falsify the original.)

This program is breath-taking in its conception and Frege's published works contain elaborate constructions which tend to convince one of the credibility of the approach. However, Frege's own version of a formal system for mathematics was proved to contain inconsistencies by one of the newly discovered contradictions in set-theoretic reasoning. Since the particular contradiction, the so-called Russell paradox, derived directly from one of the "universally valid" principles of logic, the modifications necessary to restore consistency were not easily forthcoming. A more detailed consideration of Frege's system will be the object of Chapter 3 of the present study.

A brief mention of certain philosophical questions is perhaps in order here. The notion of analytic truth, which Frege developed, had its origins in a related distinction found in Kant's *Critique of Pure Reason*. For Kant, contingent truths were called "synthetic", and logical truths "analytic". Unlike Frege, Kant considered that the truths of arithmetic, for instance, were synthetic *a priori*, that is, prior to experience but still contingent on the real world. However, Kant wrote before the separation of mathematics from its dependency on physics was generally accepted, and the logic of Kant's time was still scholastic Aristotelian logic. Consequently, Kant's notion of the analytic was much narrower than Frege's. One can only speculate as to what position Kant would have taken with respect to the question in the light of Frege's more powerful logic of relations.

In any case, the philosophical attitude often taken as a corollary to Frege's foundational approach is called *logicism*. Logicism is generally defined as the philosophical thesis that the truths of mathematics are all analytic and thus derivable from pure logic, rather than being synthetic as Kant had asserted. The main concern of the present study is on the technical and mathematical questions of foundations, rather than the various philosophical positions that are possibly related to these questions. Nevertheless, an occasional philosophic interlude may occur.

2.5. Criteria for foundations

Now that the basic ingredients of modern foundational approaches are before us, in however incomplete a form, let us state in a general way some criteria that seem necessary for any kind of foundation for mathematics. What, in short, must a foundation be, and what must it do?

(1) A foundation of mathematics must be adequate for a reasonably large portion of mathematics.

By the well-known incompleteness theorem of Gödel, which will be discussed in Chapter 6, there is not too much hope for a foundation which is adequate for mathematics in its entirety and which also satisfies some of the other criteria to be examined (in particular, the criterion of consistency). There will be some true statements unobtainable by our purely formal process of reasoning. However, some truths of mathematics are more important than others, and so concern for "lost" portions can be minimized.

(2) A foundation must derive from some intuitively natural principles.

When an axiomatic system is essentially just one among many possible ways of cataloguing a set of truths, the old notion of "self-evidence" can be discarded. Much has been made of this fact. But when foundations are concerned, the principles upon which a system rests (as distinguished from some particular axiomatization of it) must have some kind of intuitive credibility.

(3) The basic principles and primitive (undefined) notions should be as economical as possible (the "possible" being criterion 1 of this discussion).

Here we are not so much concerned that the system be used or articulated in its most economical form, since an increase in primitive notions tends to result in a practical flexibility. But theoretic economy is quite desirable, especially when it comes to comparing one foundation to another and to understanding the over-all structure of the foundation. Perfect examples are afforded by the propositional and predicate calculi of Chapter 1. It is of considerable theoretic importance that the existential quantifier be definable by means of negation and the universal quantifier, and that all the propositional sentential connectives be definable by means of one binary connective. Yet, for practical purposes, it is far better to multiply the basic signs used than to engage in laborious translations into a spare set of primitive notions.

(4) The foundation must be consistent.

If it is not consistent, then the laws of logic will yield every statement as a theorem and so the system is utterly useless.

(5) The foundation should be expressed (or expressible) as a formal system.

It is, of course, quite possible that a legitimate foundation for mathematics could be given which would not be expressible as a formal system. Indeed, the Gödel results concerning the incompleteness of formal systems have sometimes been taken to mean just that. Yet experience has shown that without criterion 5 discussions about foundations easily degenerate into quasi-philosophic and obscure debates in which no one really knows quite what the issue is. The discipline of formalizing a foundation forces one to be explicit about the principles involved. It allows for exact comparison between different approaches and often forces the discovery of unforeseen difficulties and unforeseen advantages. The consequences of even slight changes in the system can be exhaustively examined.

An excellent case in point is intuitionistic arithmetic (more will be said on intuitionism in Chapter 3). Brouwer attacked formalization and resisted attempts by others to formalize his own work. When, in spite of this, intuitionistic arithmetic was formalized, Gödel proved that the arithmetical truths provable by classical means were the same as those provable within the seemingly more restricted intuitionistic framework. Thus, on the level of arithmetic at least, the difference between intuitionistic and classical approaches is mainly philosophic. Such a clear result would never have been possible without first formalizing intuitionistic arithmetic.

(6) The construction of everyday mathematics in the system should be "natural" and "orderly".

If too many well-known principles are deduced by intuitively unnatural and grotesque means, then one loses confidence in the whole process and feels that mathematics itself represents a better basis for mathematics.

The last remark needs further comment. Since mathematics is obviously adequate for mathematics, is there any real need to choose another basis? Do we really need a foundation other than mathematics itself (whatever, indeed, mathematics itself really is)? Some mathematicians might make just such a contention, but there are strong arguments against this point of view. The clearly interrelated character of a host of mathematical notions leads one time and again toward a simplification and economy by reducing certain notions to others. Many fundamental advances in mathematics have been made by seeing that two different theories were really "the same all the time". If it can be shown that all notions follow from a few basic ones by means of some sort of reasonable process of definition and deduction, then one feels almost duty-bound to describe such a basis.

By such a process of reduction, we naturally arrive at some kind of foundation that is not the same as swallowing mathematics whole. Almost every creative mathematician goes through some similar kind of reduction process in his own thinking about mathematics. It is only when an attempt is made to systematize and render explicit this process that one can really see what is going on.

These general considerations tend to discourage any dogmatic defense of a particular system as being the only system or the best one. Most modern students of foundations do not seem to be inclined to such dogmatism. In recent years, foundational studies have come to mean primarily a comparative study of various systems and of different variants of existing systems. This is the attitude we have adopted in the present study.

Chapter 3

Frege's System and the Paradoxes

BEGINNING in 1879, G. Frege [1] and [2] elaborated a formal system intended to serve as a foundation for mathematics. We shall present Frege's system in a form quite close to the original except for extensive notational simplification.

3.1. The intuitive basis of Frege's system

The basic ideas underlying Frege's system have to do with sets and properties. If we think of a set as a collection of objects, then we can see that there are basically two ways of describing sets: (1) We can exhibit the set by giving a list of the objects we wish to regard as members of the set or (2) we can give some property that is a necessary and sufficient condition for membership in the set.

For finite sets, both methods may be employed. Thus $\{1, 2, 3, 4\}$ = the set of all integers between 0 and 5. For infinite sets, only the second method will do, since an infinite list is clearly impossible. Moreover, any finite set can be determined by some property. The set $A = \{a_1, \ldots, a_n\}$ is determined by the property $x_1 = a_1 \lor x_1 = a_1 \lor \ldots \lor x_1 = a_n$. These considerations lead us to examine closely the exact nature of the logical relationship between sets and properties.

First, we must decide what conditions we wish to regard as determining the equality of sets. The usually accepted, intuitively correct assumption is that two sets are to be considered equal when and only when they have the same elements. Though intuitive, this assumption is not trivial. Observe that, under this assumption, two equal sets can be determined by different properties. The set of all irrational integers and the set of all centaurs are the same. They are both the null set, the set with no elements. The property "to be a French city of 2,000,000 or more inhabitants" and the property "to be a capital of the French republic" are different properties both determining $\{Paris\}$. Two sets with the same elements are said to have the same "extension" or "to be coextensive".

If we accept the relation of sameness of elements as determining set equality, then we must insist that this relation between sets satisfy all the requirements of the equality relation. Reflexivity obviously holds, since every set has the same elements as itself. The other basic requirement of equality is that two equal sets must share all their properties; i.e. $x = y \supset P(x) \equiv P(y)$ where x and y are any sets and P is any property. This last requirement is just the substitutivity of identity already discussed in connection with first-order theories. Unlike reflexivity, this second characteristic of identity cannot be directly deduced from the relation of coextensiveness. Hence, if we desire coextension to characterize set identity, we must posit as an explicit hypothesis or axiom of set theory that any two coextensive sets share all their properties (and therefore are equal). This principle is known as the *principle of extensionality* or the *axiom of extensionality*.

The principle of extensionality says in effect that two sets are identical if and only if they have the same elements. This principle may seem reasonable, but it is possible to conceive of a coherent set theory in which it does not hold. We shall not have occasion in this book to consider such set theories (though in some systems considered in Chapter 4 the principle holds in a limited way). We shall generally adopt the principle of extensionality as one of the basic principles of set theory.

Once the principle of extensionality is agreed upon, a second question must be examined: does every property determine a set? Informally, properties are thought of as conditions stateable in some language. In our formal language of first-order theories, properties are represented by wffs with one or more free variables. When we ask whether every property determines a set, we mean to ask whether there exists, for any given property P, the set of precisely those objects (and no others) which satisfy the condition P.

Again it seems intuitively natural that every property should determine a set in the indicated manner. For example, the property "to be red" or "x is red" determines the set of all red things. The thesis that every property or condition does determine a set is known as the *principle of abstraction* (the set is "abstracted" from the property). As with the principle of extensionality, the principle of abstraction seems plausible. In fact, the principle is certainly used constantly in informal set theory. This is so much the case, that properties are identified with sets, and relations with sets of n-tuples, as we have done ourselves in Chapter 1. To Frege (and others of his time) the intuitive plausibility of the principle was fairly great, and Frege made it an axiom of his system, where the conditions were not any possible conditions but those stateable in Frege's formal language of set theory. When Frege's principle is formally stated, we have:

$$(Ey)\,(x)\,(x \in y \equiv A(x)),$$

where $A(x)$ is some wff of our language (to be defined) that may contain x free, and y is different from x and does not occur in $A(x)$. This asserts that "there is a y such that for all x, x is in y if and only if x satisfies the condition $A(x)$". Notice that, by the principle of extensionality, the set y whose existence is affirmed is unique for a given condition $A(x)$.

Exercise. Prove the last statement; in which all we need to assume about our language is that "$\in$" is a predicate letter of it and that extensionality and abstraction are axioms of it.

At this point, the reader may begin to feel that the principles we are examining are rather far removed from mathematical questions. What do they have to do with furnishing a foundation for mathematics? It was Frege (and others such as Dedekind) who showed that the relevance of these simple principles for constructing a foundation for mathematics is considerable. They showed, by a series of ingenious constructions, that one can recover ordinary mathematics from nothing more than first-order logic plus the two principles of extensionality and abstraction. That is, if we consider a first-order system with the "$\in$" of membership as the only primitive predicate letter (a binary one), and with the axioms of extensionality and abstraction as the only proper axioms, then it is possible, by means of certain definitions and constructions within the system, to define such sets as "Natural Numbers" and "Real Numbers", and to reproduce formally within the system the usual proofs of the known theorems about these sets and their members. Not only is it not immediately obvious that this can be done, it is surprising that such subtle mathematical notions can even be expressed in such a relatively simple language.

One can appreciate why Frege was so excited about his discoveries. Holding that the two principles of abstraction and extensionality were universally valid logical principles, Frege felt he had shown conclusively that the truths of mathematics were all universally valid logical principles.[†] We have already seen that the rules of first-order logic clearly preserve universal validity.

Let us now formalize Frege's system and carry out some of his constructions in order to understand how the reconstruction of mathematics is effected.

3.2. Frege's system

Our language $\mathbf{F}$ will contain only one binary predicate letter, which we will write as "$\in$". Instead of writing "$\in (x, y)$" we write "$(x \in y)$", writing the arguments on either side of the predicate letter. Obviously, this is useful only for binary predicate letters. We might allow no terms other than variables, and let our wffs be simply those expressions obtained by iterating quantification and sentential connectives starting with wffs of the form $(x \in y)$ as prime formulas, x and y being variables. However, we introduce a term operator of abstraction, a *vbto*.

We thus define the wffs of $\mathbf{F}$ as follows: (1) Every variable is a term. (2) If x and y are terms, then $(x \in y)$ is a wff. (3) If A is a wff, and x is a variable, then $(x)A$ and $(Ex)A$ are wffs and $\{x \mid A\}$ is a term. The variable x is, of course, bound in the term so formed. (4) If A and B are wffs, then $(\sim A)$ and $(A \vee B)$ are wffs. (5) The wffs and terms are precisely the expressions that can be obtained by application of the first four rules.[‡]

We think of $\{x \mid A(x)\}$ as standing for "the set of all x such that $A(x)$ is true". The axioms of abstraction and extensionality will be axioms of $\mathbf{F}$, and so it will immediately follow that the two schemes V.1 and V.2 of Section 1.8 are provable in $\mathbf{F}$ (verify this). We could avoid using the *vbto* of abstraction by formulating the principle of abstraction as we did in our previous discussion (by using the existential quantifier). The main disadvantage in disregarding the term operator is that we must continually use the rule *eE* in applying the principle of abstraction, and this can be technically cumbersome. It is for purely practical convenience that we introduce the abstraction operator, for the two ways of formulating $\mathbf{F}$ are essentially equivalent.

We now begin the development of our system by stating the proper axioms.

Definition 1. $(x = y)$ for $(z)(z \in x \equiv z \in y)$ where the variable z does not occur in either of the terms x or y.

[†] As we shall later discover, Frege had cause to doubt the universal validity of the principle of abstraction, and in one passage he states that he was never as sure of the principle as he had been of the other laws of logic.

[‡] We suppose, as always, the usual definitions of our other sentential connectives.

Let us recall here our convention, established in Chapter 1, Section 1.9, concerning bound variables in definitions. Technically, we should specify that z is the first variable in numerical order of subscript that does not occur in either of the terms x or y. We shall, however, not bother to add this type of specification in the statement of definitions, and we shall leave to the reader the task of doing so if he chooses.

F.1 $(x)(y)(x = y \supset A(x, x) \equiv A(x, y))$ where $A(x, y)$ is obtained from $A(x, x)$ by replacing y for zero, one, or more occurrences of x in the wff $A(x, x)$ and y is free for x in all occurrences of x which it replaces.

This is the axiom of extensionality whose intuitive meaning we have already discussed.

F.2 $(x)(x \in \{y \mid A(y)\} \equiv A(x))$, where x is free for y in $A(y)$, and $A(x)$ results from $A(y)$ upon replacing y by x in all the free occurrences of y in $A(y)$.

This is Frege's axiom of abstraction.

Both **F.1** and **F.2** are axiom schemes which give us an infinite set of axioms, one for every instance of a scheme.

We will often write "$x \neq y$" for "$\sim (x = y)$" and "$x \notin y$" for "$\sim (x \in y)$", where x and y are any terms. We now examine some elementary consequences of our axioms.

THEOREM 1. $\vdash (x_1)(x_1 = x_1)$.

Proof.   1. $x_2 \in x_1 \equiv x_2 \in x_1$ Taut
   2. $(x_2)(x_2 \in x_1 \equiv x_2 \in x_1)$ 1, $i\forall$
   3. Concl   2, $Df.$ 1, $i\forall$

Everything is equal to itself.

Definition 2. V for $\{x_1 \mid x_1 = x_1\}$.

THEOREM 2. $\vdash (x_1)(x_1 \in V)$.

Proof.   1. $(x_1)(x_1 = x_1)$ Th. 1
   2. $x_1 = x_1$ 1, $e\forall$
   3. $x_1 \in \{x_1 \mid x_1 = x_1\}$ 2, **F.2**, $e\forall$, Taut, MP
   4. $x_1 \in V$ 3, $Df.$ 2
   5. $(x_1)(x_1 \in V)$ 4, $i\forall$

V is the *universal set* containing everything. In particular, V contains itself as an element.

COROLLARY. $\vdash V \in V$.

Proof. Immediate from Theorem 2 by $e\forall$.

Definition 3. Λ for $\{x_1 \mid x_1 \neq x_1\}$.

Λ is the *null* or *empty* set, the set with no elements. We have:

THEOREM 3. $\vdash (x_1)(x_1 \notin \Lambda)$.

Proof. The reader will prove Theorem 3 as an exercise.

Notice that, because of our axiom of extensionality, the null set is the only no-element thing. There are no "individuals" in our system in the sense of different no-element things. In other words, we have:

THEOREM 4. $\vdash (x_2)((x_1)(x_1 \notin x_2) \equiv x_2 = \Lambda)$.

Proof. (1) 1. $x_2 = \Lambda$ H

  2. $x_2 = \Lambda \supset ((x_1)(x_1 \notin x_2) \equiv (x_1)(x_1 \notin \Lambda))$ **F.**1, $e\forall$

(1) 3. $(x_1)(x_1 \notin x_2) \equiv (x_1)(x_1 \notin \Lambda)$ 1, 2, MP

(1) 4. $(x_1)(x_1 \notin x_2)$ Th. 3, 3, Taut, MP

  5. $x_2 = \Lambda \supset (x_1)(x_1 \notin x_2)$ 1, 4, eH

(6) 6. $(x_1)(x_1 \notin x_2)$ H

(6) 7. $x_1 \notin x_2$ 6, $e\forall$

  8. $x_1 \notin \Lambda$ Th. 3, $e\forall$

  9. $x_1 \notin \Lambda \supset (x_1 \notin x_2 \supset x_1 \notin \Lambda)$ Taut

  10. $x_1 \notin x_2 \supset x_1 \notin \Lambda$ 8, 9, MP

(6) 11. $x_1 \notin \Lambda \supset x_1 \notin x_2$ 7, Taut, MP

(6) 12. $x_1 \notin x_2 \equiv x_1 \notin \Lambda$ 10, 11, Taut, MP

(6) 13. $x_1 \in x_2 \equiv x_1 \in \Lambda$ 12, Taut, MP

(6) 14. $(x_1)(x_1 \in x_2 \equiv x_1 \in \Lambda)$ 13, $i\forall$

(6) 15. $x_2 = \Lambda$ 14, *Df.* 1

  16. $(x_1)(x_1 \notin x_2) \supset x_2 = \Lambda$ 6, 15, eH

  17. $(x_1)(x_1 \notin x_2) \equiv x_2 = \Lambda$ 5, 16, Taut, MP

  18. Concl   17, $i\forall$

Definition 4. $\{x\}$ for $\{y \mid y = x\}$, where x is any term and y is any variable not occurring in x.

Definition 5. $\bar{x}$ for $\{y \mid y \notin x\}$ where x is any term and y is any variable not occurring in x.

$\{x\}$ is the set whose only element is x, and $\bar{x}$ is the *complement* of x, the set of objects not in x.

Definition 6. $(x \cap y)$ for $\{z \mid z \in x \wedge z \in y\}$, where x and y are any terms and z occurs in neither.

Definition 7. $(x \cup y)$ for $\{z \mid z \in x \vee z \in y\}$, where x and y are any terms and z occurs in neither.

These the reader will recognize as definitions for the notions of the *intersection* and *union* of two sets. Where there is no cause for ambiguity, we shall often omit parentheses in terms such as the term of Definition 7.

Exercise. Prove

$$\vdash(x\cap y)\cap z = x\cap(y\cap z); \quad \vdash(x\cap y) = (y\cap x); \quad \vdash(x\cup y) = (y\cup x);$$
$$\vdash(x\cup y)\cup z = x\cup(y\cup z); \quad \vdash\overline{(x\cap y)} = (\bar{x}\cup\bar{y}); \quad \vdash\overline{(x\cup y)} = (\bar{x}\cap\bar{y});$$
$$\vdash(x\cap(y\cup z)) = (x\cap y)\cup(x\cap z); \quad \vdash(x\cup(y\cap z)) = (x\cup y)\cap(x\cup z).$$

These are the usual laws governing the operations of union, intersection, and complementation. They follow directly from **F.2** and our definitions and we shall feel free to use them in the future. Another useful property is:

THEOREM 5. $\vdash(x_4)(x_3)(x_3\in x_4 \supset ((x_4\cap\overline{\{x_3\}})\cup\{x_3\}) = x_4)$

Proof. (1)   1. $x_3\in x_4$ H

(2)   2. $x_2\in((x_4\cap\overline{\{x_3\}})\cup\{x_3\})$ H

(2)   3. $x_2\in(x_4\cap\overline{\{x_3\}})\vee x_2\in\{x_3\}$ 2, Df. 7, **F.2**, $e\forall$, MP

(2)   4. $(x_2\in x_4\wedge x_2\neq x_3)\vee x_2 = x_3$ 3, Df. 4, 5, 6, 7, **F.2**, $e\forall$, Taut, MP

(1)   5. $x_2 = x_3 \supset x_2\in x_4$ 1, **F.1**, $e\forall$, Taut, MP

      6. $(x_2\in x_4\wedge x_2\neq x_3) \supset x_2\in x_4$ Taut

(1)   7. $[4 \supset x_2\in x_4]$ 5, 6, Taut, MP

(1, 2)   8. $x_2\in x_4$ 4, 7, MP

(1)   9. $[2 \supset 8]$ 2, 8, eH

(10)   10. $x_2\in x_4$ H

     11. $x_2\in x_4 \supset (x_2\in x_4\wedge(x_2\neq x_3\vee x_2 = x_3))$ Taut

(10)   12. $x_2\in x_4\wedge(x_2\neq x_3\vee x_2 = x_3)$ 10, 11, MP

(10)   13. $(x_2\in x_4\wedge x_2\neq x_3)\vee(x_2\in x_4\wedge x_2 = x_3)$ 12, Taut, MP

(1)   14. $x_2 = x_3 \supset (x_2\in x_4\wedge x_2 = x_3)$ 5, Taut, MP

     15. $(x_2\in x_4\wedge x_2 = x_3) \supset x_2 = x_3$ Taut

(1)   16. $(x_2\in x_4\wedge x_2 = x_3) \equiv (x_2 = x_3)$ 14, 15, Taut, MP, Df. $\equiv$

(1, 10) 17. $(x_2\in x_4\wedge x_2\neq x_3)\vee(x_2 = x_3)$ 13, 16, Taut, MP

(1, 10) 18. $x_2\in((x_4\cap\overline{\{x_3\}})\cup\{x_3\})$ 17, Df. 4, 5, 6, 7, **F.2**, $e\forall$, Taut, MP

(1)   19. $[10 \supset 18]$ 10, 18, eH

(1)   20. $x_2\in((x_4\cap\overline{\{x_3\}})\cup\{x_3\}) \equiv x_2\in x_4$ 9, 19, Taut, MP

(1)   21. $(x_2)[20]$ 20, $i\forall$

     22. $[1 \supset 21]$ 1, 21, eH

     23. Concl   22, Df. 1, $i\forall$

This shows that the set we get by taking out an element and putting it back in is the same set. We now begin our development of number theory in **F**.

Definition 8. 0 for $\{\varLambda\}$.

Definition 9. $S(x)$ for $\{y\mid(Ez)(z\in y\wedge y\cap\overline{\{z\}}\in x)\}$, where x is any term in which the distinct variables y and z do not occur.

Definition 10. N for $\{x_1 \mid (x_2)(0 \in x_2 \wedge (x_3)(x_3 \in x_2 \supset S(x_3) \in x_2) \supset x_1 \in x_2)\}$.

We have here the definition of zero, the successor function, and the set of natural numbers (we shall consistently use this latter term for the set of nonnegative integers in the course of this work and we shall use the term "set of positive integers" if zero is to be excluded). Call a set *inductive* if it contains the successor of every element in it. Then N is the smallest inductive set containing 0.

It was Cantor (and also Dedekind) who defined the notion of cardinal similarity (i.e. sameness of number) of two sets as being the existence of a bijective (1–1 onto) correspondence between the two sets. The Cantor definition, because of its generality, allowed the extension of the counting process to infinite sets.

But the definition of cardinal similarity leaves unanswered the question of what cardinal numbers *are*. For example, we can recognize a set with five elements as being cardinally similar to any other five-element set. But what is five? Can we consider five as an object, and if so in what way?

The question of defining the cardinal numbers themselves has been handled in several basic ways. One basic way is due to von Neumann. His idea was to define the cardinal number of a given cardinal type as being some particular set with the given type. Five, for example, will be some particular five-element set. Von Neumann gave the details of the construction of typical sets for all cardinal numbers. We shall consider his method in more detail in a later chapter, but his version of the natural numbers is easy to describe: Let 0 be the null set and define the successor of any set x as the set $x \cup \{x\}$. Now, 1 is $\{0\}$, 2 is $\{0, \{0\}\}$, and so on. Each natural number is thus defined as some particular set with the appropriate number of elements.

The second method is due to Frege and is embodied in the formal definitions we have just considered. For Frege, five was not some particular five-element set, but rather the set of *all* such five-element sets. Thus, 0 is the set of all no-element sets, or $\{\varLambda\}$. Given a natural number n defined in this manner, its successor is precisely the set of all sets that belong to n (i.e. have n elements) when one element is removed from them. This is the version of successor expressed by Definition 9 of this section. Frege likewise defined cardinal number generally as the set of all sets with the given cardinal type.

The reader may accept as intuitively credible the definition of a cardinal number, and in particular a natural number, as being the set of all sets of a given cardinal type. But does this not neglect the order relation of the natural numbers? How will we recover the usual properties of addition, multiplication, well-ordering, and the like for the set N from Definition 10? Let us begin to answer this question by proving a few theorems about N.

THEOREM 6. $\vdash 0 \in N$.

Proof.   1. $0 \in x_1 \wedge (x_2)(x_2 \in x_1 \supset S(x_2) \in x_1) \supset 0 \in x_1$ Taut

2. $(x_1)[1]$ $i\forall$

3. $0 \in N$ 2, *Df.* 10, **F**.2, $e\forall$, Taut, MP

0 is a natural number.

THEOREM 7. $\vdash (x_1)(0 \neq S(x_1))$.

Proof. (1)   1. $0 = S(x_1)$ H

(1)   2. $(x_2)(x_2 \in 0 \equiv x_2 \in S(x_1))$ 1, *Df.* 1

(1)   3. $\Lambda \in 0 \equiv \Lambda \in S(x_1)$ 2, $e\forall$

4. $\Lambda \in 0$ *Df.* 4, *Df.* 8, **F.**2, $e\forall$, Taut, MP

(1)   5. $\Lambda \in S(x_1)$ 3, 4, Taut, MP

(1)   6. $(Ex_3)(x_3 \in \Lambda \wedge \Lambda \cap \overline{\{x_3\}} \in x_1)$ 5, *Df.* 9, **F.**2, $e\forall$, Taut, MP

(1)   7. $a_1 \in \Lambda \wedge \Lambda \cap \overline{\{a_1\}} \in x_1$ 6, eE

(1)   8. $a_1 \in \Lambda$ 7, Taut, MP

(1)   9. $(Ex_3)(x_3 \in \Lambda)$ 8, iE

10. $0 = S(x_1) \supset (Ex_3)(x_3 \in \Lambda)$ 1, 9 eH

11. $(x_3)(x_3 \notin \Lambda) \supset 0 \neq S(x_1)$ 10, Taut, MP

12. $(x_3)(x_3 \notin \Lambda)$ Th. 3

13. $0 \neq S(x_1)$ 11, 12, MP

14. $(x_1)(0 \neq S(x_1))$ 13, $i\forall$

Theorem 7 shows that 0 is not the successor of any set. In particular, then, it is not the successor of any natural number.

THEOREM 8. $\vdash (x_1)(x_1 \in N \supset S(x_1) \in N)$.

Proof. (1)   1. $x_1 \in N$ H

(1)   2. $(x_2)(0 \in x_2 \wedge (x_3)(x_3 \in x_2 \supset S(x_3) \in x_2) \supset x_1 \in x_2)$
              1, *Df.* 10, **F.**2, $e\forall$, Taut, MP

(3)   3. $0 \in x_2$ H

(4)   4. $(x_3)(x_3 \in x_2 \supset S(x_3) \in x_2)$ H

(1)   5. $(0 \in x_2 \wedge (x_3)(x_3 \in x_2 \supset S(x_3) \in x_2) \supset x_1 \in x_2)$ 2, $e\forall$

(1, 3, 4) 6. $x_1 \in x_2$ 3, 4, 5, Taut, MP

(4)   7. $x_1 \in x_2 \supset S(x_1) \in x_2$ 4, $e\forall$

(1, 3, 4) 8. $S(x_1) \in x_2$ 6, 7, MP

(1)   9. $0 \in x_2 \wedge (x_3)(x_3 \in x_2 \supset S(x_3) \in x_2) \supset S(x_1) \in x_2$ 3, 4, eH, Taut, MP

(1)   10. (x_2) [9] $i\forall$

(1)   11. $S(x_1) \in N$ 10, *Df.* 10, **F.**2, $e\forall$, Taut, MP

12. $x_1 \in N \supset S(x_1) \in N$ 1, 11, eH

13. $(x_1)(x_1 \in N \supset S(x_1) \in N)$ 12, $i\forall$

The successor of any natural number is also a natural number.

Definition 11. $x \subset y$ for $(z)(z \in x \supset z \in y)$ where x and y are any terms in which z does not occur.

THEOREM 9. $\vdash (x_1)(0 \in x_1 \wedge (x_2)(x_2 \in x_1 \supset S(x_2) \in x_1) \supset N \subset x_1)$.

Proof. (1)   1. $0 \in x_1 \wedge (x_2)(x_2 \in x_1 \supset S(x_2) \in x_1)$ H

(2)   2. $x_4 \in N$ H

(2)   3. $(x_3)(0 \in x_3 \wedge (x_2)(x_2 \in x_3 \supset S(x_2) \in x_3) \supset x_4 \in x_3)$

2, *Df.* 10, **F.2**, $e\forall$, Taut, MP

(2)   4. $0 \in x_1 \wedge (x_2)(x_2 \in x_1 \supset S(x_2) \in x_1) \supset x_4 \in x_1$ 3, $e\forall$

(1, 2)   5. $x_4 \in x_1$ 1, 4, MP

(1)   6. $x_4 \in N \supset x_4 \in x_1$ 2, 5, eH

(1)   7. $(x_4)(x_4 \in N \supset x_4 \in x_1)$ 6, $i\forall$

(1)   8. $N \subset x_1$ 7, *Df.* 11

9. $[1 \supset 8]$ 1, 8, eH

10. $(x_1)[9]$ 9, $i\forall$

This tells us that N is contained in every set x_1 containing 0 and the successor of each of its elements. Theorem 9 allows us to state the following metatheorem which is a form of mathematical induction:

THEOREM 10. $\vdash A(0) \wedge (x)(A(x) \supset A(S(x)) \supset (x)(x \in N \supset A(x))$ *where* $A(x)$ *is a wff and* $A(S(x))$ *and* $A(0)$ *are the result of replacing x by $S(x)$ and 0 respectively in all the free occurrences of x in* $A(x)$.

Proof. Use $e\forall$ on Theorem 9, substituting $\{x \mid A(x)\}$ for x_1. The result follows from **F.2.**

We now use Theorem 10 to prove an even stronger principle of induction:

THEOREM 11. $\vdash (x_1)((0 \in x_1 \wedge (x_2)(x_2 \in x_1 \wedge x_2 \in N \supset S(x_2) \in x_1)) \supset N \subset x_1)$.

Proof. (1)   1. $0 \in x_1 \wedge (x_2)(x_2 \in x_1 \wedge x_2 \in N \supset S(x_2) \in x_1)$ H

(1)   2. $0 \in x_1 \wedge 0 \in N$ 1, Taut, MP, Th. 6

(3)   3. $x_3 \in x_1 \wedge x_3 \in N$ H (induction)

(1)   4. $(x_2)(x_2 \in x_1 \wedge x_2 \in N \supset S(x_2) \in x_1)$ 1, Taut, MP

(1)   5. $x_3 \in x_1 \wedge x_3 \in N \supset S(x_3) \in x_1$ 4, $e\forall$

(1, 3)   6. $S(x_3) \in x_1$ 3, 5, MP

7. $x_3 \in N \supset S(x_3) \in N$ Th. 8, $e\forall$

(3)   8. $S(x_3) \in N$ 3, 7, Taut, MP

(1, 3)   9. $S(x_3) \in x_1 \wedge S(x_3) \in N$ 6, 8, Taut, MP

(1)   10. $[3 \supset 9]$ 3, 9, eH

(1)   11. $(x_3)[10]$ 10, $i\forall$

(1)   12. $[2 \wedge 11]$ 2, 11, Taut, MP

(1)   13. $(x_3)(x_3 \in N \supset x_3 \in x_1 \wedge x_3 \in N)$ 12, Th. 10, MP

14. $(x_3 \in x_1 \wedge x_3 \in N) \supset x_3 \in x_1$ Taut, MP

(1)   15. $x_3 \in N \supset (x_3 \in x_1 \wedge x_3 \in N)$ 13, $e\forall$

(1)   16. $x_3 \in N \supset x_3 \in x_1$ 14, 15, Taut, MP

(1)   17. (x_3) [16]   16, $i\forall$

(1)   18. $N \subset x_1$ 17, $Df.$ 11

19. $[1 \supset 18]$ 1, 18, eH

20. (x_1) [19]   19, $i\forall$

Associated with Theorem 11 is another metatheorem analogous to Theorem 10:

THEOREM 12. $\vdash A(0) \wedge (x)(x \in N \wedge A(x) \supset A(S(x))) \supset (x)(x \in N \supset A(x))$, where $A(0)$ and $A(S(x))$ are the result of replacing 0 and $S(x)$ respectively in all free occurrences of x in the wff $A(x)$.

Proof. The reader will prove Theorem 12 as an exercise (analogous to proof of Theorem 10).

Theorem 11 tells us that N is a subset of any set x_1 which has 0 as an element and which has as an element the successor of every natural number x_2 that is in it. This is a stronger form of mathematical induction, since Theorem 9 requires that x_1 contain the successor of every element in it. A set may well contain the successor of each of its *natural number* elements without containing the successors of all its elements.

Of course, Theorem 11 was proved by using Theorem 9 (via a metatheorem, Theorem 10), which shows that we could always make do with the weaker principle, Theorem 9 (respectively, Theorem 10), at the price of making some proofs longer.

We now use induction to prove a useful fact about natural numbers. What we show is that if $y \cap \overline{\{w\}} \in x$, $x \in N$, and $w \in y$, then for any element $z \in y$ there obtains $y \cap \overline{\{z\}} \in x$. In other words, the z whose existence is asserted in Definition 9 can be any element of y. Again, our weaker principle, Theorem 10, will suffice.

THEOREM 13. $\vdash (x_1)(x_1 \in N \supset (x_2)(x_3)(x_4)((x_2 \in x_4 \wedge x_3 \in x_4 \wedge x_4 \cap \overline{\{x_3\}} \in x_1) \supset x_4 \cap \overline{\{x_2\}} \in x_1))$.

Proof.   (1)   1. $x_2 \in x_4 \wedge x_3 \in x_4 \wedge (x_4 \cap \overline{\{x_3\}}) \in 0$ H

(1)   2. $x_4 \cap \overline{\{x_3\}} \in 0$ 1, Taut, MP

(1)   3. $x_4 \cap \overline{\{x_3\}} = \Lambda$ 2, $Df.$ 8, **F.2**, $e\forall$, Taut, MP

4. $x_3 \in x_4 \supset (x_4 \cap \overline{\{x_3\}}) \cup \{x_3\} = x_4$ Th. 5, $e\forall$

(1)   5. $x_4 = (x_4 \cap \overline{\{x_3\}}) \cup \{x_3\}$ 1, 4, Taut, MP

(1)   6. $x_4 = \Lambda \cup \{x_3\}$ 3, **F.1**, $e\forall$, Taut, MP

7. $\Lambda \cup \{x_3\} = \{x_3\}$ $Df.$ 1, 7, **F.2**, $e\forall$, Th. 3, Taut, MP

(1)   8. $x_4 = \{x_3\}$ 6, 7, **F.1**, $e\forall$, Taut, MP

(1)   9. $x_2 \in x_4 \supset x_2 = x_3$ 8, $Df.$ 1, 4, **F.2**, $e\forall$, Taut, MP

(1)   10. $x_2 = x_3$ 1, 9, Taut, MP

(1)   11. $x_4 \cap \overline{\{x_2\}} \in 0$ 2, 10, **F.1**, $e\forall$, Taut, MP

12. $(x_2 \in x_4 \wedge x_3 \in x_4 \wedge (x_4 \cap \overline{\{x_3\}}) \in 0) \supset x_4 \cap \overline{\{x_2\}} \in 0$ 1, 11, eH

13. $(x_2)(x_3)(x_4)$ [12]   12, $i\forall$

(14)   14. $(x_2)(x_3)(x_4)((x_2 \in x_4 \land x_3 \in x_4 \land (x_4 \cap \overline{\{x_3\}}) \in x_1) \supset x_4 \cap \overline{\{x_2\}} \in x_1)$
          H (induction)

(15)   15. $x_2 \in x_4 \land x_3 \in x_4 \land x_4 \cap \overline{\{x_3\}} \in S(x_1)$ H

(15)   16. $x_4 \cap \overline{\{x_3\}} \in S(x_1)$ 15, Taut, MP

(15)   17. $(Ex_5)(x_5 \in (x_4 \cap \overline{\{x_3\}}) \land (x_4 \cap \overline{\{x_3\}}) \cap \overline{\{x_5\}} \in x_1)$ 16, $Df.\,9$, $\mathbf{F}.2$,
          $e\forall$, Taut, MP

(15)   18. $a_1 \in (x_4 \cap \overline{\{x_3\}}) \land (x_4 \cap \overline{\{x_3\}}) \cap \overline{\{a_1\}} \in x_1$ 17, eE

          19. $(x_4 \cap \overline{\{x_3\}}) \cap \overline{\{a_1\}} = (x_4 \cap \overline{\{a_1\}}) \cap \overline{\{x_3\}}$ comm., assoc. "$\cap$"

(15)   20. $(x_4 \cap \overline{\{a_1\}}) \cap \overline{\{x_3\}} \in x_1$ 18, 19, $\mathbf{F}.1$, $e\forall$ Taut, MP

(21)   21. $x_2 = a_1$ H

(15, 21) 22. $(x_4 \cap \overline{\{x_2\}}) \cap \overline{\{x_3\}} \in x_1$ 20, 21, $\mathbf{F}.1$, $e\forall$, Taut, MP

(23)   23. $x_2 \neq x_3$ H

(15)   24. $x_3 \in x_4$ 15, Taut, MP

(15, 23) 25. $x_3 \in (x_4 \cap \overline{\{x_2\}})$ 23, 24, $\mathbf{F}.2$, $e\forall$, $Df.\,4, 5, 6$, Taut, MP

(15, 21, 23) 26. $x_3 \in (x_4 \cap \overline{\{x_2\}}) \land ((x_4 \cap \overline{\{x_2\}}) \cap \overline{\{x_3\}}) \in x_1$ 22, 25, Taut, MP

(15, 21, 23) 27. $(Ex_3)(x_3 \in (x_4 \cap \overline{\{x_2\}}) \land ((x_4 \cap \overline{\{x_2\}}) \cap \overline{\{x_3\}}) \in x_1)$ 26, iE

(15, 21, 23) 28. $(x_4 \cap \overline{\{x_2\}}) \in S(x_1)$ 27, $Df.\,9$

(15, 23) 29. $x_2 = a_1 \supset (x_4 \cap \overline{\{x_2\}}) \in S(x_1)$ 21, 27, eH

(15)   30. $x_2 \in x_4$ 15, Taut, MP

(15)   31. $x_2 \neq a_1 \supset x_2 \in (x_4 \cap \overline{\{a_1\}})$ 30, $Df.\,4, 5, 6$, $\mathbf{F}.2$, $e\forall$, Taut, MP

(32)   32. $x_2 \neq a_1$ H

(15, 32) 33. $x_2 \in (x_4 \cap \overline{\{a_1\}})$ 31, 32 MP

          34. $a_1 \in (x_4 \cap \overline{\{x_3\}}) \supset a_1 \neq x_3$ $Df.\,4, 5, 6$, $\mathbf{F}.2$, $e\forall$, Taut, MP

(15)   35. $a_1 \neq x_3$ 18, 34, Taut, MP

(15)   36. $x_3 \in (x_4 \cap \overline{\{a_1\}})$ 24, 35, $Df.\,4, 5, 6$, $\mathbf{F}.2$, $e\forall$, Taut, MP

(15, 32) 37. $x_2 \in (x_4 \cap \overline{\{a_1\}}) \land x_3 \in (x_4 \cap \overline{\{a_1\}}) \land (x_4 \cap \overline{\{a_1\}}) \cap \overline{\{x_3\}} \in x_1$
          20, 33, 36, Taut, MP

(14, 15, 32) 38. $(x_4 \cap \overline{\{a_1\}}) \cap \overline{\{x_2\}} \in x_1$ 14, $e\forall$, 37, MP

(14, 15, 32) 39. $(x_4 \cap \overline{\{x_2\}}) \cap \overline{\{a_1\}} \in x_1$ 38, comm., assoc. "$\cap$", $\mathbf{F}.1$, $e\forall$, Taut, MP

(15)   40. $a_1 \in x_4$ 18, $Df.\,6$, $\mathbf{F}.2$, $e\forall$, Taut, MP

(15, 32) 41. $a_1 \in x_4 \cap \overline{\{x_2\}}$ 32, 40, $Df.\,4, 5, 6$, $\mathbf{F}.2$, $e\forall$, Taut, MP

(14, 15, 32) 42. $a_1 \in x_4 \cap \overline{\{x_2\}} \land (x_4 \cap \overline{\{x_2\}}) \cap \overline{\{a_1\}} \in x_1$ 39, 41, Taut, MP

(14, 15, 32) 43. $(x_4 \cap \overline{\{x_2\}}) \in S(x_1)$ 42, iE, $Df.\,9$, $\mathbf{F}.2$, $e\forall$, Taut, MP

(14, 15) 44. $x_2 \neq a_1 \supset (x_4 \cap \overline{\{x_2\}}) \in S(x_1)$ 32, 43, eH

(14, 15, 32) 45. $(x_2 = a_1 \lor x_2 \neq a_1) \supset (x_4 \cap \overline{\{x_2\}}) \in S(x_1)$ 29, 44, Taut, MP

          46. $x_2 = a_1 \lor x_2 \neq a_1$ Taut

(14, 15, 23)   47. $(x_4 \cap \overline{\{x_2\}}) \in S(x_1)$ 45, 46, MP

(14, 23)   48. $[15 \supset 47]$ 15, 47 eH

(14)   49. $x_2 \neq x_3 \supset [48]$ 23, 48, eH

                50. $(x_2 \in x_4 \wedge x_2 \in x_4 \wedge (x_4 \cap \overline{\{x_2\}}) \in S(x_1)) \supset (x_4 \cap \overline{\{x_2\}}) \in S(x_1)$ Taut

                51. $x_2 = x_3 \supset ((x_2 \in x_4 \wedge x_3 \in x_4 \wedge (x_4 \cap \overline{\{x_3\}}) \in S(x_1))$
                $\supset (x_4 \cap \overline{\{x_2\}}) \in S(x_1))$ 50, **F.**1, e∀, Taut, MP

(14)   52. $(x_2 \neq x_3 \vee x_2 = x_3) \supset [48]$ 49, 51, Taut, MP

                53. $(x_2 \neq x_3 \vee x_2 = x_3)$ Taut

(14)   54. $(x_2 \in x_4 \wedge x_3 \in x_4 \wedge (x_4 \cap \overline{\{x_3\}}) \in S(x_1)) \supset (x_4 \cap \overline{\{x_2\}}) \in S(x_1)$
                52, 53, MP

(14)   55. $(x_2)(x_3)(x_4) [54]$ 54, i∀

                56. $[14 \supset 55]$ 14, 55 eH

                57. $(x_1) [56]$ 56, i∀

                58. $[13 \wedge 57]$ 13, 57, Taut, MP

                59. Concl 58, Th. 10, MP

Proofs by induction such as the above are tedious, but they involve no essential difficulties. Part of the tedium results from the necessity of treating the various cases such as $x_2 = x_3$; $x_2 \neq x_3$ and $a_1 = x_2$; $a_1 \neq x_2$, and the like, but the proof for each case is simple. For an exercise, the reader should prove the following:

THEOREM 14. $\vdash (x_2)(x_2 \in N \supset (x_1)(x_3)((x_1 \in x_2 \wedge x_3 \in x_2 \wedge x_1 \subset x_3) \supset x_1 = x_3))$.

Proof. The reader will prove Theorem 14 as an exercise. (*Hint:* Proof by induction; use Theorem 13.)

Definition 12. 1 for $S(0)$.

Definition 13. 2 for $S(1)$.

Definition 12 and 13 serve to give names to the first three natural numbers. We can continue in the usual way to name as many as we please.

Definition 14. Fin (x) for $(Ez)(z \in N \wedge x \in z)$ where x is any term in which z does not occur.

A set x is finite if it is an element of a natural number. Since a natural number, as we have defined the notion, consists intuitively of the set of all sets of a given finite cardinality, this definition is justified.

Definition 15. Inf (x) for $\sim$ Fin (x) where x is any term.

An infinite set is one that is not finite.

Theorem 14 says, in effect, that any two sets of the same finite cardinality and with one of the sets contained in the other are equal. In other words, a finite set cannot have the same cardinality as a proper subset of itself.

Notice that we have not yet proved that there are any infinite sets.

The reader may well wonder how it is that we have proved a theorem of mathematical induction when this is often taken as a primitive axiom in axiomatic approaches to the number system. In fact, we have proved not only the principle of induction but most of the other Peano postulates as well. Theorem 6, Theorem 7 (where we restrict x_1 to range over natural numbers), Theorem 8, and Theorem 11 each express one of the five Peano postulates. The only Peano postulate remaining to be proved is that the successor function is 1–1; i.e. $x \in N \wedge y \in N \wedge S(x) = S(y) \supset x = y$. The proof of the last postulate will be forthcoming.

Landau [1] shows how to construct analysis by starting with the Peano postulates as axioms. However, the logic used for the construction involves extensive use of set-theoretic concepts. Mere quantification theory plus the Peano postulates does not suffice for the construction of analysis. In Chapter 1, we considered the first-order theory which has as axioms the Peano postulates plus the recursive definitions of addition and multiplication. This is the system **S** of first-order arithmetic. First-order arithmetic is also generally inadequate for the development of mathematics. For the development of mathematics, we need a system strong enough to express set-theoretic concepts (such as the principle of abstraction of Frege's system or some similar general principles).

When the Peano postulates are used as a starting point, as in Landau [1], the usual operations of addition, multiplication, and the like on natural numbers are introduced by recursive definitions. For example, $x + y$ is defined by the two recursion equations $x + 0 = x$ and

$$x + S(y) = S(x + y).$$

From these equations we can see that

$$x + 1 = x + S(0) = S(x + 0) = S(x),$$

the last being already well defined. Similarly, $x + 2 = S(S(x))$, and so on, $x + n$ being simply n iterations of the successor function starting with x.

The difficulty with this approach is that one must somehow demonstrate that there really is a function from $N \times N$ into N defined by such recursion equations. Of course this is no problem in first-order arithmetic, since "$+$", "$\cdot$", and "′" are primitive function letters. But in set theory, in which these are introduced by definitions, some justification is needed. What is needed is a theorem of *primitive recursion* which proves the following: Given two number-theoretic functions $g(x_1, \ldots, x_n)$ and $h(x_1, \ldots, x_n, y, z)$, there is a function $f(x_1, \ldots, x_n, x_{n+1})$ such that $f(x_1, \ldots, x_n, 0) = g(x_1, \ldots, x_n)$ and, for all $y \in N$,

$$f(x_1, \ldots, x_n, S(y)) = h(x_1, \ldots, x_n, y, f(x_1, \ldots, x_n, y)).$$

(It is easy to prove by induction that there can be at most one such f.) The proof of the primitive recursion theorem involves the use of general notions of set theory; it cannot be proved from the Peano axioms alone. As it turns out, once we have axioms strong enough to prove such a theorem of primitive recursion, we have axioms strong enough to develop arithmetic and to prove the Peano postulates as theorems (as we have begun to do in our system **F**). Hence, the Peano postulates tend to lose their place as an ultimate set of axioms, since they can be proved as theorems within a more general theory, the latter being necessary in any case to make any real use of the Peano postulates themselves.

However, there is another aspect of the development of mathematics within a given foundational system that keeps the Peano postulates from losing their importance completely. It is
the aspect of simplicity or economy. Definitions such as Definition 10 may have intuitive appeal,
but deductions from them can be cumbersome. In proving most of our theorems thus far, we
have had to return to such definitions and grind out the result. However, once we have obtained
sufficiently many properties of the natural numbers to characterize them completely, we can
simply dispense with our definition and proceed on the basis of the proved properties. It is
known that, within a set theory such as ours, the Peano postulates do characterize the natural
numbers completely; i.e. any two sets that satisfy these properties will be isomorphic. (It should
be noted here that this is not so if we take as our underlying logic quantification theory
rather than set theory. Skolem [2] has shown that there exist nonisomorphic models of first-order
arithmetic. Using a method different from that of Skolem, we show in Chapter 6 that there are
at least $2^{\aleph_0}$ nonisomorphic models of first-order arithmetic.)

Hence, once we have proved the fifth Peano postulate, it will suffice for us to know about
the set N that Theorem 4, Theorem 5, Theorem 6, and Theorem 11, and the fifth postulate are
true of it. Definition 10 will never be needed again. We can then prove a theorem of primitive
recursion and deduce the usual properties of the natural numbers by defining addition, multiplication and other number-theoretic functions by recursive methods. The integers can be introduced as equivalence classes of ordered pairs of natural numbers, and the rational numbers as
certain equivalence classes of ordered pairs of integers; the real numbers can be constructed from
the rational numbers by any one of the various known means. Analysis can be carried on in the
system. For examples of these constructions see Bernays [1] and Birkhoff and Maclane [1]
and Section 5.7 of the present work.

In the case of each of these number systems, sets of characterizing properties can be proved,
and so the process of "throwing away" the construction can be repeated on each level. The virtue of this is that it removes the necessity of laborious retranslations of statements into unabbreviated, definition-free form in order to prove theorems. The reader can well appreciate
that even a simple statement of analysis such as $(x_1)(x_1 \in R \supset |x_1| \geq 0)$, "the absolute value of
every real number is nonnegative", would run to great length if all definitional abbreviations
were removed from it.

Exercise. A form of recursion even simpler than primitive recursion is *simple recursion*:
Given any set X, any element $a \in X$, and any function f from X to X, there exists a unique function g from N to X such that $g(0) = a$ and $g(S(n)) = f(g(n))$, where N is the set of natural numbers, 0 is zero, and S is the successor function. Such a theorem will be proved for the natural
numbers in a later system (see Chapter 5, Section 5.3, Theorem 29), and it is provable within
most axiomatic set theories (and certainly in intuitive set theory) on the basis of the Peano pos-
tulates. Prove, in intuitive set theory, that if $\langle N, 0, S \rangle$ and $\langle N', 0', S' \rangle$ are two systems satisfying
the Peano postulates (and thus the theorem of simple recursion) then there is an isomorphism
between them. (In the system of Chapter 8, we shall assume simple recursion as a basic axiom,
and prove the Peano postulates. Thus we shall reverse the usual procedure of proving the theorem of simple recursion from the Peano postulates.)

Having now appreciated the virtue of proving a complete set of properties for the natural
numbers, let us turn to the problem of proving the last Peano postulate in **F**. It will turn out that

proving the remaining property is considerably more complicated than proving the other four. One reason for this is that the fifth postulate, together with the others already proved, implies the existence of infinite sets. In particular, the set V of all sets and the set N of all natural numbers will be infinite sets.

As it now stands, nothing would exclude the possibility that we have only a finite number of natural numbers, and that all the successors of the last one are simply the empty set, Λ. This would mean that $x = S(x)$ for some natural number, but we have not yet disproved this either. Proving the remaining Peano postulate amounts to proving that no natural number is empty. In other words, every natural number n has a representative n-element set. The representatives we shall choose are precisely the von Neumann natural numbers mentioned in our previous discussion (though our development in **F** will be quite different from von Neumann's, which will be treated in a later system).

3.3. The theorem of infinity

The actual technique involved in proving the last Peano postulate uses an idea due to Frege, and which we now explain. Suppose that some relation $R(x, y)$ is given (we can think of this simply as some wff of our system with two free variables). Now let some set z be given. By the *closure* of z under the relation R we mean the smallest set that includes z and that contains as elements all those objects which can be obtained by starting with z and iterating the application of the relation R.

In less pretentious language, we start with z, and with R we generate a set from z in the following manner: if $x \in z$ and further if $R(x, y)$ then y is in our new set. That is, we throw in all things that are paired under R with some element of z. We now consider the newly obtained set and repeat the process, throwing in those objects paired under R with objects of the enlarged set. We repeat this operation indefinitely and the union of all the sets we get in this way is the closure of z under the R relation.

But how do we define this notion precisely? Frege's answer was the following:

$$Cl(x_1) = \{x_2 \mid (x_3)(x_1 \subset x_3 \wedge (x_5)(x_4)(x_4 \in x_3 \wedge R(x_4, x_5) \supset x_5 \in x_3) \supset x_2 \in x_3)\}.$$

The closure of x_1 is the set of all objects x_2 that are elements of every set x_3 which includes x_1 as a subset and which contains all the objects paired under R with any given object of x_3. A moment's reflection will afford the observation that we have already used this idea. We have defined the set N as the closure of the set $\{0\}$ under the relation of successor. Let $x_1 = \{0\}$ in the preceding definition, and write $S(x_4) = x_5$ for $R(x_4, x_5)$. We obtain an equivalent of Definition 10.

Let us take a more intuitive nonmathematical example. Let x_1 be the set whose only element is Abraham, and let the relation R be the relation "x is the parent of y" (parenthood). Then the closure of x_1 is the set consisting of Abraham and all his descendants. Thus, the natural numbers can be thought of as 0 together with the "descendants" of 0 under the successor relation.

Now, what we intend do to is to define a set ω which will be the set containing Λ and all the descendants of Λ under the relation $x \cup \{x\} = y$. Intuitively such a set ω has exactly one representative from each natural number. We shall establish this fact.

It is important to realize that it is this method of defining the closure of a set under a given relation (and the fact that the definition of closure can be expressed in our formal language) which allows us to prove a theorem of mathematical induction, a point whose importance we have already emphasized. In fact, Peano's postulate of induction really says that the natural numbers are the closure of the set $\{0\}$ under successor. The way in which Theorem 9 was proved can actually be generalized to the case of any relation where "N" stands for the closure of some set under the given relation.

Thus, it was Frege who saw that the idea underlying Peano's postulate of induction could be generalized to any relation and any given set. We now use these ideas in the following development.

Definition 16. ω for $\{y \mid (z)(\varLambda \in z \wedge (x)(x \in z \supset x \cup \{x\} \in z) \supset y \in z)\}$ where x, y, z are all distinct variables.

THEOREM 15. $\vdash \varLambda \in \omega$.

Proof. The reader will prove Theorem 15 as an exercise (see proof of Theorem 6).

THEOREM 16. $\vdash (x_1)(\varLambda \neq x_1 \cup \{x_1\})$.

Proof. The reader will prove this theorem as an exercise.

THEOREM 17. $\vdash (x_1)(x_1 \in \omega \supset x_1 \cup \{x_1\} \in \omega)$.

Proof. The reader will prove this theorem as an exercise (see proof of Theorem 8).

THEOREM 18. $\vdash (x_1)(\varLambda \in x_1 \wedge (x_2)(x_2 \in x_1 \supset x_2 \cup \{x_2\} \in x_1) \supset \omega \subset x_1)$.

Proof. The reader will prove this theorem as an exercise (see proof of Theorem 9).

THEOREM 19. $\vdash A(\varLambda) \wedge (x)(A(x) \supset A(x \cup \{x\})) \supset (x)(x \in \omega \supset A(x))$, where $A(x)$ is any wff and $A(x \cup \{x\})$ and $A(\varLambda)$ are obtained from $A(x)$ by substituting at all free occurrences of x in $A(x)$.

Proof. See proof of Theorem 10.

Let us use this metatheorem to prove:

THEOREM 20. $\vdash (x_1)(x_1 \in \omega \supset (x_2)(x_2 \in x_1 \supset x_2 \subset x_1))$.

Proof.
1. $(x_1)(x_1 \notin \varLambda)$ Th. 3
2. $x_2 \notin \varLambda$ 1, $e\forall$
3. $(x_2)(x_2 \in \varLambda \supset x_2 \subset \varLambda)$ 2, Taut, MP, $i\forall$
(4)   4. $(x_2)(x_2 \in x_1 \supset x_2 \subset x_1)$ H
(5)   5. $x_2 \in x_1 \cup \{x_1\}$ H

(5)   6. $x_2 \in x_1 \lor x_2 = x_1$ 5, *Df.* 4, 7, **F.**2, $e\forall$, MP

7. $x_2 = x_1 \supset x_2 \subset x_1 \cup \{x_1\}$ *Df.* 7, 11, **F.**1, $e\forall$, Taut, MP

(4)   8. $x_2 \in x_1 \supset x_2 \subset x_1$ 4, $e\forall$

9. $x_2 \subset x_1 \supset x_2 \subset x_1 \cup \{x_1\}$ *Df.* 11, 7, **F.**2, $e\forall$, Taut, MP

(4)   10. $x_2 \in x_1 \supset x_2 \subset x_1 \cup \{x_1\}$ 8, 9, Taut, MP

(4)   11. $x_2 \in x_1 \lor x_2 = x_1 \supset x_2 \subset x_1 \cup \{x_1\}$ 7, 10, Taut, MP

(4, 5) 12. $x_2 \subset x_1 \cup \{x_1\}$ 6, 11, MP

(4)   13. $[5 \supset 12]$ 5, 12, eH

(4)   14. $(x_2)[13]$ 13, $i\forall$

15. $[4 \supset 14]$ 4, 14, eH

16. $(x_1)[15]$ 15, $i\forall$

17. Concl   3, 16, Th. 19, Taut, MP

In the following we shall need the stronger principle:

THEOREM 21. $\vdash (x_1)(\varLambda \in x_1 \land (x_2)(x_2 \in \omega \land x_2 \in x_1 \supset x_2 \cup \{x_2\} \in x_1) \supset \omega \subset x_1)$.

Proof. By strict analogy with the proof of Theorem 11, Theorem 21 bears the same relation to Theorem 18 as does Theorem 11 to Theorem 9. Finally, we have the metatheorem:

THEOREM 22. $\vdash A(\varLambda) \land (x)(x \in \omega \land A(x) \supset A(x \cup \{x\})) \supset (x)(x \in \omega \supset A(x))$, *where $A(x)$ and the other formulas have the usual meaning.*

Now let us prove:

THEOREM 23. $\vdash (x_1)(x_1 \in \omega \supset (x_1 \notin x_1))$.

Proof.   1. $(x_1)(x_1 \notin \varLambda)$ Th. 3

2. $\varLambda \notin \varLambda$ 1, $e\forall$

(3)   3. $x_1 \in \omega \land x_1 \notin x_1$ H

(4)   4. $x_1 = x_1 \cup \{x_1\}$ H

5. $x_1 \in x_1 \cup \{x_1\}$ *Df.* 4, 7, **F.**2, $e\forall$, Taut, MP

(4)   6. $x_1 \in x_1$ 4, 5, *Df.* 1, $e\forall$ Taut, MP

7. $x_1 = x_1 \cup \{x_1\} \supset x_1 \in x_1$ 4, 6, eH

(8)   8. $(x_1 \cup \{x_1\}) \in x_1$ H

(3)   9. $(x_1 \cup \{x_1\}) \in x_1 \supset (x_1 \cup \{x_1\}) \subset x_1$ Th. 20, 3, Taut, MP, $e\forall$

(3, 8) 10. $x_1 \cup \{x_1\} \subset x_1$ 8, 9, MP

(3, 8) 11. $x_1 \in x_1$ 5, 10, *Df.* 11, $e\forall$, Taut, MP

(3)   12. $(x_1 \cup \{x_1\}) \in x_1 \supset x_1 \in x_1$ 8, 11, eH

(3)   13. $(x_1 = (x_1 \cup \{x_1\}) \lor (x_1 \cup \{x_1\}) \in x_1) \supset x_1 \in x_1$ 7, 12, Taut, MP

(3)   14. $x_1 \notin x_1 \supset (x_1 \cup \{x_1\}) \notin (x_1 \cup \{x_1\})$ 13, Taut, MP, *Df.* 4, 7, **F.**2, $e\forall$

(3)   15. $(x_1 \cup \{x_1\}) \notin (x_1 \cup \{x_1\})$ 3, Taut, MP

16. $[3 \supset 15]$ 3, 15, eH

17. $(x_1)[16]$ 16, $i\forall$

18. Concl   2, 17, Th. 22, Taut, MP

We use Theorem 23 to prove that every natural number has a member of ω as an element (and is thus nonempty). First we prove:

THEOREM 24. $\vdash (x_1)(x_2)(x_1 \in \omega \wedge x_2 \in N \wedge x_1 \in x_2 \supset x_1 \cup \{x_1\} \in S(x_2))$.

Proof.   (1)   1. $x_1 \in \omega \wedge x_2 \in N \wedge x_1 \in x_2$ H

2. $x_1 \notin x_1 \supset ((x_1 \cup \{x_1\}) \cap \overline{\{x_1\}}) = x_1$ Df. 1, 4, 5, 6, 7, **F**.2, $e\forall$, Taut, MP

(1)   3. $x_1 \notin x_1$ 1, Th. 23, $e\forall$, Taut, MP

(1)   4. $(x_1 \cup \{x_1\}) \cap \overline{\{x_1\}} = x_1$ 2, 3, MP

(1)   5. $(x_1 \cup \{x_1\}) \cap \overline{\{x_1\}} \in x_2$ 1, 4, **F**.1, $e\forall$, Taut, MP

6. $x_1 \in (x_1 \cup \{x_1\})$ Df. 4, 7, **F**.2, $e\forall$, Taut, MP

(1)   7. $(Ex_1)(x_1 \in (x_1 \cup \{x_1\}) \wedge ((x_1 \cup \{x_1\}) \cap \overline{\{x_1\}}) \in x_2)$ 5, 6, Taut, MP, iE

(1)   8. $(x_1 \cup \{x_1\}) \in S(x_2)$ 7, Df. 9, **F**.2, $e\forall$, MP

9. $[1 \supset 8]$ 1, 8, eH

10. $(x_1)(x_2)[9]$ 9, $i\forall$

Now we can establish the following theorem:

THEOREM 25. $\vdash (x_1)(x_1 \in N \supset (Ex_2)(x_2 \in \omega \wedge x_2 \in x_1))$.

Proof. Exercise for the reader (proof by induction on x_1).

Now we can prove a theorem of infinity (i.e. that V is infinite). First we have the following theorem:

THEOREM 26. $\vdash \varLambda \notin N$.

Proof.   Exercise for the reader (follows immediately from Theorem 25).

It will also simplify things to prove:

THEOREM 27. $\vdash (x_1)(x_1 \subset V)$.

Proof.   1. $x_2 \in V$ Th. 2, $e\forall$

2. $x_2 \in x_1 \supset x_2 \in V$ 1, Taut, MP

3. $(x_1)(x_2)[2]$ $i\forall$

4. Concl   3, Df. 11

Now we prove our theorem of infinity:

THEOREM 28. $\vdash (x_1)(\sim(V \in x_1 \wedge x_1 \in N))$.

Proof. (1)   1. $V \in x_1 \wedge x_1 \in N$ H

(2)   2. $x_2 \in S(x_1)$ H

(2)   3. $(Ex_3)(x_3 \in x_2 \wedge x_2 \cap \overline{\{x_3\}} \in x_1)$ 2, *Df.* 9, **F.**2, Taut, MP

(2)   4. $a_1 \in x_2 \wedge x_2 \cap \overline{\{a_1\}} \in x_1$ 3, *eE*

         5. $x_1 \in N \supset (x_4)(x_5)(x_4 \in x_1 \wedge x_5 \in x_1 \wedge x_4 \subset x_5 \supset x_4 = x_5)$ Th. 14, *e*∀,
         changing the names of several bound variables.

(1)   6. $V \in x_1 \wedge x_2 \cap \overline{\{a_1\}} \in x_1 \wedge x_2 \cap \overline{\{a_1\}} \subset V \supset x_2 \cap \overline{\{a_1\}} = V$
         5, 1, Taut, MP, *e*∀

         7. $x_2 \cap \overline{\{a_1\}} \subset V$ Th. 27, *e*∀

(1, 2)   8. $x_2 \cap \overline{\{a_1\}} = V$ 6, 1, 4, 7, Taut, MP

(1, 2)   9. $a_1 \notin V$ 8, *Df.* 1, 4, 5, 6, **F.**2, *e*∀, Taut, MP

(1, 2) 10. $(Ex_3)(x_3 \notin V)$ 9, *iE*

(1)   11. $x_2 \in S(x_1) \supset (Ex_3)(x_3 \notin V)$ 2, 10, *eH*

(1)   12. $(x_3)(x_3 \in V) \supset \sim (x_2 \in S(x_1))$ 11, Taut, MP

(1)   13. $x_2 \notin S(x_1)$ 12, Th. 2, MP

(1)   14. $(x_2)(x_2 \notin S(x_1))$ *i*∀

(1)   15. $S(x_1) = \varLambda$ 14, *Df.* 1, Th. 4, *e*∀, Taut, MP

(1)   16. $S(x_1) \in N$ 1, Th. 8, *e*∀, MP

(1)   17. $\varLambda \in N$ 15, 16, **F.**1, *e*∀, Taut, MP

        18. $[1 \supset 17]$ 1, 17, *eH*

        19. $\varLambda \notin N \supset \sim (V \in x_1 \wedge x_1 \in N)$ 18, Taut, MP

        20. $\sim (V \in x_1 \wedge x_1 \in N)$ Th. 26, 19, MP

        21. $(x_1)[20]$ 20, *i*∀

COROLLARY. $\vdash (x_1)(x_2)(x_1 \in x_2 \wedge x_2 \in N \supset (Ex_3)(x_3 \notin x_1))$.

Proof. (1)   1. $x_1 \in x_2 \wedge x_2 \in N$ H

(2)   2. $(x_3)(x_3 \in x_1)$ H

(2)   3. $x_1 = V$ 2, Th. 2, *e*∀, *i*∀, *Df.* 1, Taut, MP

(1, 2) 4. $V \in x_2 \wedge x_2 \in N$ 1, 3, **F.**1, *e*∀, Taut, MP

(1)   5. $(x_3)(x_3 \in x_1) \supset V \in x_2 \wedge x_2 \in N$ 2, 4, *eH*

(1)   6. $\sim (V \in x_2 \wedge x_2 \in N) \supset (Ex_3)(x_3 \notin x_1)$ 5, Taut, MP

(1)   7. $(Ex_3)(x_3 \notin x_1)$ 6, Th. 28, *e*∀, MP

        8. Concl   1, 7, *eH*, *i*∀

We use the foregoing corollary in proving:

THEOREM 29. $\vdash (x_1)(x_2)((x_1 \in N \wedge x_2 \in N \wedge S(x_1) \subset S(x_2)) \supset x_1 \subset x_2)$.

Proof. (1)   1. $x_1 \in N \wedge x_2 \in N \wedge S(x_1) \subset S(x_2)$ H

(2)   2. $x_3 \in x_1$ H

$(1, 2)$ 3. $(Ex_4) (x_4 \notin x_3)$ 1, 2, Corollary to Th. 28 (with a change of name of bound variables), Taut, MP

$(1, 2)$ 4. $a_1 \notin x_3$ 3, eE

$(1, 2)$ 5. $x_3 = (x_3 \cup \{a_1\}) \cap \overline{\{a_1\}}$ 4, Df. 4, 5, 6, 7, **F**.2, $e\forall$, Taut, MP

$(1, 2)$ 6. $(x_3 \cup \{a_1\}) \cap \overline{\{a_1\}} \in x_1$ 2, 5, **F**.1, $e\forall$, Taut, MP

$(1, 2)$ 7. $x_3 \cup \{a_1\} \in S(x_1)$ 6, Df. 4, 5, 6, 7, 9, iE, **F**.2, $e\forall$, Taut, MP

$(1, 2)$ 8. $x_3 \cup \{a_1\} \in S(x_2)$ 1, 7, Df. 11, $e\forall$, Taut, MP

$(1, 2)$ 9. $(Ex_4) (x_4 \in (x_3 \cup \{a_1\}) \wedge (x_3 \cup \{a_1\}) \cap \overline{\{x_4\}} \in x_2)$
$\qquad$ 8, Df. 9, **F**.2, $e\forall$, Taut, MP

$(1, 2)$ 10. $a_2 \in (x_3 \cup \{a_1\}) \wedge (x_3 \cup \{a_1\}) \cap \overline{\{a_2\}} \in x_2$ 9, eE

$\qquad$ 11. $x_2 \in N \supset (x_5) (x_6) (x_7) (x_5 \in x_7 \wedge x_6 \in x_7 \wedge x_7 \cap \overline{\{x_6\}} \in x_2 \supset x_7 \cap \overline{\{x_5\}} \in x_2)$
$\qquad$ Th. 13 (with a change in the name of bound variables), $e\forall$

(1) 12. $a_1 \in (x_3 \cup \{a_1\}) \wedge a_2 \in (x_3 \cup \{a_1\}) \wedge (x_3 \cup \{a_1\}) \cap \overline{\{a_2\}} \in x_2 \supset (x_3 \cup \{a_1\})$
$\qquad$ $\cap \overline{\{a_1\}} \in x_2$ 11, 1, MP, $e\forall$

$(1, 2)$ 13. $(x_3 \cup \{a_1\}) \cap \overline{\{a_1\}} \in x_2$ 10, 12, Df. 4, 7, **F**.2, $e\forall$, Taut, MP

$(1, 2)$ 14. $x_3 \in x_2$ 5, 13, **F**.1, $e\forall$, Taut, MP

(1) 15. $x_3 \in x_1 \supset x_3 \in x_2$ 2, 14, eH

(1) 16. $(x_3) (x_3 \in x_1 \supset x_3 \in x_2)$ 15, $i\forall$

(1) 17. $x_1 \subset x_2$ 16, Df. 11

$\qquad$ 18. $[1 \supset 17]$ 1, 17, eH

$\qquad$ 19. Concl   18, $i\forall$

From this, we immediately obtain:

THEOREM 30. $\vdash (x_1) (x_2) (x_1 \in N \wedge x_2 \in N \wedge S(x_1) = S(x_2) \supset x_1 = x_2)$.

Proof. The reader will prove Theorem 30 as an exercise (immediate from Theorem 29). This establishes our last Peano postulate.

3.4. Criticisms of Frege's system

We have already indicated in our discussion the way in which we could continue the development of analysis in the system **F**. Let us now pause to reflect on the system itself and its merits or demerits. As we worked in the system, the reader may have become convinced of its viability. Certainly it seems to meet most of the five criteria stated at the end of Chapter 2. The constructions certainly do seem intuitively natural. Moreover, the axioms seem to do nothing more nor less than express formally some intuitively sensible truths of set theory, namely that every condition gives rise to a set, and that sets with the same elements are equal. However, the system **F** does not meet criterion number 4. It is not consistent. It was Russell who first pointed this out in a letter to Frege. Let us deduce "Russell's paradox" within the system **F**.

THEOREM 31. $\vdash \{x_1 \mid x_1 \notin x_1\} \notin \{x_1 \mid x_1 \notin x_1\}$.

Proof. (1) 1. $\{x_1 \mid x_1 \notin x_1\} \in \{x_1 \mid x_1 \notin x_1\}$ *H*

     2. $\{x_1 \mid x_1 \notin x_1\} \in \{x_1 \mid x_1 \notin x_1\} \equiv \{x_1 \mid x_1 \notin x_1\} \notin \{x_1 \mid x_1 \notin x_1\}$ **F.2**, $e\forall$

 (1) 3. $\{x_1 \mid x_1 \notin x_1\} \notin \{x_1 \mid x_1 \notin x_1\}$ 1, 2, Taut, MP

     4. $[1 \supset 3]$ 1, 3 *eH*

     5. $\{x_1 \mid x_1 \notin x_1\} \notin \{x_1 \mid x_1 \notin x_1\}$ 4, Taut, MP

THEOREM 32. $\vdash \{x_1 \mid x_1 \notin x_1\} \in \{x_1 \mid x_1 \notin x_1\}$.

Proof. 1. $\{x_1 \mid x_1 \notin x_1\} \notin \{x_1 \mid x_1 \notin x_1\}$ Th. 31

     2. $\{x_1 \mid x_1 \notin x_1\} \notin \{x_1 \mid x_1 \notin x_1\} \equiv \{x_1 \mid x_1 \notin x_1\} \in \{x_1 \mid x_1 \notin x_1\}$ **F.2**, $e\forall$

     3. Concl   1, 2, Taut, MP

These last two theorems together give us our contradiction. Intuitively, Russell's paradox can be stated as follows: If every condition determines a set, then consider the set y determined by the condition $x \notin x$. That is, y is the set of all sets that are not elements of themselves. Presumably y is a large set, since most sets do not contain themselves as elements (the set of cats is not a cat, the set of real numbers is not a real number, and so on). Now, does y have itself as an element? By the rules of sentential logic, either it does or it does not. If it does, then $y \in \{x \mid x \notin x\}$ and so y must satisfy the defining condition of the set y; i.e. it must not belong to itself. On the other hand, if y does not belong to itself, then y satisfies the defining condition of y and is thus an element of itself.

Exercise. Suppose we formulate **F** without the term operator of abstraction. "$\in$" is our only predicate letter and variables are our only terms. The axiom of abstraction is stated: $(Ey)(x)(x \in y \equiv A(x))$, where x may be free in the wff $A(x)$, which does not contain an occurrence of y. Give a formal derivation of Russell's paradox in this formulation of **F** (the axiom of extensionality is the same).

In 1902 Russell communicated this contradiction to Frege just as the last volume of his work was going to print (see van Heijenoort [1], p. 124). Frege appended an acknowledgment of the contradiction to this work which reads in part:

> Hardly anything more unwelcome can befall a scientific writer than that one of the foundations of his edifice be shaken after the work is finished. I have been placed in this position by a letter of Mr. Bertrand Russell just as the printing of this [second] volume was nearing completion. It is a matter of my Basic Law V [our principle of abstraction F.2]. I have never concealed from myself its lack of the self-evidence which the others possess, and which must properly be demanded of a law of logic. .... (Translation by M. Furth, see Frege [4], p. 127.)

Frege continued his appendix by exhibiting the formal deduction within his system of the Russell paradox. Thus, Frege's system **F** is inconsistent and we lose the distinction between truth and falsehood. Everything is provable within the system and it cannot serve as a foundation for mathematics.

3.5. The paradoxes

Actually, Russell's contradiction was only one of many contradictions which began to appear in set-theoretic reasoning about the turn of the century. Had it not been for these contradictions, quite probably mathematicians would have considered that the question of a foundation for mathematics was solved. In particular, the inconsistency of Frege's system showed that certain intuitively natural principles, such as the principle of abstraction, were false. There was thus a certain emotional shock which accompanied the appearance of the contradictions. Just what, after all, was wrong with the intuitive reasoning underlying set theory?

Let us first note that, though Frege's system **F** is contradictory, the way in which Frege actually constructs analysis within his system may not be contradictory. Russell's contradiction, which is deduced in a simple and straightforward way from **F.2**, does not involve any of the detailed constructions that we have deduced in **F**. Russell's contradiction certainly invalidates the general principle of abstraction, but it does not necessarily invalidate certain arguments using *particular* applications of the principle of abstraction. One might pose the question arising from Russell's paradox as follows: How can we set up a system that allows us to develop something like Frege's constructions of mathematics but which is not contradictory?

Of course, it is conceivable that our constructions are also contradictory. It may be that mathematics itself is contradictory and that no formal consistent system exists which is capable of reproducing mathematics (or a large portion thereof). The point to realize is that such a drastic conclusion does not follow from Russell's contradiction. What Russell's contradiction shows is that we cannot allow arbitrary conditions to determine sets and then indiscriminately consider the sets so formed to be members of other sets. The key words here are "arbitrary" and "indiscriminately". How do we, in view of the possibility of contradiction, discriminate between improper and proper uses of sets and properties? The attempt to solve this problem has been the genesis of most of the systems of foundations that have been propounded until the present. Each foundational system may be thought of as representing one answer to the question, "What are the permissible rules for operating with sets and properties?"

Now, clearly the answer one gives to the question of the legitimate use of sets and properties rests on what one considers to be the source of contradictions in the first place. At the turn of the century, different mathematicians gave different explanations for the source of the contradictions. We shall see how each of these different analyses gave rise to a different foundational system. Since it is not just Russell's, but other contradictions as well that must be avoided, let us first consider several other well-known paradoxes.

Cantor's paradox has to do with the notion of cardinal number. We say that two sets have the same cardinality if there is a 1–1 correspondence between them. The set A has cardinality greater than or equal to the cardinality of the set B (in symbols card(A) $\geqslant$ card(B)) if there is a 1–1 correspondence between B and a subset of A. Finally, A has greater cardinality than B, card(A) $>$ card(B), if card(A) $\geqslant$ card(B) and card(A) $\neq$ card(B). Cantor proved, by contradiction, that the set of all subsets (the power set) of a given set must have greater cardinality than the set itself. (See our discussion of Cantor's theorem in Chapter 7, Section 7.2.) Thus, we can obtain an unending hierarchy of infinite cardinalities by starting with the set of all natural numbers and iterating the operation of power set. Now what about the set of all sets V? The power set of V is just V itself. And, by Cantor's theorem, the power set of

V must have higher cardinality than V. Yet, V (the power set of V) cannot have higher cardinality than itself (the identity mapping gives a 1–1 correspondence). Hence, we have a contradiction.

Cantor's contradiction can also be deduced in Frege's system without passing through Russell's paradox, but the deduction is more complicated than the deduction of Russell's contradiction alone.

Berry's paradox is the following one: Consider all sentences of the English language with less than 200 letters. There are only finitely many such sentences. Thus, the set of all the natural numbers nameable by sentences of less than 200 letters must be finite. Its complement must be nonvoid and consequently must have a least element. This least element will be the least number not nameable by a sentence with less than 200 letters. But this number has just been named by a sentence with less than 200 letters.

The liar paradox: "The sentence in quotation marks and which occurs immediately after the words 'The liar paradox' on page 98 of Chapter 3 of *The Logical Foundations of Mathematics* by W. S. Hatcher is false." This last sentence says of itself that it is false and so it is true if and only if it is false.

There are many other forms of these paradoxes. The liar paradox in particular is subject to seemingly infinite reformulation. Envisage, for example, a card with two sides A and B. On side A is written: "The sentence on side B of this card is false." On side B is written: "The sentence on side A of this card is true." Now it is easy to see that the sentence on side A is both true and false, an impossible situation.

Each of these paradoxes results from methods of reasoning and ways of thinking that do not appear to be intuitively unjustifiable. Thus, it is our intuition itself which is shown to be faulty and which must be revised and rendered more sensitive. But we must first decide exactly what the true source of the paradoxes is.

3.6. Brouwer and intuitionism

One of the earliest explanations of the contradictions was given by L. E. J. Brouwer, a Dutch mathematician. Brouwer's basic contention was that the fault lies in our assumption of the existence of infinite sets and our unjustified application of certain principles, true for finite sets, to such infinite sets. For Brouwer, there is no such thing as a logic that is true apart from mathematics. For him, mathematics itself is simply the precise part of our thinking, and no general logical principles, independent of mathematical content, can be found. In particular, the so-called "law of the excluded middle" (represented by $(P \lor (\sim P))$) is true for finite sets, but not for infinite sets.

We can admit that some sets, such as the natural numbers, are "potentially infinite" in the sense that we can think of no highest natural number. But we do not have the right to "collect" the natural numbers into one packet called the *set of natural numbers* and then, in turn, to reason with this set by allowing it to be a member of other sets (such as its own power set), and so on.

Brouwer's position has become known as *intuitionism*. The term derives from his conviction that our intuition of the natural numbers, beginning with an object and allowing each next one to be determined by a simple operation, is the most basic possible intuition on which to found mathematics. For intuitionists, this intuition of the natural numbers is much more basic than our intuition of any "logical laws" such as the law of excluded middle. In particular, Brouwer held that any attempt to construct natural numbers from some other intuition (as Frege and Dedekind had done) was basically mistaken.

Thus, Brouwer avoids the paradoxes essentially by restricting the logical tools used in reasoning about mathematical objects, and also by requiring that mathematical objects be defined or generated in a certain, more restrictive manner.

To show that a set has members, it is not enough, for Brouwer, to show that the assumption that it is empty leads to a contradiction. One must give a procedure which is in some sense constructive and which allows us to exhibit at least one member of the set. Thus, much of Brouwer's criticism can be formulated in terms of the conditions of existence in mathematics. For intuitionists, freedom from contradiction is not enough to assert existence.

There is a strong rebuttal to Brouwer's argument about existence and it is contained in the completeness theorem of first-order logic. This theorem shows us that any consistent system has a model. In other words, whenever we are talking consistently, we are talking *about* something. There is no freedom from contradiction without the existence of structures satisfying the noncontradictory ideas.

Needless to say, Brouwer would refuse to admit the completeness theorem as valid, since its proof uses general set-theoretic ideas which are not wholly constructive. But for the mathematician who believes that intuitive set theory is consistent, Brouwer's argument about existence has little weight.

It should be pointed out here that some constructivists had taken their essential position with regard to the nonexistence of infinite sets before Russell's paradox was well known. In the light of this observation, intuitionism and constructivism (see our discussion in Chapter 2, Section 2.3) should be considered as positive viewpoints on the nature of mathematics rather than as reactions to the paradoxes. Of course, the appearance of the paradoxes was certainly taken by constructivists as a confirmation of their viewpoint.

For most mathematicians, intuitionism is much too radical to be acceptable. Not only are the paradoxes avoided, but considerable portions of modern mathematics are also excluded. Cantor's hierarchy of transfinite cardinals is rejected and so are certain basic properties of the real numbers. The completeness of the reals cannot be established by constructive methods, for example. Over the years, most students of intuitionism have worked on the problem of seeing just how much mathematics can be reproduced from a purely constructive viewpoint. Generally speaking, the results show that most of elementary number theory can, but that considerable portions of analysis cannot be recaptured. Even those portions of analysis which can be reproduced are extremely complicated from the intuitionist viewpoint.

Any discussion of constructivism should always point out that constructive proofs are interesting even to the mathematician who does not accept a philosophy of constructivism, because proofs which satisfy constructive conditions often furnish additional useful information. If such a proof tells us how to exhibit a member of a set, then this gives us more information than if we simply know that there is something in the set.

We have discussed intuitionism at some length here, because we do not intend to develop it in any detail. In particular, it does not satisfy criterion number 1 of Chapter 2. It does not allow for the development of a portion of mathematics which we regard as reasonably large and thus cannot be considered as having solved the problem of giving a foundation *for mathematics*. But intuitionism does represent a philosophical position concerning the nature of mathematics which cannot be refuted in any simple way. Let us again remind ourselves that we have, as yet, no proof that the mathematics for which we are trying to give a foundation is itself consistent.

3.7. Poincaré's notion of impredicative definition

Henri Poincaré held that the source of the paradoxes was a certain type of definition of sets which Poincaré regarded as illegitimate. What impressed Poincaré about the paradoxes was the aspect of "circularity" or self-reference inherent in them. For example, in Cantor's paradox we consider the set V of all sets and then we consider V as an element of itself. Now if we think of a set as being well defined only when each of its members has been specified, then V cannot be defined until each of its members is specified. Hence, if V is an element of V, then V is not well defined until V has been specified. This is circular. Definitions involving this type of circularity, in which an object is defined with reference to sets whose existence in turn depend on the object to be defined, were called *impredicative* definitions by Poincaré. Let us quote Poincaré's own statement of the matter: "[Impredicative definitions are] definitions by a relation between the object to be defined and *all* the objects of a certain kind of which the object to be defined is itself supposed to be a part (or at least some objects which depend for their definition on the object to be defined)." (See Poincaré [2], p. 7.)

Poincaré never attempted to pursue his idea and elaborate a system for mathematics based on his notion. The reason for this is that Poincaré had a distaste similar to Brouwer's for attempts to construct the natural numbers from other axioms or principles. As did Brouwer, Poincaré regarded the natural numbers as the most basic possible structure with which to begin mathematics. In several articles, he engaged in various polemics with logicians of his day and defended in particular the law of mathematical induction as an irreducible logical principle.

In a first attempt to evaluate the notion of impredicative definition as a source of paradox, let us observe that no obvious logical flaw is involved, especially if we view sets as independent, pre-existing objects. Viewing sets in this way, a definition does not have to show us how to construct a set. Rather, it only has to pick out some particular set from our pre-existing realm of sets. Suppose, for example, that we wish to argue on this basis that the set V of all sets is legitimate. We can take the position that, after examining our pre-existing realm of sets, we find that there is among them a particular one which happens to contain as elements all the other sets and which, in fact, has itself as an element. We call this set V. True, a set is determined by the objects which are its elements (principle of extensionality), and this is so with V. V is one of the objects which help to determine V. When we define V as $\{x \mid x = x\}$ or some other such definition, we are not constructing it, but only picking it out from among other sets.

On the other hand, when we view the definition of the set V of all sets as illegitimate, we seem to have in mind some sort of constructive or progressive generation of sets by which sets are generated at any stage only from those sets already existing. Thus, V would be illegitimate, since it is not generated exclusively from sets that exist *before* (notice the temporal or iterative implication) V exists. Thus, implicit in Poincaré's analysis of the paradoxes is a constructivism not unlike Brouwer's.

It can be inferred from much of what Poincaré said on the subject that he believed that all mathematics could be constructed from the integers without ever using impredicative definitions. As it turns out, however, this is not possible. Hence, we can only speculate about what Poincaré would have done when forced to choose between his own notion, with its abandonment of certain parts of mathematics, or rejection of his notion and acceptance of the "nonconstructive" or impredicative parts of mathematics.

Let us see briefly how impredicative procedures are involved in the classic Frege–Dedekind construction of the real numbers. Given the rational numbers, constructed as equivalence classes of ordered pairs of integers, we can construct the real numbers by the method known as Dedekind cuts. First we define a *cut* as a nonempty set of rational numbers that contains no largest rational, that does not contain all rationals, and that contains every rational q less than a rational r in the cut. We define an algebra of cuts by adding and multiplying them in a way that makes the set of cuts satisfy the axioms of an ordered field. We then undertake to prove that the field of cuts is complete, and thus make it a model for the real number system. It is at this point that an impredicative procedure enters into consideration.

Let a bounded set of real numbers (each real number is now a cut, a certain set of rational numbers) S be given. We want to prove that S has a least upper bound. The proof proceeds by considering the set B of rationals which is the union of all the real numbers (considered as cuts) in S. It is easy to verify that this set B is itself a cut and that it is the least upper bound of the set S. (The reader should consult some book in analysis in order to supply the details from our sketch.) But is it legitimate to define B the way we have? Not if we refuse impredicative definitions, for the totality of cuts (real numbers) is defined only when all members of the set (all cuts) are specified. Thus, the totality of all real numbers is defined only after B is specified. But then we define B itself by a relation over all real numbers, for we must take each real number, decide whether it is in S or not, and if it is, contribute its members to the union of all the cuts in S. In particular, B itself is a real number which is either in S or not, and so we must decide whether or not B is part of the union. And yet B is defined by the very procedure that ultimately depends on the existence of B itself.

Again we can see that no obvious logical flaw is involved, but the procedure is impredicative. So those who would refuse impredicative procedures must refuse this construction. No way is known to construct analysis without recourse to impredicative definitions. Hermann Weyl, in his book *Das Kontinuum*, concluded from this that analysis was not well founded and that part of classical mathematics should be sacrificed.

Later on, we shall consider a formal system propounded by Bertrand Russell in which the notion of impredicativity is made quite precise. This system, the theory of types, was influenced by Poincaré's notion of impredicativity. Thus, although Poincaré did not himself elaborate a system, one can see some of this thought reflected in the analysis of the paradoxes given by Russell and incorporated in his famous theory of logical types.

3.8. Russell's principle of vicious circle

Poincaré's notion of impredicativity was formulated by Russell as the principle of vicious circle. To avoid the paradoxes, said Russell, one must be careful to consider the *range of significance* of a given open sentence. To quote Russell: "Every proposition containing *all* asserts that some propositional function [open sentence] is always true; and this means that all values of the said function are true, not that the function is true for all arguments, since there are arguments for which any given function is meaningless, i.e., has no value. Hence we can speak of *all* of a collection when and only when the collection forms part or the whole of the *range of significance* of some propositional function, the range of significance being defined as the collection of those arguments for which the function in question is significant, that is, has a value." (This statement is contained in the paper Russell [1] and in the book van Heijenoort [1], p. 163, where Russell [1] is reprinted.)

The principle of vicious circle says that the values which are meaningless for a given open sentence must be those involving the type of supposedly illegitimate self-reference previously discussed. "The principle which enables us to avoid illegitimate totalities may be stated as follows: 'Whatever involves all of a collection must not be one of the collection'; or, conversely: 'If, provided a certain collection had a total, it would have members only definable in terms of that total, then the said collection has no total.'" (Russell and Whitehead [1], p. 37.)

In the next chapter, we shall examine Russell's system in detail.

3.9. The logical paradoxes and the semantic paradoxes

Ramsey [1] pointed out that the paradoxes were essentially of two different kinds. Paradoxes which involve notions directly expressible within a set-theoretic language such as **F** are called *logical* paradoxes. Russell's paradox and Cantor's paradox are logical paradoxes. On the other hand, the *semantic* paradoxes involve notions such as "truth" or "definability". These notions cannot be *directly* formulated within a formal language of the kind usually sufficient for the expression of mathematical notions. Berry's paradox and the liar paradox are semantic paradoxes.

In Chapter 6 we shall study Gödel's arithmetization of formal languages. This technique consists of assigning natural numbers to signs and expressions of a given language. It then becomes possible to talk about the language by talking about the corresponding numbers of the signs and expressions. Since, as we have seen, number theory is derivable in a sufficiently rich set-theoretic language, there is a sense in which certain semantic notions can be expressed in such a set-theoretic language after all.

Let us recall, for example, that in Chapter 1 we gave a purely set-theoretic definition of truth for first-order theories. Of course this was in intuitive set theory, but such a definition is expressible in sufficiently rich formalized set theories by means of our arithmetization process (if not otherwise). Tarski has shown (see Tarski [1], pp. 152 ff.) that if such a language F is so rich that the concept "true in F" is definable in F, then we can derive contradictions in F by a reasoning quite analogous to that of the semantic paradoxes.

Ramsey's distinction serves, however, to emphasize again the usefulness of formalization in clarifying the exact degree of expressiveness possible for a given axiomatic system stated within a given language.

Chapter 4

The Theory of Types

BEGINNING in 1908, Russell proposed a system called the *theory of types*, which was to serve as a foundation for mathematics. It represented Russell's answer to the paradoxes and was founded on Russell's vicious circle principle which we discussed in Chapter 3. The form of Russell's theory presented in the famous *Principia Mathematica* written in collaboration with A. N. Whitehead is very complex and contains some partially conflicting tendencies which will be explained later. We shall develop first a theory **PT** of *predicative type theory*, weaker than the system **PM** of *Principia Mathematica*, but like it in form and spirit. We shall then proceed to *type theory* **TT**, a system essentially equivalent to the system **PM** but simpler in form. Finally, we shall present a simplified theory **ST** of *simple types*, which is a reformulation by Gödel and Tarski of type theory. During the course of the discussion we shall consider briefly a system **RT** of ramified types.

4.1. Quantifying predicate letters

Fundamentally, type theory is a generalization of the first-order predicate calculus by allowing, within the language, quantification of predicate letters as well as of individual variables. The different versions of type theory result primarily from different choices of rules and axioms for operating within this generalized language, though there will be some differences in grammar also.[†]

It might appear at first that such a generalization of first-order logic would be straightforward, but this turns out not to be so. In fact, the different versions of type theory result from different ways of dealing with the difficulties which do arise.

In order to appreciate the nature of these difficulties, let us begin by formulating the most straightforward possible generalization of first-order logic, admitting quantification over predicate letters. Our alphabet consists of the alphabet of the pure predicate calculus plus the following: (1) all individual constant letters; (2) all symbols B_j^i with $i, j \geqslant 1$. The latter are called *constant predicate letters* while the predicate letters A_j^i are now called *variable* predicate letters. The set of wffs is defined as follows: (1) All variables and constants are terms. (2) $A_i^n(t_1, \ldots, t_n)$ and $B_k^n(t_1, \ldots, t_n)$ are wffs where the t_i are any terms. These are *prime* formulas.

[†] There is an interesting anachronism in this otherwise accurate description of type theory: historically, type theory preceded first-order logic which is thus more properly viewed as a specialization of type theory. However, the subsequent development of first-order logic amply justifies the central place we have given it in this study. Cf. also the last section of this chapter.

(3) If Y is a formula and X is any variable, then $(X)Y$ and $(EX)Y$ are formulas in which X is bound. (4) Iteration with our sentential connectives in the usual way yields formulas when applied to formulas. (5) Nothing else is a formula or a bound variable except by the above.

Thus, we now have as formulas such expressions as $A_1^2(A_1^1, x_1)$, $A_1^1(A_2^1)$ and $(EA_1^1)\, A_1^2(A_1^1, x_2)$.

The rules for deduction within this language may now be taken to be the obvious generalization of our natural deduction rules for first-order logic. The constant letters are used for dummy constants in applications of rule eE. The only restriction is the obvious one on substitution: For any rule involving substitution of terms for variables, any term substituted for a variable must have the same superscript as the variable. Otherwise, the rules of Section 1.6 all apply immediately to our generalized language.

The restriction on substitution is necessary because of clause (2) in the definition of the formulas which requires that the number of terms involved in the formation of a prime formula correspond to the argument number of the (constant or variable) predicate letter. We call the argument number of a predicate letter its *degree* and individual variables are said to have degree 0. The restriction thus amounts to requiring that substitution must take place only among terms of the same degree. Failure to observe the degree restriction on substitution would mean that substitution applied to a wff could yield a non-wff expression as a result, a clearly unacceptable situation.

Let us now think of what interpretation we might give to our generalized predicate calculus. As in the case of the first-order predicate calculus, we may start with a nonempty domain D of *individuals* which are thought of as values for the individual variables x_i. Also, as before, the predicate letters such as A_1^2 will be thought of as representing relations, but not in the same way: for A_1^2 is now a quantifiable variable, and so cannot be given a fixed interpretation. It has all possible binary relations (in whatever universe we are considering) as values. Moreover, we cannot even think of these binary relations as holding only between individuals (members of D) since expressions like $A_1^2(A_1^1, x_1)$ and $A_1^2(A_1^1, A_2^1)$ are now wffs. Thus, we now have relations between individuals and sets, sets and sets, and even between relations and relations.

Thus, in order to define an interpretation, we must not only specify the domain D of individuals, we must specify, for each degree n, an appropriate domain D_n of n-ary relations which serves as a universe of discourse for the variables of degree n. For $n = 0$, $D_n = D$ and we will speak of the individuals as 0-ary relations.

We can see, then, that our generalized language has increased immensely in expressive power. Indeed, given any sequence of domains D_n of n-ary relations, and without any further interpretation whatever, every wff involving only variables now expresses a relation. For example, the wff $A_1^2(A_1^1, x_1)$ expresses the ternary relation "to be a binary relation between a singulary relation (i.e. a set) and an individual". Or, $A_1^1(A_1^1)$ expresses the singulary relation "to be an element of itself".

It is natural to want to have some device within our language for naming such relations expressed by our wffs so that we can talk about them in our language. We now introduce such a device due to Russell. It amounts to introducing for each $n \geqslant 1$ a restricted n-ary *vbto* of abstraction into the language.

Let $A(z_1, \ldots, z_n)$ be any wff having exactly the distinct variables z_i free in *alphabetic order*. For our generalized language, alphabetic order means that we order first according to degree in increasing order, then according to subscript number among the variables having the same degree. We now form a term of degree n by placing a circumflex "^" over each free variable in A and enclosing the result in parentheses. We represent this by writing $(A(\widehat{z_1}, \ldots, \widehat{z_n}))$. The variables thus circumflexed are bound, and the usual restrictions concerning bound variables apply to these terms which we call *abstracts*.

We now extend our language by allowing formulas to be also of the form $(A(\widehat{z_1}, \ldots, \widehat{z_n}))(t_1, \ldots, t_n)$ where each term t_i has the same degree as z_i. We extend our rules by allowing substitution of abstracts of degree n for variables of degree n, as well as the following special axiom scheme:

$$\vdash (A(\widehat{z_1}, \ldots, \widehat{z_n}))(y_1, \ldots, y_n) \equiv A(y_1, \ldots, y_n),$$

where the y_i are variables having the same degree as the corresponding z_i, where each y_i is free for z_i in $A(z_1, \ldots, z_n)$, and where $A(y_1, \ldots, y_n)$ is the result of replacing in the formula $A(z_1, \ldots, z_n)$ each free occurrence of z_i by y_i.

The first difficulty now appears: our generalized logic is contradictory, for we can deduce Russell's paradox within it. We proceed to do so.

We begin with the wff $A_1^1(A_2^1) \equiv A_1^1(A_2^1)$ which is a tautology and thus a theorem of our system. Applying the rule $i\forall$ we obtain

$$\vdash (A_2^1)(A_1^1(A_2^1) \equiv A_1^1(A_2^1)).$$

Applying now iE, we can obtain

$$\vdash (EA_3^1)(A_2^1)(A_3^1(A_2^1) \equiv A_1^1(A_2^1)).$$

Finally, we apply $i\forall$ again to obtain

$$\vdash (A_1^1)(EA_3^1)(A_2^1)(A_3^1(A_2^1) \equiv A_1^1(A_2^1)).$$

We now apply $e\forall$ substituting the term $((\sim \widehat{A_4^1}(\widehat{A_4^1})))$ to obtain

$$\vdash (EA_3^1)(A_2^1)(A_3^1(A_2^1) \equiv ((\sim \widehat{A_4^1}(\widehat{A_4^1})))(A_2^1))$$

which gives us $\vdash (EA_3^1)(A_2^1)(A_3^1(A_2^1) \equiv \sim A_2^1(A_2^1))$ by our axiom for operating with terms formed by circumflexion. Applying eE we obtain

$$\vdash (A_2^1)(B_1^1(A_2^1) \equiv \sim A_2^1(A_2^1))$$

(where B_1^1 is some dummy constant letter). Finally, we employ $e\forall$ to obtain $B_1^1(B_1^1) \equiv \sim B_1^1(B_1^1)$ which is a contradiction.

Our generalized logic is, essentially, an alternative formulation of Frege's system **F** since the notion "$x \in y$" of **F** can be rendered here by a wff of the form $y(x)$, and since our abstracts formed by circumflexion play the role of the abstraction operator in **F**. In fact, the grammar of the present system is much closer to that of Frege's original one than is the grammar of **F**.

Having seen how the straightforward generalization of the predicate calculus leads to difficulties, we now want to see exactly how Russell's type theory deals with these problems.

4.2. Predicative type theory

Let us recall that Russell's principle for overcoming the paradoxes is the vicious-circle principle which hinges on restricting the "range of significance" (domain of values) for predicate variables. It is not easy, at first, to see how to apply this principle in the present case since the domain of values we assign to variables of a certain type depends on the choice of an interpretation whereas the deduction of Russell's paradox is a purely formal, syntactical affair which does not depend on interpretations in any way. Clearly, if we are to avoid contradiction we must modify the language or the deductive system or both.

We must therefore carefully separate the semantic and syntactic aspects of the situation.

In fact, Russell's exposition of type theory, especially in *Principia Mathematica*, did not make this clear distinction. As a result his treatment is sometimes confused and often difficult to follow. In order to avoid similar difficulties here, then, we must keep in mind from now on that type theory has two basic, distinct aspects: (1) the language and deductive system, and (2) an intuitively-given semantic interpretation. Since the way that the language and deductive system are modified is motivated by the modifications of the semantic interpretation, we begin our discussion with the semantic aspect of type theory.

The key idea in modifying the semantics of our generalized logic is the observation that the domains D_n, which served as domains for the variables of degree n in the previous section, had no relationship to each other. They were totally arbitrary except that their elements all had to be n-ary relations. This arbitrariness was reflected in the language by the fact that we could express very complicated relations such as relations applying to themselves, etc. This over-expressiveness resulted in contradiction: in fact we could express *everything* in the language, both truth and falsehood! In formulating our new system we will restrict this over-liberal expressiveness of the language by restricting the way predicate variables can have arguments. In particular, variables will not be able to have themselves as arguments.

On the semantic level this means that we must introduce more structure into our interpretations. We will define a *type hierarchy* in such a way that it is completely determined by the set D of individuals. The basic idea is that all the other relations in the hierarchy are built up from D by iterating the two basic operations of taking cartesian products and forming subsets. The hierarchy is thus progressively generated in such a way that relations at one level are all strictly built up from relations of a lower level, with the individuals at the bottom level. However, because we iterate with two operations, natural numerals will not suffice as type symbols.

As a result of the modifications of the formal system on the one hand and of the structuring of our interpretations on the other, we will obtain interpretations which are actually models of our system. None of the interpretations D_n of Section 4.1 could have served as models for our generalized predicate calculus of that section since its contradictory character precluded its having any models.

We now begin to make all of this precise by defining what we mean by a type symbol:

Definition 1. A *type symbol* is any expression which satisfies the following recursive definition: (i) "o" is a type symbol. (ii) If $t_1, \ldots, t_n$ are all type symbols, then the expression $(t_1, \ldots, t_n)$ is also a type symbol. (iii) These are the only type symbols.

Examples of type symbols are (o) and $(o, (o, o))$. The usefulness of type symbols in defining a type hierarchy is made immediately clear by the following definitions:

Definition 2. Given any nonempty set D, *a typed relation based on D of type t* or simply a *typed relation on D of type t* is defined as follows: (i) Elements of D, also called *individuals*, are typed relations of type o. (ii) A typed relation of type $t = (t_1, \ldots, t_n)$ is a set R of n-tuples such that $\langle r_1, \ldots, r_n \rangle \in R$ only if r_i is a typed relation of type t_i, for each $1 \leqslant i \leqslant n$. (iii) The only typed relations on D are those that have types according to the preceding.

Definition 3. By the *type hierarchy based on* any nonempty set D, we mean the collection consisting of all the typed relations on D. We also say that D *generates* the type hierarchy based on D. By *a type hierarchy*, we mean the type hierarchy based on some nonempty set D.

It is the type hierarchies which will be the interpretations of our new language, called the *language of type theory*. Of course, these definitions involve general set theory just as the semantics for first-order logic involved general set theory. Nevertheless, we will subsequently see that these semantic notions can be rendered syntactically by means of *translations* from one theory into another. Such translations will be purely formal associations between two languages which define the semantics of one language in the other. For the moment, we will simply work in intuitive set theory. As with first-order theories, we will make no appeal to general set theory in defining the formal system of type theory. Indeed, our main purpose in considering type hierarchies here is to furnish an intuitive basis for understanding formal type theory.

Exercise. Let D and D' be two nonempty sets with the same cardinality. Prove, in intuitive set theory, that the type hierarchies based on D and D' are isomorphic in a natural way. This shows that a type hierarchy is essentially determined by the cardinality of the set D of individuals of the hierarchy.

For a type hierarchy based on any nonempty set D, we call the elements of D individuals, but nothing excludes the possibility that the elements of D are, themselves, sets with elements. However, the foregoing exercise shows that, for the purposes of type theory, the nature of the individuals is not of any essential importance. In particular, our formal language of type theory will be so formulated that it will not be possible either to affirm or deny within the language that an individual has elements or is otherwise structured. This neutrality regarding the nature of the individuals allows us to regard them as "uncomposed" or "elemental" as Russell did, but it does not force us to do so.

One thing we have not yet made precise is the notion of "level" in a type hierarchy. The term used by Russell was "order" and so we maintain this usage in the following definition:

Definition 4. The *order* of a type symbol is given by the following recursive definition: (i) The type symbol "o" has order 0. (ii) A type symbol $(t_1, \ldots, t_k)$ has order $n+1$ if the highest order of the type symbols $t_1, \ldots, t_k$ is n.

It is easy to give a parenthesis-counting algorithm for determining the order of a type symbol. Begin counting parentheses from left to right, adding $+1$ for each left parenthesis and -1 for each right parenthesis. Then the order of the type symbol is the highest (necessarily nonnegative) integer obtained in the counting process. For example: The order of o is zero, the order of $(o, o, (o), ((o), (o, o)), (o))$ is three, and the order of (o, o, o) is one. (The process always ends with a count of zero, but we are interested only in the highest integer obtained in the counting process.)

Definition 5. By the *order* of a typed relation in a type hierarchy we mean the order of the type symbol which is its type.

Thus, a relation is "further up" than another one in a type hierarchy if its order is higher than the order of the other. Only individuals have type 0 and so are on the lowest level. The progressive generation of a type hierarchy is represented by the fact that the components r_i of any n-tuple $\langle r_1, \ldots, r_n \rangle$ in a typed relation R all have lower order than R.

Notice that a type hierarchy is limited not only in that all the relations in it are built up from the set D of individuals, but also by the fact that not all relations that can be built up from D are typed relations. The essential feature of a typed relation is that each of its arguments can have as values only relations of some fixed type. For example, a relation R of type $((o))$ can have only sets of individuals as elements. Thus, if x is an individual, and if x is not a set of individuals, then no relation of type $((o))$ can have both x and $\{x\}$ as elements. Yet, the set $\{x, \{x\}\}$ is built up strictly from individuals by forming only sets whose elements comprise relations of lower order.

When we construct our formal language of type theory, we shall have continually in mind the interpretation of some nonempty set D and the type hierarchy based on D. All the variables in our language will have type indices, and will be thought of as ranging only over relations of the indicated type. We shall also form terms by circumflexion (abstraction), and every term so formed will also have an unambiguously designated type symbol associated with it. This symbol will be called the type of the term. A closed term (an abstract with no free variables or a constant) will be thought of as naming some entity of the indicated type. Quantification will be applied only in the name of typed variables, and substitution of terms for variables will be allowed only for terms of the same type as the variable for which substitution is being made. Predicate variables can have only terms of the indicated type as arguments. There will be no function letters, and so the only terms are variables, constants, or abstracts.

In such a system, an expression of the form $x(x)$, where x is of a given type, will not even be well formed, since whatever the type t of the variable x may be, the type of any variable which has x as argument must be (t). It follows that x cannot have itself as an argument, and the bothersome property $\sim x(x)$ is not expressible in our system. This point will be made clearer after our formal language is given.

We have already mentioned that Russell did not clearly separate the syntactical and semantic aspects of type theory as we have done here. For example, in *Principia Mathematica* the distinction between a relation in the type hierarchy and a term or variable of the formal language is continually confused. Russell defined a type in these words: "A 'type' is defined as the range of significance of some function." (Whitehead and Russell [1], p. 161.) By a "function" or "propositional function", Russell sometimes meant a relation and sometimes a wff with free variables (which would have a relation as its truth set). Quine [4] discusses some of the unfortunate results of this confusion.

We now define the language of type theory. The system we are now going to set up is called the system **PT** of *predicative type theory*. Our alphabet contains the small italic letter x with any positive natural number (numeral) subscript and any type symbol superscript. These are called *variables*. We have *constants* consisting of the small italic a with any positive natural number (numeral) subscript and any type symbol superscript. We also have parentheses, commas, the circumflex, and the signs "$\forall$" and "E". (We shall introduce in this system a special symbol "$\forall$" for universal quantification, because our parentheses are becoming dangerously overworked. Instead of the monstrosity $(x_1^o) \, x_1^{(o)}(x_1^o)$ we shall have the slightly more readable monstrosity $(\forall x_1^o) \, x_1^{(o)}(x_1^o)$. The subscripts are to distinguish variables or constants and the superscripts indicate the type. A variable or constant will be said to be *of the type* indicated by the type symbol it has as superscript.

Before defining the set of wffs of the system **PT**, we need to establish a convention concerning the order of variables in any expression. Given any expression A in the alphabet of type theory, and given any list $y_1, \ldots, y_k$ of distinct variables occurring in the expression, the *canonical order* of these variables *relative to A* is determined as follows: We first order the variables according to increasing order of subscript (alphabetical order). If there are then two or more distinct variables among the y_i having the same subscript, these variables must all be of different types (i.e. their superscripts are all different). We order these variables (for each case of distinct variables having identical subscripts) among themselves according to the order of first occurrence from left to right in the expression *(typographical order)*. It is understood that the typographically ordered variables are inserted as a segment into the alphabetical ordering at the precise position where their common subscript first appeared. The variables in the list $y_1, \ldots, y_n$ are now totally ordered. This total ordering is the canonical ordering (relative to the given expression). The reason for giving absolute priority to alphabetical ordering over typographical ordering in defining the canonical order will shortly appear.

We now define the wffs of **PT**: (1) A variable or constant of a given type is a term of that type. (2) Where A is a term of type $(t_1, \ldots, t_n)$ and where $y_1, \ldots, y_n$ are terms, each y_i of type t_i respectively, then $A(y_1, \ldots, y_n)$ is a wff of our theory. (3) Where A is any wff and y is some variable, then $(\forall y)A$ and $(Ey)A$ are wffs. As usual, quantified variables are considered bound. (4) If A and B are wffs, then $(\sim A)$ and $(A \lor B)$ are wffs. (5) Let A be a wff in which the variables $y_1, \ldots, y_n$ $(n > 0)$ are distinct free variables in canonical order relative to A (it is not necessary that these be all the free variables of A). Let t_i be the type of y_i. Then the expression obtained from A by placing a circumflex over *all occurrences* of the free variables in the list $y_1, \ldots, y_n$ and enclosing the result in parentheses is a term of type $(t_1, \ldots, t_n)$, the variables y_i being bound in the term so formed.[†] Moreover, if there are in A two or more variables of the same type, at least one of which is bound in forming the given term, then the variables of that type which are circumflexed must be written as subscripts on the term (in canonical order among the subscripted variables). (6) These are the only terms and wffs of **PT**.

The terms defined by circumflexion under part (5) of our definition are called *abstracts*. The reason for subscripting variables under certain conditions as prescribed in part (5) is to avoid ambiguities that would otherwise result. Consider, for example, the wff $(x_1^{(o)}(x_1^o) \supset x_1^{(o)}(x_2^o))$. Now, let us form an abstract by circumflexing the variable x_1^o. We obtain $((x_1^{(o)}(\widehat{x_1^o}) \supset x_1^{(o)}(x_2^o)))_{x_1^o}$, which is of type (o). Now, taking this term, we can form a wff $((x_1^{(o)}(\widehat{x_1^o}) \supset x_1^{(o)}(x_2^o)))_{x_1^o}(x_3^o)$.

We then form an abstract by circumflexing this time x_2^o. We obtain $(((x_1^{(o)}(\widehat{x_1^o}) \supset x_1^{(o)}(\widehat{x_2^o})))_{x_1^o}(x_3^o))_{x_2^o}$, again of type (o). Suppose, however, that we suppress the subscripts on the abstracts. We obtain in this case $(((x_1^{(o)}(\widehat{x_1^o}) \supset x_1^{(o)}(\widehat{x_2^o})))(x_3^o))$ which is ambiguous, for we do not know which of the two variables x_1^o and x_2^o was the first one circumflexed. This ambiguity cannot be tolerated, since there is considerable difference in meaning associated (or to be associated) with the two ways of reading the foregoing formula.

Notice, now, that by the way we have defined the canonical order of variables, any wff of our language can be simply transformed into a structurally similar one in which the canonical

---

† As before, we use the expression $(A(\widehat{y_1}, \ldots, \widehat{y_n}))$ as a metalinguistic name for this term.

order of all variables in it will be the alphabetical order. We simply replace (in any way we choose) every distinct variable of a given type which happens to have a subscript identical to a variable of another type by a completely new variable of the same type. In any case, substitution of terms for variables will only take place within a given type, and the canonical order is just the alphabetical order among variables of the same type.

In *Principia Mathematica*, Russell and Whitehead refer only to alphabetical order in forming abstracts (see page 200 of Whitehead and Russell [1]). However, they practiced what they called "typical ambiguity", namely the suppression of all type indices.[†] Thus, their language was never completely defined in a rigorous way as a formal system. In particular, then, the problem of defining alphabetical order among type symbols never arises in their study, by way of imprecision. Since there is no natural alphabetic ordering of the type symbols, we have resorted here to the canonical ordering which is the alphabetical ordering among variables of the same type.[‡]

As we already remarked in Section 4.1, the circumflex notation for forming abstracts is due to Russell. In his treatment, Russell made no attempt to relieve the ambiguity inherent in unrestricted circumflexion, and this represents a further ambiguity in *Principia Mathematica*. The present method of the limited subscripting of circumflexed variables is our own modification of Russell's notation. Notice that we did not need subscripts in the system of Section 4.1 since we there required all free variables of a formula to be circumflexed in forming an abstract. The present way of forming abstracts is thus more general in that the choice of free variables to be circumflexed is unrestricted.

Clearly, the abstracts formed by circumflexion are closely related to terms defined by the *n*-ary *vbtos* of Chapter 1: Given a wff A or our present language and n distinct variables $y_1, \ldots, y_n$ in *any order*, of types $t_1, \ldots, t_n$ respectively, we could form the term $((\widehat{y_1}, \ldots, \widehat{y_n})A)$ of type $(t_1, \ldots, t_n)$ in which the variables y_i would be bound. This would amount to adding an *n*-ary *vbto* to our language for each $n \geqslant 1$.

Terms of the form $((\widehat{y_1}, \ldots, \widehat{y_n})A)$, formed by the use of *vbtos*, are more general than the abstracts of our present language on two counts. First, the variables y_i do not necessarily have to appear in A for terms formed by *vbtos*, whereas in the case of abstracts they do. Secondly, in the case of terms formed by *vbtos*, we can choose any one of the $n!$ possible orders of the prefixed variables $y_1, \ldots, y_n$, while for abstracts the order is fixed once and for all as the canonical one. The terms formed by abstracts thus constitute a special case of terms formed by *vbtos*, namely the subset of those *vbto*-formed terms in which the prefixed variables all occur in the formula and such that the typographical order of the prefixed variables is the canonical order of these variables relative to the formula.

In *Principia Mathematica*, Russell and Whitehead did, in fact, introduce *n*-ary *vbtos* as in the above for $n = 1$ and $n = 2$, by means of so-called "contextual definitions" (which means roughly that the *vbtos* are not explicitly added to the alphabet but that symbols involving their use can be defined in certain contexts). Their approach was motivated by their desire to distinguish between general and extensional uses of terms, the former being rendered by abstracts, and the latter by terms involving *vbtos*. As we have no need of this distinction, we shall work directly with

---

[†] More will be said about typical ambiguity further on.

[‡] The usual way of ordering expressions over a finite alphabet is lexicographic ordering among each (necessarily finite) set of expressions having the same length. We have rejected this as a way of ordering the type symbols since it is not very natural for them and since it does not generalize to the order-type symbols of Section 4.5 for which the alphabet is infinite.

the abstracts as we have defined them. An alternative approach would be to use only terms defined by *n*-ary *vbto*s and avoid abstracts formed by circumflexion altogether. We have rejected
this approach primarily out of historical considerations. The spirit of our treatment here is to
maintain the closest reasonable approximation to Russell's original symbolism and presentation while correcting the errors and avoiding the ambiguities inherent in them.

Returning now to the development of the system **PT**, our next task is to define its axioms and
rules of inference. To do this, we need to introduce one further notion, namely that of the *order*
of a term.

Definition 6. The *order* of a variable or a constant is the order of the type symbol that is its
superscript. For an abstract *t*, let *n* be the highest order of all its variables (bound or free) and
its constants. If there is a bound variable *y* of order *n* in *t*, then *t* is of *order* $n+1$. Otherwise, the
order of *t* is *n* (in this case *t* must have some free variable or constant of order *n*).

Notice that abstracts may not have the same order as the order of their type. If *A* is an abstract whose type is of order *n*, *A* will not have order *n* if *A* contains bound variables of order
n or higher, or free variables or constants of order $n+1$ or higher. For example, the term

$$((\forall x_1^{(o)}) \, x_1^{(o)}(\widehat{x_1^o}))$$

is a term of type (*o*), and this type symbol has order 1. However, the term has order 2, since it
contains a quantified variable of order 1. Obviously the order of an abstract is at least as high
as the order of its type.

Let us see what this means in terms of our semantic interpretations. An interpretation of our
language is any type hierarchy. Let us choose an arbitrary but fixed interpretation. We thus
think of the variables of type *t* as having for values the relations of type *t* in this type hierarchy.
Now, a closed abstract (one without free variables) is thought of as defining or naming an entity
of its type. If a closed abstract *A* of type *t* does not have the order of its type, then it must contain bound variables whose type has order greater than or equal to the order of the type of *A*,
or else constants whose type has order greater than the type of *A*. In either case, the abstract
defines an entity of type *t* by referring to entities whose existence in turn depends on the existence of entities of the type we are defining. In short, we are involved in the kind of impredicative definition discussed informally and without reference to a type hierarchy in Chapter 3.

For an example, consider the abstract $((\forall x_1^{(o)}) \, x_1^{(o)} \, (\widehat{x_1^o}))$ of two paragraphs ago. The type
of this term is (*o*), and so this abstract defines an entity of type (*o*). But involved in the abstract
is the quantifier $(\forall x_1^{(o)})$, which quantifies over *all* entities of type (*o*), and thus the very entity we
are defining by the given abstract. The impredicative nature of this abstract is reflected in the
fact that it does not have the same order as the order of its type symbol.

On the other hand, the abstracts $(x_1^{(o)}(\widehat{x_1^o}))$ and $(a_1^{(o)}(\widehat{x_1^o}))$ are both of order 1, the same as the
order of their type.

These considerations lead us to define in a purely formal way the notions of *predicative* and
impredicative abstracts. An abstract is predicative if and only if the abstract has the same order
as the order of its type. It is impredicative otherwise; that is, if it has higher order than the order
of its type.

Since an abstract is predicative if and only if it has the same order as its type, and since we
have defined the order of a variable or constant to be the order of its type, it would follow

that such variables and constants should stand not for any arbitrary entity of the indicated type, but only for predicatively defined entities. When we formulate our rules of inference for **PT** we shall have precisely this interpretation in mind, and we shall allow substitution only of predicative terms for a variable of a given type. We thus declare all variables and constants to be predicative, for they have the same order as their type.

In the system of *Principia Mathematica* which we call **PM**, Russell did things in a different and more complicated manner. Variables were thought of as being differentiated not only by type, but further by order. To incorporate this into our formal system, we would have to add further indices to each variable to indicate the order of the variable and the order of each argument places of its type. For example, $x_1^{((o),\ (o,\ o))/5/(3,\ 2)}$ would be a variable of type $((o), (o, o))$, of order five, whose first argument was of type (o) and order 3 and whose second argument was of type (o, o) and order 2. (The 2 here has nothing to do with the fact that (o, o) has two argument places.)

In such a system, the order of a variable is no longer the order of its type, but the indicated order, such as "5" in the preceding example. Of course, the order of a variable cannot be less than the order of its type. For example, the order of any variable of type $((o), (o, o))$ must be 2 or more, since this type symbol has order 2. Similarly, the first and second argument places in the preceding example must be at least of order 1.

Another criterion of a well-formed term would be that the order of a term must be higher than the order of any of its arguments. This requirement does not otherwise follow from the already stated requirement that the order of a term cannot be lower than the order of its type. Thus, a term of type $((o), (o))$ whose arguments are of order $(3, 2)$ must be of order 4 at least, even though the order of the type $((o), (o))$ is only a. Again, this case is seen as part of our example.

As an example of an abstract that can be substituted for the variable in the above paragraph, we have

$$(((\forall x_1^{((o))/4/3})\,(x_1^{((o))/4/3}(\widehat{x_1^{(o)/3/0}}))\supset \widehat{x_1^{(o,\ o)/2/(0,\ 0)}}(x_1^{o/0},\ x_2^{o/0}))).$$

This abstract has type $((o), (o, o))$. The highest order of any variable in the abstract is 4 and that variable is bound. Thus, the abstract has order 5. The order of the first circumflexed variable is 3 and of the second 2. As might be inferred from this example, all variables of type o will have order 0.

In setting up rules for a system with order indices, we would not require that the orders of a term and its arguments be the same as those of a variable for which the term is substituted, only the same or lower, provided, of course, that the particular instance of substitution was acceptable according to our various other restrictions. This is intuitively justified by the understanding that the order of a term indicates now high up in the type hierarchy we have to go to define it. However, terms would still be required to have the same type as variables for which they are substituted.

The reader may begin to wonder at a system with such formidable complications. He will wonder even more if he glances at *Principia Mathematica* and sees no type indices whatever (with the exception of one or two instances). Russell and Whitehead did not present their work in a carefully defined formal system and all the complications of types and orders within types and variables with type and order indices were to be understood. The reader could put in the indices in any way consistent with the understood conventions.

In a later section, after we have given the rules and axioms for our simpler system **PT** of predicative types, we shall indicate what changes should be made to formulate a system with variables with order indices. This latter system is called the *ramified theory of types* **RT**. The system **PM** of *Principia Mathematica* was not exactly the system **RT**, but rather **RT** plus another axiom called the *axiom of reducibility*. This axiom, which will be explained later, actually had the effect of nullifying the order concept, thus making the system **PM** roughly equivalent to the system we shall call **TT** for type theory. We now return to complete our formulation of **PT**.

Let A be a wff with at least one free variable and let $y_1, \ldots, y_n$ be *all* the free variables of A in canonical order relative to A. Let $(A\widehat{y_1}, \ldots, \widehat{y_n}))$ be the abstract formed by circumflexing *all* the free variables of A, and let t be the type of this abstract. Then we define the *order* of the wff A to be the order of the abstract $(A(\widehat{y_1}, \ldots, \widehat{y_n}))$. We say that the wff A is *predicative* if the order of A is the same as the order of the type symbol t, which is the type of the abstract formed from A. Otherwise, it is impredicative.

With this definition, we can speak of the order of an open wff, and also determine whether or not an open wff is predicative. We could also define the order of a closed wff A to be $n+1$ where the term of highest order in A is a bound variable of order n or a constant of order $n+1$. We shall, however, never need to make use of the notion of the order of a closed wff. Whenever we speak of the order of a wff, we presume the wff to be open unless there is a specific mention to the contrary.

Exercise 1. Determine the order of the following type symbols:

$$(o, (o, (o, (o), ((o))))); \qquad (((o))); \qquad ((o), (o), (o), (o)).$$

Exercise 2. Give two examples of an abstract of each of the types of Exercise 1, one predicative and one impredicative. Do the same for wffs.

Exercise 3. Give an example of an abstract of order 7 and of type (o). The point here is that abstracts of a given type may have arbitrarily high order depending on the other variables which may occur in them. Give an example of a predicative wff of order 3; an impredicative wff of order 3.

Exercise 4. Formulate an algorithm for determining whether a given abstract is predicative or not.

Another useful concept is that of a wff $A(x)$ being predicative *relative* to a free variable x in $A(x)$. Given a wff $A(x)$, we say that $A(x)$ is predicative relative to the variable x if the term $(A(\hat{x}))$ formed by circumflexing x is predicative. We also say that $A(x)$ is predicative *with respect to* x in this case. If $A(x)$ is predicative relative to x, then the term $(A(\hat{x}))$ will have order $n+1$ where the order of x is n. This is clear, since the type of $(A(\hat{x}))$ is (t) where t is the type of x.

We are now in a position to state the axioms and rules of inference of the system **PT** of predicative type theory. Our rules of inference are our rules of natural deduction described in Chapter 1, which we modify in the following way: In rule eE, the dummy constants are just the constants of **PT** already defined. They need be new only in the particular proof and must have the same type as the variables they replace. In the rule $e\forall$, the term substituted for the variable must have the same type and order as the type of the variable originally quantified.

A similar remark applies for the rule iE: the term t in $A(t)$ must be predicative and of the same type as x in $(Ex) A(x)$.

For handling abstracts, we posit as an axiom a type-theoretic analogue of the axiom used to handle abstracts in our contradictory general predicate calculus:

PT.1. $(A(\widehat{y_1}, \ldots, \widehat{y_n})) (z_1, \ldots, z_n) \equiv A(z_1, \ldots, z_n)$ is an axiom where y_i are distinct free variables of the wff $A(y_1, \ldots, y_n)$ and $y_1, \ldots, y_n$ is their canonical order relative to it. The abstract is predicative, and for each i, z_i has the same type and order as y_i. $A(z_1, \ldots, z_n)$ is the result of replacing the term z_i for all free occurrences of y_i in $A(y_1, \ldots, y_n)$. In each case, z_i is free for y_i.

The requirement that terms substituted for variables be predicative and that the abstract of **PT**.1 be predicative is a characteristic feature of **PT**. We call this feature of **PT** the *predicativity restriction* of **PT**. It effectively excludes impredicative abstracts from the system, since we can never substitute an impredicative abstract for any variable nor apply **PT**.1 to them. (We could have formulated **PT** by excluding impredicative abstracts from the grammar of the system.) **PT**.1 will be the only axiom dealing with abstracts.

We shall state another axiom for **PT** which amounts to an axiom of extensionality. Russell actually treated the question of identity in a somewhat different manner which we have thought better to avoid here. While not requiring that all properties and relations satisfy the principle of extensionality, he worked only with those which did. Thus, our approach here involves no essential departure from Russell's theory.

Definition 7. $x = y$ for $(\forall z^{(t)}) (z^{(t)}(x) \equiv z^{(t)}(y))$, where x and y are any two terms of type t in which $z^{(t)}$ does not occur.

According to this definition, two terms are equal if they satisfy the same properties. Notice that equality is defined only between entities of the same type. Definition 1 is different from the definition used in the system **F** of Chapter 3. In our notation, that definition would be stated

$$(\forall z) (x(z) \equiv y(z)),$$

two terms are equal if the same arguments satisfy both of them. The trouble is that this definition is only valid for entities x and y which can have arguments, and this excludes individuals. Formally, we can see that for this alternative definition to be well formed, x and y must have type (t) where t is the type of z. Thus, x and y can never have type o.

Let us also recall that one consequence of our definition in Chapter 3 was that any two no-element things were equal; they were both the null set Λ. There was thus no possibility of different individuals; that is, different no-element things. Our present way of treating equality is motivated by our desire to allow for the possibility of distinct "atomic" individuals.

We now posit, as a final axiom scheme for **PT**, the following axiom of extensionality:

PT.2. $(\forall z_1) (\forall z_2) \ldots (\forall z_n) (x^{(t_1, \ldots, t_n)}(z_1, \ldots, z_n)$
$$\equiv y^{(t_1, \ldots, t_n)}(z_1, \ldots, z_n)) \supset x^{(t_1, \ldots, t_n)} = y^{(t_1, \ldots, t_n)}$$

where the z_i are of type t_i, and are all distinct variables that do not occur in the terms $x^{(t_1, \ldots, t_n)}$ and $y^{(t_1, \ldots, t_n)}$ of indicated type.

This axiom tells us that any two relations which are satisfied by the same arguments are equal. Our intuition tells us that the converse should also hold; that is, two equal relations should be satisfied by the same arguments. For an illustration, we prove this fact for *sets*, that is, for terms of type (t) where t is any type.

THEOREM 1. $\vdash x^{(t)} = y^{(t)} \equiv (\forall z^t)(x^{(t)}(z^t) \equiv y^{(t)}(z^t))$ *where t is any type and $x^{(t)}$ and $y^{(t)}$ are any terms of type (t), and z^t is a variable of type t not occurring in $x^{(t)}$ or $y^{(t)}$.*

Proof. (1) 1. $x^{(t)} = y^{(t)}$ H
      (1) 2. $(\forall w^{((t))})(w^{((t))}(x^{(t)}) \equiv w^{((t))}(y^{(t)}))$ 1, *Df.* 7
      (1) 3. $((\widehat{x^{(t)}})(z^t))(x^{(t)}) \equiv (\widehat{x^{(t)}}(z^t))(y^{(t)})$ 2, $e\forall$, the abstract is of type $((t))$, and it is
             predicative, z^t is new.
      (1) 4. $x^{(t)}(z^t) \equiv y^{(t)}(z^t)$ 3, **PT**.1, Taut, MP, the abstract is predicative and so
             PT.1 applies.
      (1) 5. $(\forall z^t)(x^{(t)}(z^t) \equiv y^{(t)}(z^t))$ 4, $i\forall$, z^t is not in (1)
          6. $x^{(t)} = y^{(t)} \supset (\forall z^t)(x^{(t)}(z^t) \equiv y^{(t)}(z^t))$ 1, 5, *eH*

This is one-half of the biconditional. The other half follows directly from the axiom of extensionality, and thus our theorem is established.

Exercise. Prove the generalization of Theorem 1 to the case in which x and y have any types, except type o in which we cannot state the theorem. The proof is by analogy with the proof of Theorem 1.

In the future, when we speak of Theorem 1, we shall presume the results of this exercise to be a part of it.

The importance of Theorem 1 and the foregoing exercise is that they show that we can take the right side of Theorem 1 as a definition of equality for all terms except those of type o. The relevance of this lies in the fact that the right side of Theorem 1 is predicative with respect to $x^{(t)}$ whereas Definition 7 is *not* predicative relative to x. We shall have occasion in the future to form abstracts involving equality, and we therefore always suppose that equality is given by the right side of Theorem 1 (unless, of course, we are dealing with equality between individuals).

The reflexivity of equality follows from Definition 1 and the fact that tautologies are theorems of our system. Definition 7 is, in itself, the principle of the substitutivity of equality. By our axiom **PT**.1, any predicative abstract of the appropriate type can be substituted for the variable $z^{(t)}$ in an application of $e\forall$. Applying **PT**.1 we obtain the general principle of the substitutivity of identity only for wffs that are predicative with respect to the variables x and y. We state the theorem:

THEOREM 2. $\vdash (\forall x)(\forall y)(x = y \supset (A(x, x) \equiv A(x, y)))$ *where $A(x, x)$ is predicative with respect to x, and y replaces x in A(x, x) in zero, one, or more occurrences, y being free for x in all those occurrences that it replaces. y has the same type as x.*

Proof. Let $A(x, x)$ be predicative with respect to x. Replace x by a new variable w in all places where we wish to replace x by y and obtain $A(x, w)$. w is of the same type as that of x and y.

Thus, $(A(x, \hat{w}))$ is of type (t) where t is the common type of x and y, and $(A(x, \hat{w})$ is predicative, since $A(x, x)$ is predicative with respect to x. Now, starting with

$$(\forall z^{(t)})(z^{(t)}(x) \equiv z^{(t)}(y)),$$

we obtain by $e\forall$ $(A(x, \hat{w}))(x) \equiv (A(x, \hat{w}))(y)$ and, by **PT.1**,

$$A(x, x) \equiv A(x, y)$$

which was to be proved. We can apply **PT.1**, since $(A(x, \hat{w}))$ is predicative.

It should be clear that if we removed the predicativity restriction from **PT** we would obtain immediately the general substitutivity of identity by the same reasoning as in the proof of Theorem 2 since the worry of the predicativity relative to the variables x and y would no longer be present.

Having now set up **PT**, we need to see what can be done with it and how it avoids the paradoxes.

4.3. The development of mathematics in PT

Let t be any type. By reasoning strictly analogous to that considered at the beginning of Section 4.1, we can deduce:

$$(Ex^{(t)})(\forall x^t)(x^{(t)}(x^t) \equiv A(x^t))$$

where $A(x^t)$ is any wff predicative relative to x^t. Briefly, starting with $x^{(t)}(x^t) \equiv x^{(t)}(x^t)$, which is a tautology and thus a theorem, we proceed by $i\forall$ to $(\forall x^t)(x^{(t)}(x^t) \equiv x^{(t)}(x^t))$ and by iE and $i\forall$ to

$$(\forall x^{(t)})(Ey^{(t)})(\forall x^t)(y^{(t)}(x^t) \equiv x^{(t)}(x^t)).$$

By applying $e\forall$ to this wff in the name of $(A(\widehat{x^t}))$, we obtain the desired result by using **PT.1**. $(A(\widehat{x^t}))$ is predicative, since $A(x^t)$ is predicative relative to x^t.

By applying **PT.1** directly, we obtain with greater facility the following metatheorem:

THEOREM 3. $\vdash (\forall x^t)((A(\widehat{y^t}))(x^t) \equiv A(x^t))$ where t is any type, $A(y^t)$ contains y^t free and is predicative relative to y^t, and x^t is free for y^t; $A(x^t)$ results from $A(y^t)$ by substituting x^t for y^t in all the latter's free occurrences.

Proof. This is an instance of **PT.1** to which we apply $i\forall$.

Obviously, $A(x^t)$ is predicative relative to x^t in Theorem 3.

Definition 1. $(x^t \in x^{(t)})$ for $x^{(t)}(x^t)$ where t is any type, and the terms are of indicated type.

Definition 2. $\{y \mid A(y)\}$ for $A(\hat{y})$ where $A(y)$ is any wff containing y free.

We do not intend to use this notation in **PT**, but it allows us to reformulate Theorem 3 in a way that may aid the reader to compare the latter with Frege's axiom of abstraction in the inconsistent system **F**.

COROLLARY. $\vdash (\forall x^t) (x^t \in \{y^t \mid A(y^t)\} \equiv A(x^t))$, *where the variables and wffs are subject to the same restrictions as in Theorem 3.*

Proof. Restatement of Theorem 3 in the notation of Definitions 1 and 2.

The differences between the foregoing corollary and Frege's contradictory axiom of abstraction are: (1) In **PT** our wffs are defined differently and in a more restrictive manner than in **F** (or in our contradictory general predicate calculus). Certain wffs expressible in Frege's system are not expressible in **PT**. (2) The wff $A(y^t)$ in Theorem 3 is predicative relative to y^t, whereas no such restriction is put on the wff of Frege's axiom of abstraction. Indeed, there is no clear way to define the notion of predicativity in **F**, since our definition in **PT** has depended on the type indices that are not present in **F**.

How do these differences allow us to avoid Russell's paradox? The contradictory property $\sim \hat{x}(\hat{x})$, "not an attribute of itself", cannot even be expressed in **PT**. Consider any wff of the form $y(x)$ and let t be the type of x. Then (t) must be the type of y and so y cannot be equal to x, since sameness of type is necessary to the equality of two terms. Thus, **PT** excludes Russell's paradox by the very way in which its wffs are defined.

Notice that the restriction of predicativity in Theorem 3, due ultimately to our predicativity restriction for **PT**, is not used in avoiding Russell's paradox. The paradox is avoided by excluding from our language certain kinds of expressions that give rise to contradiction. Of course Poincaré's notion of impredicative and predicative definitions certainly influenced Russell in his formulation of the vicious-circle principle, and in his conception of the type hierarchy. The restriction of predicativity in **PT** owes its existence primarily to our intuition of the constructive nature of a type hierarchy. We "feel" intuitively that it is wrong to define an entity of lower order by quantifying over things of higher order whose existence may depend on the entity we are defining. But, as it is often pointed out, this intuition breaks down if we view sets (even the sets of a type hierarchy) as preexisting, independently given entities. In this case, definition by an abstract serves only to pick out a preexisting one from those already present. Later we shall see that predicativity is a restriction which prevents us from developing all mathematics within **PT**, and we must ultimately choose between mathematics in its classic form and our love for predicativity.

Let us now turn to the question of how mathematics can be developed within **PT**. First, let us see how Frege's constructions fare in **PT**. We have a form of the theorem of abstraction and we have the principle of extensionality. We go a few steps of the way and give definitions and theorems in **PT** primed numerals corresponding to the same definitions and theorems in Chapter 3. The reader is invited to compare them step by step.

Definition 2'. $V^{((o))}$ for $(\widehat{x_1^{(o)}} = \widehat{x_1^{(o)}})$.

We must indicate the type of defined terms by attaching an appropriate type symbol, since we cannot tell the type directly unless we are dealing with the original defining abstract. We choose to work here with type (o), since the abstract for equality is predicative in this (and higher) types but not in type o. This has already been explained in our remarks following Theorem 1 of Section 4.2.

As we define terms by circumflexion, we suppose all such abstracts to be predicative unless explicit mention to the contrary is made. The reader may take it as a continuing exercise to verify this for each definition. The point here is that we may thus apply **PT.1** to our defined terms.

THEOREM 2'. $\vdash (\forall x_1^{(o)})\, V^{((o))}(x_1^{(o)})$.

Proof. Follows immediately from Theorem 1, Section 4.2 and Definition 2'.

This tells us that $V^{((o))}$ is satisfied by every argument of type (o). $V^{((o))}$ is thus the universal set of all things of type (o).

Definition 3'. $\Lambda^{((o))}$ for $(\widehat{x_1^{(o)}} \neq \widehat{x_1^{(o)}})$.

THEOREM 3'. $\vdash (\forall x_1^{(o)})\, (\sim \Lambda_1^{((o))}(x_1^{(o)}))$.

Proof. The reader will prove this theorem as an exercise. (*Hint*: strictly analogous to proof of Theorem 3, Chapter 3.)

THEOREM 4'. $\vdash (\forall x_2^{((o))})\, ((\forall x_1^{(o)})\, (\sim x_2^{((o))}(x_1^{(o)})) \equiv (x_2^{((o))} = \Lambda^{((o))}))$.

Proof. Strictly analogous to Theorem 4, Chapter 3, using Theorem 1 and Theorem 2, Section 4.2, to handle equality.

Although our definition of equality allows for the possibility of distinct individuals, the axiom of extensionality **PT.2**, resulting as it does in Theorem 1, Section 4.2, does not allow for the possibility of different no-element things of types other than type o. This is shown by Theorem 4'. However, the null sets of different types are different, since equality is defined only between entities of the same type.

Definition 4'. $\{x^{(t)}\}^{((t))}$ for $(\widehat{y^{(t)}} = x^{(t)})$ where t is any type and $x^{(t)}$ is any term of type (t) in which the variable $y^{(t)}$ of type (t) does not occur.

Definition 5'. $(\overline{x^{(t)}})^{(t)}$ for $(\sim x^{(t)}(\widehat{x^t}))$ where $x^{(t)}$ is any term of type (t) in which the variable x^t of type t does not occur.

Definition 6'. $(x^{(t)} \cap y^{(t)})^{(t)}$ for $(x^{(t)}(\widehat{z^t}) \wedge y^{(t)}(\widehat{z^t}))$ where z^t is a variable of type t and $x^{(t)}$ and $y^{(t)}$ are any terms of type (t) in which z^t does not occur.

Definition 7'. $(x^{(t)} \cup y^{(t)})^{(t)}$ for $(x^{(t)}(\widehat{z^t}) \vee y^{(t)}(\widehat{z^t}))$ where z^t is a variable of type t and $x^{(t)}$ and $y^{(t)}$ are any terms of type (t) in which z^t does not occur.

The complement of an entity of a given type is an entity of that type. Similarly, the union and intersection of two sets of a given type are again of that type. Moreover, union and intersection are defined only within a given type. The complement of a set of type (t) is not the set of all other things not in the set, but rather the set of all other things of type t not in the set.

On the other hand, the formation of the unit set of a given set, by Definition 4', raises the order of the term. In this way, we see that the union "$x^{(t)} \cup \{x^{(t)}\}^{((t))}$" is not definable in **PT**, since the two summands are of different types.

Definition 8'. $0^{(((o)))}$ for $\{\varLambda^{((o))}\}^{(((o)))}$.

This is the definition of zero. Notice that this zero has a certain type.

Definition 9'. $S(x^{((((t)))}))^{(((o)))}$ for $(Ez^{(t)}) \, \widehat{(y^{((t))}}(z^{(t)}) \wedge x^{(((o)))}\widehat{(y^{((t))}} \cap \overline{\{z^{(t)}\}^{((t))}})^{(\,t)})$

where the variables and terms are of indicated type, the variables $z^{(t)}$ and $y^{((t))}$ are distinct and do not occur in the term $x^{((((t))))}$.

Notice that the successor of a term of a given type is of the same type. Notice also that the terms defined by the definitions 4' to 7' and by Definition 9' are all predicative if the terms contained in them are predicative to begin with.

Definition 10'. $N^{(((((o)))))}$ for $((\forall x_1^{(((((o))))})) \, (x_1^{(((((o))))}}(0^{(((o))}})) \wedge (\forall x_2^{((((o)))}}) \, (x_1^{(((((o))))}}(x_2^{((((o)))}})$

$$\supset x_1^{(((((o))))}}(S(x_2^{((((o)))}})^{(((o))}})) \supset x_1^{(((((o))))}}\widehat{(x_3^{((((o)))}})))).$$

We now encounter the first difficulty in developing mathematics in **PT**, for $N^{(((((o)))))}$ is impredicative. This prevents us from applying Theorem 3 of this section to N in the manner of Frege's abstraction axiom, because of the predicativity restriction in Theorem 3.

Suppose, for a moment, that we remove the predicativity restriction from **PT** and thus from Theorem 3. We can then proceed, by strict analogy with the method of system **F** of Chapter 3, to prove the analogues Theorems 6', 7', 8', 9', and 11' of Theorems 6, 7, 8, 9, and 11 respectively of Chapter 3. In order to complete the theory of natural numbers, we would still need to prove a theorem of infinity analogous to Theorem 30 of Chapter 3, the fifth Peano postulate. This would imply that $N^{(((((o)))))}$ and thus $V^{(((((o)))))}$ (which contains $N^{(((((o)))))}$ as a subset) is infinite. However, even without the predicativity restriction in **PT**, our method of proving a theorem of infinity in Chapter 3 will not work here. In Chapter 3 we used the operation of "self-adjunction", that is, forming the set $x \cup \{x\}$, in order to construct the set ω of representatives of the natural numbers. But the union $x \cup \{x\}$ is not definable in **PT**, since x and $\{x\}$ will never have the same type, and the union operation is defined only within a given type as we have previously observed. This difficulty is not involved in any way with the predicativity restriction in **PT**. It is a feature of the way we have defined our wffs by limiting the way a term can have arguments.

The fact is that it is impossible to prove a theorem of infinity in **PT** even without the predicativity restriction.

We can clearly see the impossibility of proving a theorem of infinity in **PT** if we think again of our type hierarchies. We never specified that the domain of individuals had to be infinite. Suppose the set D of individuals, expressed in our system by $V^{(o)}$, is finite.[†] Then any relation or set of individuals will also be finite. Inductively, it follows that every set and relation in our type hierarchy will be finite. (The intuitively conceived class of *all* of these is infinite, but this class is not a relation or set of our type hierarchy. This class is, in fact, the

---

† $V^{(o)}$ (respectively $\varLambda^{(o)}$) can be defined predicatively by taking any universally valid (respectively universally false) wff that yields a predicative abstract: e.g. $(x_1^{(o)}\widehat{(x_1^{o})} \equiv x_1^{(o)}\widehat{(x_1^{o})})$ (respectively $(x_1^{(o)}\widehat{(x_1^{o})} \wedge \sim x_1^{(o)}\widehat{(x_1^{o})}))$.

type hierarchy based on D as we have defined it.) A finite domain of individuals entails the finitude of every set and relation in our type hierarchy. Although a type hierarchy always has an infinity of relations in it, we shall nonetheless speak of a *finite* type hierarchy whenever the domain D of individuals is finite.

Since no axiom of our system **PT** forces the domain of individuals to be infinite, it follows that if **PT** has any type hierarchy as a model, then any finite type hierarchy is a model and no theorem of infinity will be provable in any type.

We can see, then, that if we wish to develop mathematics within **PT**, we will need a further axiom, an axiom of infinity. There are many ways to posit an axiom of infinity for **PT**. We choose the following:

PT.3. $(Ex_1^{(o,\,o)})\,((\forall x_1^o)\,(\sim x_1^{(o,\,o)}(x_1^o,\,x_1^o))\wedge(\forall x_1^o)\,(Ex_2^o)\,(x_1^{(o,\,o)}(x_1^o,\,x_2^o))$

$\wedge(\forall x_1^o)\,(\forall x_2^o)\,(\forall x_3^o)\,(x_1^{(o,\,o)}(x_1^o,\,x_2^o)\wedge x_1^{(o,\,o)}(x_2^o,\,x_3^o)\supset x_1^{(o,\,o)}(x_1^o,\,x_3^o)))$

PT.3 asserts the existence of an irreflexive, transitive, strongly connected relation on individuals and there are no finite models for such a relation. Thus, with **PT.3** as an axiom, no type hierarchy with a finite domain D of individuals is a model for **PT**. The domain of individuals must now be at least denumerably infinite in any model for **PT** with **PT.3** added.

Now that we have an axiom of infinity in **PT** we can begin to develop infinitistic mathematics, at least in a certain form. We have already seen, however, that the predicativity restriction is a substantial one and we would therefore suspect that further difficulties may be forthcoming. We examine this question in the following section.

4.4. The system TT

The system **PT** without the axiom of infinity **PT.3** is clearly consistent, since we have a model for it in which the domain of individuals is finite. We have not rigorously proved that some finite type hierarchy is a model, but the careful way we have defined our wffs and rules of inference should convince one that it would certainly be possible. Of course, such a constructive model for **PT** would hold equally well if we dropped the predicativity restriction, and we have already seen that predicativity is not strictly needed to avoid Russell's paradox in **PT**. *Predicativity is irrelevant if we omit* **PT.3**.

With **PT.3**, however, we now have an infinite set, and we can easily define the operation of power set and begin working with infinitistic mathematics. Here our intuition is not wholly trustworthy. The restriction of predicativity thus becomes more meaningful. We knew from the beginning that we would have to add an axiom of infinity and this is why we countenanced the complication of orders within types and the predicativity restriction.

From now on, **PT** will mean **PT** with infinity unless otherwise stated. The sad fact is that even with the axiom of infinity, **PT** is not sufficient for the full development of analysis. To see this, let us return to our discussion of the completeness of the real numbers, which was begun informally at the end of Chapter 3.

In Section 3.7, Chapter 3, we saw intuitively that the Dedekind cut construction of least upper bounds was impredicative. We will now see this precisely. Suppose G is some set of real numbers bounded above (each real number being a set of rationals). Then, the least upper bound B of G is defined to be the set of all rationals y that are elements of some element of G.

Formally, we define B for $((Ex)(G(x) \wedge x(\hat{y})))$. Let y be of type t and let the order of t be n. Thus, for the abstract to be well formed, x must be of type (t) and G of type $((t))$. The abstract B has type (t), and thus would have to be of order $n+1$ to be predicative. But B contains a free variable G of order $n+2$ and is of order $n+2$. B is impredicative, since it does not have the same order as the order of its type. The wff $(Ex)(G(x) \wedge x(y))$ is thus impredicative with respect to y, and we cannot use Theorem 3 to obtain the existence of the desired set.

More precisely, the theorem we desire is the wff

$$(Ez^{(t)})(\forall y)(z^{(t)}(y) \equiv (Ex)G(x) \wedge x(y)).$$

We have the general theorem $(\forall w^{(t)})(Ez^{(t)})(\forall y)(z^{(t)}(y) \equiv w^{(t)}(y))$ in our system, but we can apply $e\forall$ only for predicative terms and thus not for B. Nor can we use **PT**.1 to obtain $B(y) \equiv (Ex)(G(x) \wedge x(y))$.

Though we shall not show it here, predicativity also prevents the deduction of Cantor's theorem that the cardinality of the power set of a given set is greater than the cardinality of the set. (The interested reader can obtain this impossibility from a careful analysis of our discussion of Cantor's theorem in Chapter 7 if he defines predicativity in the obvious way for the system of that chapter.)

These two examples show that predicativity is a powerful restriction on infinite mathematics in **PT**. We cannot carry on analysis in its fullness nor can we carry on the general theory of cardinal numbers and higher infinities. On the other hand, these parts of mathematics can be obtained if we maintain the same wffs, rules, and axioms of **PT**, but ignore predicativity altogether. It is by effecting exactly these modifications in **PT** that we now obtain the more general system **TT** of *type theory*.

Precisely, **TT** has the same wffs as **PT**. The rules of inference and axioms are the same except that we remove the predicativity restriction. Substitution of any term of a given type for a variable of that type is permitted. The order of a term does not have to be that of its type symbol. In fact, there is no need even to define the concept of order to develop **TT**. The variables are thought of as ranging over all entities of the indicated type regardless of how these entities may be defined. Axiom **TT**.1 is the same as **PT**.1 except that we exclude the stipulation that the abstract and the terms z_i be predicative. Axioms **TT**.2 and **TT**.3 are the same as **PT**.2 and **PT**.3. This completes the description of **TT**.

Among the consequences of this generalization of **PT** is the general substitutivity of equality, which is in line with our remark following Theorem 2, Section 4.2. The Russell paradox is excluded in exactly the same way, since predicativity was never necessary for avoiding it.

In the preceding section, we saw that if we dropped the predicativity restriction from **PT** we could develop part of the Frege theory of the natural numbers, proving all of the Peano Postulates except the theorem of infinity. Since the predicativity restriction is dropped in **TT**, we immediately obtain in **TT** the theorems 6′, 7′, 8′, 9′ and 11′ referred to in the preceding section. These theorems express analogues in **TT** of the first four Peano postulates. Since we have also an axiom of infinity in **TT**, we can prove the fifth Peano postulate, which will be a type-theoretic analogue of Theorem 30 of Chapter 3 (the method of proof will not be exactly the same, of course).

Since much of the Frege construction carries over from **F** to **TT**, we could envisage dropping **TT**.3 from **TT** and posing the fifth Peano postulate as an alternative axiom of infinity. This axiom has the form:

TT.3*. $(\forall x_1^{(((o)))}) (\forall x_2^{(((o)))}) ((N^{((((o))))}(x_1^{(((o)))}) \wedge N^{((((o))))}(x_2^{(((o)))})$
$\wedge S(x_1^{(((o)))})^{((((o))))} = S(x_2^{(((o)))})^{((((o))))}) \supset x_1^{(((o)))} = x_2^{(((o)))}).$

TT.3*, in conjunction with the other four Peano postulates, forces $N^{((((o))))}$ and thus the class $V^{((((o))))}$ of all things of type $(((o)))$ to be at least denumerably infinite. Thus, **TT.3*** excludes finite type hierarchies as models, since the finitude of the domain D of individuals of a type hierarchy would entail the finitude of every set in the hierarchy and thus the set which is the interpretation of $V^{((((o))))}$.

TT.3 rather than **TT.3*** will be our form of the axiom of infinity in **TT** unless otherwise indicated.

In **TT**, Cantor's theorem is provable. Moreover, we have least upper bounds for bounded sets of real numbers, for now we can apply substitution and **TT.1** to our impredicative abstract $((Ex) (G(x) \wedge x(\hat{y})))$. In short, we have a foundation for mathematics.

An intuitive model for **TT** is, of course, any type hierarchy where the individuals are infinite in number. In particular, the type hierarchy with a denumerably infinite set of individuals is a model for **TT**. We can take this type hierarchy to be the standard model for **TT**. (Remember that the only thing that distinguishes one type hierarchy from another in any essential way is the cardinality of the domain D of individuals.) If we regard this intuitive model as sufficient proof of consistency, then **TT** is consistent. However, we cannot *prove* that our type hierarchy, with an infinite domain of individuals, is a model unless we appeal to strongly nonconstructive principles of set theory. Moreover, our constructive approach of defining all higher types from lower ones is now gone, for we have abandoned predicativity.

TT is a generalization of **PT** in the strict sense of the term: every wff and theorem of **PT** is a wff and theorem of **TT**, but not conversely.

4.5. Criticisms of type theory as a foundation for mathematics

The theory of types does furnish a foundation for mathematics. If we accept the type hierarchy with a denumerably infinite domain of individuals as a model of **TT**, then **TT** is consistent. Of course, there is already here a certain conflict of intuition. If we accept the Poincaré–Russell analysis, which says that the basic source of the paradoxes is impredicative definition, then type theory can be criticized because it admits impredicative definition. On the other hand, predicative type theory is not adequate as a foundation for mathematics. We must ultimately choose between adequacy and predicativity.

Some (see Weyl [1]) have clearly chosen predicativity. Most mathematicians, however, would obviously choose adequacy and renounce predicativity. However, we must then find a different explanation for the source of the paradoxes. Of course, the type hierarchy itself excludes certain kinds of vicious circle, and so we could take the position that Frege's basic fault was his failure to recognize the necessity of a type hierarchy in dealing with quantification of predicate letters.

From the point of view of the working mathematician, a basic criticism of type theory is its preoccupation with purely logical and linguistic questions. Even though the complication of orders is omitted, **TT** involves a complicated symbolism that is surely distasteful to the mathematician who is concerned with the structures he is defining.

Coupled with this criticism is another one. In **TT**, arithmetic is constructed by Frege's method. In our system, 0 has type $(((o)))$ and the set N of natural numbers type $((((o))))$. But we could have begun the construction with any type t whatsoever in the place of o. This important point should be clearly understood. Our Theorem 3 of class abstraction was obtained relative to any type t. Our definitions of union, intersection, and the like were relative to any type. We started by defining the null set of type $((o))$, but this, we recall, was only because of predicativity considerations for the abstract defining the null set. Now that these considerations are no longer present, we can define $\Lambda^{(t)}$ for $(\widehat{x^t \neq x^t})$ for any type t, and $0^{((t))}$ as $\{\Lambda^{(t)}\}^{((t))}$, again for any type t. Thus the set N of natural numbers would be of type $(((t)))$ relative to the type t.

This all means that there is definable in **TT** an *infinity of arithmetics*, one relative to each type! Arithmetic, indeed all the mathematics we construct, is constructible relative to any type t. The natural numbers will always be of type $(((t)))$ relative to such a type. This shows that, with an axiom of infinity, it is only *relative types* that matter.

Foreseeing this intuitively repugnant result, Russell and Whitehead employed a device of *typical ambiguity* to which we have already referred (cf. Section 4.2). Type indices never appear (except once or twice in the latter portions of the work), and the reader is to imagine the definitions and constructions performed relative to any type, and in any way in which a restoration of type indices would conform to a well-formed construction. An example would be our abstract $((Ex)(G(x) \wedge x(\hat{y})))$ obtained from the wff $(Ex)(G(x) \wedge x(y))$. No type indices are present, but whatever the type t of y is, x must be of type (t) and G of type $((t))$. Even with typical ambiguity, we must always be sure that the expressions we write permit us to restore type indices in some consistent way.

Exercise. Restore relative type indices to the following forms or prove that it is impossible to do so (thus the expression in question cannot be a wff):

(a) $(Ey)(y(x) \vee x(y))$;   (b) $x(y, z)$;   (c) $x(x)$;

(d) $x(y, (\hat{y} = x))$;   (e) $x(y, (\hat{y} = z))$.

Notice that in cases in which there are two or more arguments, there may be more than one consistent way of restoring relative types. This is not, however, the case for monadic (single-argument) wffs.

Though typical ambiguity may lessen the emotional reaction to an infinite hierarchy of arithmetics, it does not change the basic theory. In fact, it serves to obscure the theory if a careful definition of the wffs is not given. Such a formal presentation of the expressions of the language was not given in *Principia Mathematica*, and this fact can be made the basis of further criticism of the system of that book (though it does not apply to **TT**).

One escape from the dilemma of relative types is to add explicit axioms (such as the Peano postulates), which require the individuals to be the natural numbers. Then, the types of higher order can be viewed as sets and relations over sets and relations over, and so on down to the natural numbers. Such a form of type theory was used by Gödel in Gödel [2], and it is perhaps one of the most satisfactory forms of type theory for practical use as a foundation.

Russell argued that type theory was intuitively natural (see Whitehead and Russell [1], p. 37). Let us recall the philosophical thesis defended by Frege and Russell: that all mathematics is deducible from universally valid logical principles. The universally valid wffs of **TT** will be those wffs true in all type hierarchies. However, the axiom of infinity is not true in type hierarchies with a finite domain of individuals. It is thus not a universally valid principle, and so cannot be viewed as a truth of logic. Thus, even if we grant Russell's contention that type theory represents a consistent, intuitively correct reconstruction of Frege's general predicate calculus, it still does not follow that mathematics is deducible from logic, since there are mathematical principles in this reconstructed calculus which are not universally valid.

To sum up, Frege failed to substantiate the logistic thesis because his system was contradictory. It does not follow from the inconsistency of his system that his constructions are contradictory. But Russell's reformulation of general logic, though consistent without an axiom of infinity, is too weak to allow for all of Frege's constructions, and so fails to substantiate the logistic thesis because mathematics is not deducible within it. A foundation for mathematics is obtained if we add an axiom of infinity, but then we can no longer claim that our axioms are universally validlogical principles.

The original system of *Principia Mathematica* is subject to yet another criticism raised by Leon Chwistek [1]. To examine this criticism, we now formulate the system **RT** of ramified types described briefly in Section 4.2. Besides the usual parentheses, commas, the circumflex, and the like, the alphabet of **RT** contains variables and constants which we now describe. A *variable* is the lower-case italic x with a subscript and a superscript. The subscript is a numeral for a positive integer. (Subscripts are to distinguish among variables as usual.) The superscript is either the symbol $o/0$ or else a symbol of the form $(t_1, \ldots, t_n)/k/(y_1, \ldots, y_n)$ where the t_i are type symbols, and k and the y_i are all natural numbers subject to the restrictions that the number y_i is for each i greater than or equal to the order of the corresponding type t_i, and the number k is greater than the greatest value of the y_i. Further, if t_i is o, then y_i must be 0.

The intuitive reason for requiring all variables of type o to be of order 0 is that no term of type o can be defined by circumflexion. There is no way, then, to name an individual by binding variables of higher type than o. We can only name a property of an individual and this must have type (o) and at least order 1.

The constants are defined in exactly the same way except that the lower-case italic a is used in place of x.

A variable or constant has the type indicated by its type symbol, and the order indicated by the number immediately following its type symbol in its superscript. The whole superscript symbol is called the *order-type* of the variable or constant, and the superscript symbols are *order-type symbols.*[†] We define the wffs as follows: (1) Variables and constants are terms of the indicated order-type. (2) Let x be a term of order-type $(t_1, \ldots, t_n)/k/(y_1, \ldots, y_n)$, and let $z_1, \ldots, z_n$ be terms of type t_i respectively, and of order less than or equal to y_i respectively. Then $x(z_1, \ldots, z_n)$ is a wff. (3) Expressions obtained from wffs by iterating statement connectives and quantifiers are wffs. Quantified variables are considered bound as usual. (4) Let $A(z_1, \ldots, z_n)$ be a wff in which the free variables $z_1, \ldots, z_n$ occur in canonical order. Let t_i be the type of z_i and y_i the order of z_i. Let k be the highest order of *all* variables or constants in the wff. Then $(A(\widehat{z_1}, \ldots, \widehat{z_n}))$, formed by circumflexing all free occurrences of these variables and enclosing the result in parentheses, is a term of order-type $(t_1, \ldots, t_n)/r/(y_1, \ldots, y_n)$ where r is k if the term has no bound variables of order k, and is $k+1$ otherwise. The circumflexed variables are bound in the term so formed. We require as before that circumflexed variables must be subscripted if there are different variables of the same order-type, at least one of which is circumflexed. (5) These are the only wffs and terms of **RT**.

[†] The order-type of a term of **RT** as we have here defined it has nothing whatsoever to do with the order type of an ordered set. The latter means the isomorphism class of the given ordered set under order isomorphism.

Our rules of inference consist of our natural deduction rules modified in the obvious ways: substitution is to take place within a given type, and for orders less than or equal to the orders of the variable and its argument places for which substitution is being made, *provided* that the expression resulting from such substitution is a wff. An example will be helpful here.

$(\forall x_1^{((o))/5/3}) x_1^{((o))/5/3}(x_1^{(o)/3/0})$ is a wff. By $e\forall$ we can infer the wff $x_1^{((o))/4/3}(x_1^{(o)/3/0})$, but *not* $x_1^{((o))/4/2}(x_1^{(o)/3/0})$ which is not a wff. But, if our original wff were $(\forall x_1^{((o))/5/3}) x_1^{((o))/5/3}(x_1^{(o)/2/0})$, for example, then we could obtain $x_1^{((o))/4/2}(x_1^{(o)/2/0})$ by $e\forall$, since this latter is a wff. Notice that by substitution using $e\forall$ we can never go up in order. We can remain only in the same order, or go down, subject to the restrictions of well-formedness.

As axioms for **RT**, we shall have **RT.1**, **RT.2**, and **RT.3** corresponding to the same numbered axioms for **TT**. However, care must be taken in formulating these. To do this, we define predicativity in **RT**.

A term (variable, constant, or abstract) of order-type

$$(t_1, \ldots, t_n)/k/(y_1, \ldots, y_n)$$

is *predicative* if and only if k is $y+1$ where y is the maximum value of the orders $y_1, \ldots, y_n$. All variables and constants of order-type $o/0$ are predicative.

If, in the above definition, each y_i is the order of the type t_i then k will be the same as the order of the type $(t_1, \ldots, t_n)$ and so we obtain our definition of predicativity in **PT** as a special case of our general definition in **RT**.

This shows that if we exclude from **RT** all terms except those of order-type $o/0$ and $(t_1, \ldots, t_n)/k/(y_1, \ldots, y_n)$ where y_i is the same as the order of t_i and k is the same as the order of $(t_1, \ldots, t_n)$, then the wffs of **RT** that can be formulated with this restricted set of terms represent wffs of **PT** in an obvious way (from which impredicative abstracts will be excluded). Moreover, it will be true that the rules and axioms of **RT**, when applied to this restricted set, yield a system which is the same as **PT** (with impredicative abstracts excluded from the grammar of the system). **PT** is a subsystem of **RT**.

For **RT.1**, we shall have the same axiom as **TT.1**, except that the wffs and abstracts contained in the statement must satisfy the criteria of well-formed expressions for **RT**.

We define equality in **RT** by: $x = y$ for $(\forall z)(z(x) \equiv z(y))$ where x and y are terms of the same order-type. Where the common type of x and y is t and the common order is n, z is of order-type $(t)/n+1/n$, and does not occur in the terms x and y.

Given a term x of type t and order n, we say that another term y is *predicative with respect to x* if y is of order-type $(t)/n+1/n$. Our definition of equality says that two terms of a given order-type are equal if they satisfy the same properties z that are predicative with respect to them. Notice that if z is predicative with respect to x, it is predicative. Russell used an exclamation point ! in the metalanguage to indicate a predicative term. Thus, our definition would be stated: $x = y$ for $(\forall z)(z!(x) \equiv z!(y))$, if we used Russell's symbolic device.[†]

The axiom **RT.2** of extensionality can be stated as follows:

$$(\forall z_1) \ldots (\forall z_n)(x(z_1, \ldots, z_n) \equiv y(z_1, \ldots, z_n)) \supset x = y,$$

where the order-type of x and y is $(t_1, \ldots, t_n)/k/(y_1, \ldots, y_n)$ and the z_i are each of type t_i *and* order y_i. None of the variables z occurs in the term x or y.

[†] This particular use of "!" should not be confused with our notation "$(E!)$" of Chapter 1. The present use will be confined to the discussion of this section.

We state an axiom scheme of infinity for **RT** which corresponds to **TT**.3.

RT.3. $(Ex_1^{(o,\, o)/k/(0,\, 0)})\,((\forall x_1^{o/0})\,(\sim x_1^{(o,\, o)/k/(0,\, 0)}(x_1^{o/0},\, x_1^{o/0}))$

$\wedge\,(\forall x_1^{o/0})\,(Ex_2^{o/0})\,(x_1^{(o,\, o)/k/(0,\, 0)}(x_1^{o/0},\, x_2^{o/0}))$

$\wedge\,(\forall x_1^{o/0})\,(\forall x_2^{o/0})\,(\forall x_3^{o/0})\,(x_1^{(o,\, o)/k/(0,\, 0)}(x_1^{o/0},\, x_2^{o/0})$

$\wedge\, x_1^{(o,\, o)/k/(0,\, 0)}(x_2^{o/0},\, x_3^{o/0}) \supset x_1^{(o,\, o)/k/(0,\, 0)}(x_1^{o/0},\, x_3^{o/0}))),$

where k can take any value greater than or equal to one.

RT.3 is an axiom scheme, since we have a different proper axiom for each permissible value of k. **RT**.3 thus posits the existence of an irreflexive, strongly connected, transitive relation on individuals, one relation for each possible order.

This completes the description of **RT**.

Though highly complicated and more flexible than **PT**, **RT** is actually much closer to **PT** than **TT** in its power as a foundation. Technically it is, of course, intermediate between them, containing **PT** as a subsystem but not as strong as **TT** because of the presence of the various order restrictions. **RT** has difficulties of its own with such notions as least upper bounds. Let us examine this briefly.

Again, we let G be some bounded set of real numbers, and we define the least upper bound B to be the set of all rationals y which are elements of some element of G. B is $((Ex)\,(G(x)\wedge x(\hat{y})))$. Let t be the type of y and n its order. We can suppose x of $(n+1)$st order and G of $(n+2)$nd order. These are the minimal orders consistent with the wff, and the difficulty we are going to encounter holds just as well if we suppose higher orders for these variables. Thus, B is of order-type $(t)/n+2/n$. B is thus defined in **RT**, but it is still not predicative.

Now let G be some predicatively specified bounded class of real numbers; G thus has order-type $((t))/n+2/n+1$. (Remember, it is the rational numbers which have type t in our example, whereas the reals are certain sets of rationals.) Now, according to intuitive real analysis and set theory, the least upper bound B will be an element either of G or of its complement. But we cannot even express the notion "B is an element of G", for $G(B)$ is not well formed. The arguments of G must have order less than or equal to $n+1$, but B has order $n+2$ and thus does not qualify. The complement of G will have the same order-type as that of G (see Definition 5' for **PT**), and so B will not be an element of that set either. B, though it exists, is not an element either of G or of its complement. Such a higher-order least upper bound does not enable us to develop the usual analysis of real numbers.

RT will also be subject to limitations similar to those of **PT** concerning the substitutivity of identity as follows from our definition of equality.

Russell treated this problem, in the first edition of *Principia Mathematica*, by adding an *axiom of reducibility*. The effect of this addition was that to every property there corresponds a predicative property satisfied by the same arguments. In Russell's symbolism, we would have:

$$(Ey)\,(\forall x)\,(y!\,(x) \equiv A(x))$$

where $A(x)$ is any wff containing x free. More formally, we posit:

$$(\forall x)\,(Ew)\,(\forall z_1) \ldots (\forall z_n)\,(w(z_1,\, \ldots,\, z_n) \equiv x(z_1,\, \ldots,\, z_n))$$

where x is of order-type $(t_1,\, \ldots,\, t_n)/k/(y_1,\, \ldots,\, y_n)$ and w is *predicative* of order-type $(t_1,\, \ldots,\, t_n)/r/(y_1,\, \ldots,\, y_n)$, r less than or equal to k. We obtain a generalization of Russell's form of the axiom by applying $e\forall$ to x in the name of an appropriate abstract $(A(\widehat{z_1}, \ldots, \widehat{z_n}))$. Russell also assumed the axiom for binary relations, but did not bother to do so for relations in general, since it was not strictly necessary for his intended usage of the principle.

The axiom of reducibility lets us reduce the order of any term to the lowest possible order consistent with well-formedness (which is thus predicative by definition). Observe now that, by the axiom of reducibility, $\vdash(Ew)(\forall y)(w!(y) \equiv (Ex)(G(x)\wedge x(y)))$. Applying eE we obtain $(\forall y)(a!(y) \equiv (Ex)(G(x)\wedge x(y)))$ where $a!$ is some constant of the same order-type as w. We can then define $B!$ as $a!(\hat{y})$ which is of order-type $(t)/n+1/n$ instead of $(t)/n+2/n$. $B!$ is a predicative least upper bound, which we can handle in the usual way without fear of difficulty.

Notice that it might have been possible to find an equivalent least upper bound of lower order in some special cases without use of the axiom of reducibility. It would depend on the details of how the set G was specified. But the axiom of reducibility tells us that we can always reduce the order, regardless of the special mode of definition of the case under consideration.

The system **PM** of *Principia Mathematica* may be thought of as **RT** plus the axiom of reducibility. Now, Chwistek's criticism of **PM** was the following observation: **RT** with the axiom of reducibility is roughly the same system as **TT** while being much more complicated. The axiom of reducibility makes the distinction between various orders rather unnecessary, since we can always reduce the order and work only with predicative order-types. Thus, reasoned Chwistek, the only real choice is between type theory without orders, the system **TT**, or **RT**. **PM** is an unfortunate hybrid.

In the second edition of *Principia Mathematica*, Russell envisioned dropping the axiom of reducibility, but of course he found no truly satisfactory way of avoiding the difficulties of the orders in **RT**.

Finally, let us mention that an axiom of choice is also needed to assure a full development of mathematics in type theory. We shall delay discussion of this axiom until Chapter 5 when we take up Zermelo's system.

4.6. The system ST

The system we have called **TT** is often called "simplified type theory", meaning type theory without the use of the notion of orders. The system we shall call **ST**, *simplified types* or *simple types*, will be a version of **TT** which uses a further simplifying device due to Wiener and Kuratowski.

In our formulations of type theory, we have dealt with relations of degree n for any finite $n > 0$. Such a relation of degree n has type $(t_1, \ldots, t_n)$. We have used the term "set" for relations of degree one, relations of type (t) for some t. (Individuals have degree zero as in Section 4.1. The degree of a relation is simply the number of its argument places and an individual has no argument places.) We have used binary relations, for example, to state an axiom of infinity for **RT**. The reflective reader might wonder why we did not use relations other than sets in our development of our system **F** of Chapter 3.

The fact is that we would have needed relations of degree greater than one for anything like a full development of mathematics in **F**. As it happens, we can develop a certain portion of the theory of natural numbers (in particular Frege's constructions and the Peano postulates) without using relations other than sets. But the whole theory of functions, relations, cardinal numbers, and so on cannot be done without relations of degree greater than one. As we have already remarked, the original form of Frege's system was a general predicate calculus which countenanced relations of all finite degrees, very much like the contradictory general predicate calculus considered at the beginning of this chapter.

The presence of relations other than sets in our type theory has obviously complicated our notation considerably, especially our type symbols. We needed multi-argument type symbols for variables whose values were multi-argument relations. In the system **RT**, the symbolism was even more elaborate and complicated. We can avoid the complications of multi-argument type symbols in type theory if we can define generally relations in terms of sets. This is already accomplished when we consider relations as sets of n-tuples (as we have done throughout this work), *provided*, of course, we can successfully define the notion of an n-tuple purely in terms of sets. This we shall now proceed to do.

Given two objects x and y, we define the ordered pair $\langle x, y \rangle$ to be the set $\{\{x\}, \{x, y\}\}$. In type theory, x and y must be of the same type, for otherwise $\langle x, y \rangle$ is not well formed. If t is the common type of x and y, then $\langle x, y \rangle$ is of type $((t))$. (See Definitions 4' and 7' for **PT**. $\{x, y\}$ is $\{x\} \cup \{y\}$.)

We can define an ordered triple $\langle x, y, z \rangle$ as the ordered pair $\langle \langle x, y \rangle, z \rangle$. Inductively, we can define an ordered n-tuple $\langle y_1, \ldots, y_n \rangle$ as $\langle \langle y_1, \ldots, y_{n-1} \rangle, y_n \rangle$. Given an ordered pair $\langle x, y \rangle$, we speak of x as the first component and y as the second, and more generally for n-tuples, of the first, second, third, and so on.

Exercise. Prove $\vdash \langle x^t, y^t \rangle^{((t))} = \langle z^t, w^t \rangle^{((t))} \equiv (x^t = z^t \wedge y^t = w^t)$ in **TT**. (*Hint*: use the axiom of extensionality.)

The ordered pair differs from the set $\{x, y\}$ primarily in the identity condition reflected in the foregoing exercise. Two ordered pairs are equal if and only if they are equal component by component, but two sets are equal if they have the same elements regardless of order.

We have defined a term of **TT** to be a set if it is of type (t) for some type t. Notice that an ordered pair of two objects (and thus an ordered n-tuple) is a set. With ordered pairs, we can dispense with a primitive notion for relations of type $(t_1, \ldots, t_n)$, since we can now view such relations as sets of n-tuples, and thus as sets of sets. Our type symbols can be defined as follows: o is a type symbol, and $(\ldots (o) \ldots)$ with n pairs of parentheses is a type symbol. These are the only type symbols. We can, as a matter of fact, use natural numbers as type superscripts, letting 0 be type o, and n be the type $(\ldots (o) \ldots)$ where n pairs of parentheses are used. This is the procedure we shall use.

The type hierarchies that are possible models of our system can also be simplified. Let some nonempty set T_0 be given. The elements of T_0 are the entities of type 0. The entities of type 1, based on T_0, are subsets of T_0; that is, elements of $\mathcal{P}(T_0)$. We let T_1 designate the set $\mathcal{P}(T_0)$ of all entities of type 1. Inductively, the set T_{n+1} is defined as the set $\mathcal{P}(T_n)$. In other words, T_n is $\mathcal{P}(\ldots \mathcal{P}(T_0) \ldots)$ with n iterations of the power set operator $\mathcal{P}$. We define an entity of type n to be an element of T_n. The hierarchy we obtain in this way, starting with some nonempty set T_0 and iterating the power set operator any finite number of times, is called the *simplified type hierarchy* (or the *simple type hierarchy*) based on T_0. We shall, in the future, often speak of a "type hierarchy" where it is clear that we mean a simplified type hierarchy.

Notice that, for a fixed set $T_0 = D$, the simplified type hierarchy based on T_0 is a subcollection of the type hierarchy based on T_0 as previously defined. Moreover, we stress again that a simple type hierarchy is determined by the cardinality of the set T_0, since there is, as before, a natural isomorphism between two type hierarchies based on sets with the same cardinality.

Since we shall posit an axiom of infinity for **ST**, as we did for **TT**, the domain T_0 of individuals of any type hierarchy which is a model for **ST** will have to be at least denumerably infinite.

If two terms x and y are of type n and m respectively, then we cannot form the ordered pair $\langle x, y \rangle$ unless $n = m$. However, if $n \neq m$, let n be greater than m and let k equal $n-m$. Then the term $\{\dots \{y\} \dots\}$ to k brackets is of type n and we can form the ordered pair $\langle x, \{\dots\{y\}\dots\}\rangle$. This will be a set of type $n+2$. Finally, a relation will be a set of such pairs, and thus a term of type $n+3$. In our system **TT**, such a relation would have been of type (n, m) (where n is the type of x and m the type of y). In **ST**, it will be of type $n+3$. For example, a relation of type $(o, ((o)))$ with our notation in **TT** will have type 5 in **ST**. The highest type of the two arguments is two, and so the type of relation will be two plus three.

Exercise. In each case, give the type of the relation under our new scheme of defining relations as sets of ordered pairs and using natural number type indices. (a) $(((o)))$; (b) $((o, o), (((o))), o)$; (c) $(o, (o), ((o)))$.

We now formulate explicitly the system **ST** of simple types. We have variables, consisting of the small italic "x" with natural number (numeral) superscripts and positive natural number (numeral) subscripts. Constants, necessary mainly for the rule eE, consist of the small italic letter "a" followed by a natural number superscript and a positive natural number subscript. We have parentheses and the circumflex (but the comma is no longer necessary).

We define the wffs and terms of **ST** as follows: (1) Variables and constants are terms of the indicated type. (2) If y is a term of type n, and x a term of type $n+1$, then $x(y)$ is a wff. (3) If X and Y are wffs, then so are $(\sim X)$ and $(X \lor Y)$. (4) If Y is any wff and x is any variable, then $(\forall x)Y$ and $(Ex)Y$ are wffs. (5) If $A(y)$ is a wff containing y free, y being a variable of type n, then $(A(\hat{y}))$ is a term of type $n+1$, where the variable y is subscripted if there is another variable of type n in A. The quantified and circumflexed variables in (4) and (5) respectively are considered bound as usual. These are the only wffs and terms.

Our rules of inference are our natural deduction rules with the restriction that substitution of variables and terms is to take place only within a given type. A term must be of the same type as any variable for which it is substituted.

As axioms, we have **ST**.1, **ST**.2, and **ST**.3 corresponding to the same numbered axioms for **TT**. **ST**.1 is: $(A(\hat{y}))(x) \equiv A(x)$ where x is free for y in $A(y)$. Equality is defined by $x = y$ for $(\forall z)(z(x) \equiv z(y))$, x and y of type n and z of type $n+1$. **ST**.2 is then stated:

$$(\forall z^n)(x^{n+1}(z^n) \equiv y^{n+1}(y^n)) \supset x^{n+1} = y^{n+1}.$$

ST.3 is the same as **TT**.3 except for the difference in the way we write our type superscripts. We state:

ST.3. $(Ex_1^3)((\forall x_1^0)(\sim x_1^3(\langle x_1^0, x_1^0 \rangle)) \land (\forall x_1^0)(Ex_2^0) x_1^3(\langle x_1^0, x_2^0 \rangle)$
$$\land (\forall x_1^0)(\forall x_2^0)(\forall x_3^0)(x_1^3(\langle x_1^0, x_2^0 \rangle)$$
$$\land x_1^3(\langle x_2^0, x_3^0 \rangle) \supset x_1^3(\langle x_1^0, x_3^0 \rangle)))).$$

We can write $x_1^3(x_1^0, x_2^0)$ instead of $x_1^3(\langle x_1^0, x_2^0 \rangle)$. However, other definitions will serve to put **ST** in a form more directly comparable to **F** (and to systems to be considered).

Definition 1. $(x^n \in x^{n+1})$ for $x^{n+1}(x^n)$.

Definition 2. $\{y \mid A(y)\}$ for $(A(\hat{y}))$, where y is of type n.

ST.1 now becomes: $x \in \{y \mid A(y)\} \equiv A(x)$ where x is free for y, y of type n. **ST**.2 yields the substitutivity of identity in the same way as for **TT**.

The modern preference in formulating **ST** is to use exclusively the notation of Definitions 1 and 2, rather than the notation with which we have set up the system. In the future we shall suppose this notation to be the standard one when we speak of **ST**. In this standard notation, $x(\langle y, z \rangle)$ becomes $\langle y, z \rangle \in x$.

In formulating **ST**, we can choose to avoid the use of a primitive term operator of abstraction. In this formulation, call it **ST***, the variables are the same as for **ST** and they are the only terms. Prime wffs are of the form $(x_i^n \in x_j^{n+1})$ where the variables are of indicated type. The wffs are those expressions obtainable from the prime wffs by iterating quantification and our sentential connectives. The axiom of abstraction is stated:

ST*.1. $(Ex_i^{n+1})(\forall x_j^n)(x_j^n \in x_i^{n+1} \equiv A(x_j^n))$ where x_i^{n+1} does not occur in the wff $A(x_j^n)$ which contains the variable x_j^n free.

The definition of equality and the axiom of extensionality are the same as for **ST**, except for using the "$\in$" notation.

ST*.3. $(Ex_1^3)((\forall x_1^0)(\langle x_1^0, x_1^0 \rangle \notin x_1^3) \wedge (\forall x_1^0)(Ex_2^0)(\langle x_1^0, x_2^0 \rangle \in x_1^3)$

$\wedge (\forall x_1^0)(\forall x_2^0)(\forall x_3^0)(\langle x_1^0, x_2^0 \rangle \in x_1^3 \wedge \langle x_2^0, x_3^0 \rangle \in x_1^3 \supset \langle x_1^0, x_3^0 \rangle \in x_1^3)).$

For use as a foundation, the term operator of abstraction is useful, since it avoids continual appeal to the rule eE in applications of **ST***.1. (For dummy constants we use the constants a_m^n as defined for **ST**.) But for metamathematical considerations of **ST**, the form which avoids the primitive term operator is easier to work with, since models of **ST** in this form are easier to describe. This form is equivalent to that involving the term operator of abstraction, given the axiom of extensionality.

Exercise. Using the formulation **ST*** of type theory, *prove* that any finite simple type hierarchy T is a model for **ST*** without the axiom of infinity. (*Hint*: A finite simple type hierarchy is one whose domain T_0 of individuals is finite and nonempty. Let the interpretation of the predicate letter "$\in$" of **ST*** be the membership relation in the set hierarchy T, and prove that the axioms of **ST** are true under this interpretation. For **ST***.1, use induction on the quantifiers and sentence connectives in $A(x_j^n)$.)

We will now see that **ST** can even be formulated as a first-order theory.

4.7. Type theory and first-order logic

We began our discussion of type theory in this chapter by generalizing our first-order logic of Chapter 1 and allowing quantification over predicate letters. Contradictions in this general predicate calculus led us to impose the restrictions of a type hierarchy. We wish to see now in what sense first-order logic is a special case of type theory.

Let F be a first-order theory and $\langle D, g \rangle$ an interpretation of F. The individual variables of F are thought of as ranging over D, and each predicate letter of F has an interpretation as a relation over D. Now, consider the type hierarchy with D as its domain of individuals. Since we can bind only individual variables in a first-order theory, every relation considered in F will be of type $(o, o, \ldots, o)$ in the type hierarchy starting with D, and the order of such a type symbol is one. Hence the designation "first-order".

Of course we are here dealing with the order of the type symbol and not the ramified theory **RT** of different orders within a given type. Recognizing this fact, Church [3] uses the term "level" where we have used the term "order" in dealing with a partial formulation of **RT**. We have preferred to maintain Russell's original terminology in dealing with **RT**.

Starting with the system **TT**, suppose we restrict ourselves to those wffs which can be constructed by using as terms only variables whose type is of order 0 or variables whose type is of order one. We further suppose that only individual variables (those whose type is order 0) can be quantified, and we disregard all our axioms of **TT**. We have left only our rules of inference applied to the indicated set of wffs. What we have is a formulation of the pure predicate calculus of first-order.

More generally, by the *pure predicate calculus of order n*, n greater than or equal to 1, we mean the system obtained by excluding from **TT** all constants, all variables whose type is of order $n+1$ or higher, and all abstracts whose type is of order n or higher. (We presume, as always, that dummy constants are available for our rule eE.) We also exclude the axiom of extensionality and infinity from **TT**, and we restrict quantification of variables to variables of type $n-1$ or less. In this formulation of the pure predicate calculus of order n, the use of abstracts replaces a substitution rule for predicate variables. In the case of $n = 1$, there are no abstracts whose type has order 0, and so we do not need abstracts, since only individual variables are quantifiable.

By a theory of order n, n greater than or equal to 1, we mean a system obtained by adding proper axioms and possibly some constants to the pure predicate calculus of that order. We may also exclude any number of variables of order n as long as there is included at least one variable of that order. The variables of order n in a theory of order n are not bindable, just like the predicate letters of a first-order theory. By specialization of this definition to the case $n = 1$ we obtain a definition of a first-order theory equivalent to our original one. (We have made no provision for function symbols, but these can always be treated by taking predicate letters in a theory with equality.)

TT may be thought of as a theory of order ω (the first infinite ordinal). Yet **TT** can actually be given a first-order formulation. As an example of the method, let us describe a first-order formulation of the system **ST*** of simple type theory without a primitive term operator of abstraction.

We take a first-order theory **ST'** with one binary predicate letter "$\in$" and a denumerable number of singulary predicate letters $K_0, K_1, \ldots$. Intuitively, $K_n(x)$ will mean "x is of type n". Of course the wffs of **ST'** are all the formulas which are well formed in the usual sense for a first-order theory, i.e. the expressions which can be obtained from prime formulas such as $(x \in y)$ and $K_n(x)$ by iterating quantification and the sentential connectives. However, the axioms of **ST'** will be (with one minor exception) the first-order "translations" of the axioms of **ST*** and so the extra wffs now available will not really give an essentially stronger theory (unless we envisage adding new kinds of axioms and getting a different theory from **ST**).

Given a wff X of **ST'**, we will say that X is *restricted to types* if every subformula $(x)B$ of X is of the form $(x)(K_n(x) \supset C)$, for some predicate letter K_n and some wff C, and if every subformula $(Ex)B$ of X is of the form $(Ex)(K_n(x) \wedge C)$, again for some K_n and C. Every closed wff of **ST*** is translatable into a unique wff X of **ST'** restricted to types. Given a closed wff A of **ST***, we begin from left to right with the smallest quantified subformulas of A, replacing each occurrence of $(\forall x_m^n)B$ by $(x_j)(K_n(x_j) \supset B')$ and each occurrence of $(Ex_m^n)B$ by $(Ex_j)(K_n(x_j) \wedge B')$, continuing until the transformation is complete. In each case, x_j is the first unused variable available at the given stage of the transformation, starting with x_1, and B' is the result of replacing everywhere x_m^n by x_j in B.[†] This purely formal process transforms every closed wff of **ST*** into a closed wff of **ST'**. A proper axiom of **ST'** is now defined to be a wff X', which is the transform of the universal closure X of some proper axiom of **ST***, or else the wff $(Ex_1)K_0(x_1)$. These are the only proper axioms of **ST'**.

Notice that our transformation mapping is not 1–1 on closed wffs of **ST***, since different formulas can give rise to the same transform. However, two closed wffs of **ST*** with the same transform will be equivalent, by the rule of change of name of bound variables. Notice also that there are wffs of **ST'** which are closed and restricted to types but which are not transforms of wffs of **ST*** or variants thereof by change of bound variable. The wff $(x_1)(K_0(x_1) \supset x_1 \in x_1)$ is an example.

Exercise. Formulate an algorithm for determining whether or not a wff of **ST'** is an axiom.

In thinking about what the axioms of **ST'** look like, let us recall that the axiom of **ST*.2** of extensionality involved an abbreviational definition for equality when we stated it. The transformation process must be applied only to unabbreviated wffs. Thus, axioms of the kind **ST'.2** will be of the form:

$$(x)(y)(K_{n+1}(x) \wedge K_{n+1}(y) \supset ((z)(K_n(z) \supset (z \in x \equiv z \in y))$$
$$\supset (v)(K_{n+2}(v) \supset (x \in v \equiv y \in v))))$$

where the variables are all distinct (here we have rearranged the initial quantifiers slightly as the reader can check by formally transforming the closure of an axiom of the kind **ST*.2**). An axiom of the form **ST'.1** will be an appropriate restriction to types of the universal closure of a wff of the form $(Ey)(K_{n+1}(y) \wedge (x)(K_n(x) \supset (x \in y \equiv A(x))))$, where y does not occur in the wff $A(x)$ of **ST'**, and $A(x)$ contains x free and is restricted to types. $A(x)$ may well contain other free variables and these will be restricted to types in an appropriate way when the full universal closure of the axiom of **ST*** is transformed.

Exercise. Give explicitly the axiom **ST'.3** of infinity by transforming the axiom **ST*.3**.

We intend that our way of setting up **ST'** allow us to translate every proof of a theorem in **ST*** into a proof in **ST'**. In **ST*** we had the usual rules of the predicate calculus except that quantification and substitution applied only within a given type. In **ST'**, we have the usual

---

[†] For example, the transform of $(\forall x_1^2)(Ex_1^1)(x_1^1 \in x_1^2 \supset (Ex_1^0)(x_1^0 \in x_1^1))$ is $(x_3)(K_2(x_3) \supset (Ex_2)(K_1(x_2) \wedge (x_2 \in x_3 \supset (Ex_1)(K_0(x_1) \wedge x_1 \in x_2))))$. The transform of $(\forall x_1^1)((Ex_1^0)(Ex_1^1)(x_1^0 \in x_1^1) \supset (Ex_1^1)(x_1^0 \in x_1^1))$ is $(x_4)(K_0(x_4) \supset ((Ex_3)(K_0(x_3) \wedge (Ex_1)(K_1(x_1) \wedge x_3 \in x_1)) \supset (Ex_2)(K_1(x_2) \wedge x_4 \in x_2)))$. These examples illustrate the process of proceeding from left to right and of handling all of the smaller quantified subformulas before handling the larger ones.

rules of the predicate calculus with no restrictions, and it is easy to check that these allow us to operate with wffs which are transforms of wffs of **ST*** with the same effect as if the variables had type indices. The one exception to this is where we deduce in **ST*** a wff $(Ex_m^n) A(x_m^n)$ from the wff $(\forall x_n^m) A(x_m^n)$. This deduction is justified by our quantification rules in **ST***, since we have only to apply $e\forall$ and then iE. However, the transform of $(\forall x_m^n) A(x_m^n)$ is $(x_j) (K_n(x_j) \supset A'(x_j))$, and we can obtain $(Ex_j) (K_n(x_j) \wedge A'(x_j))$ from this only if we can establish $(Ex_j) K_n(x_j)$; i.e. only if we can establish that there is something of type n. This is the reason for our extra axiom $(Ex_1) K_0(x_1)$, which tells us that there is something of type 0. The more general principle $(Ex_1) K_n(x_1)$ follows from our special axiom in conjunction with the axioms of abstraction **ST′**.1.

We can now see how to translate a proof from **ST*** into **ST′**. To obtain the translation in **ST′** of an axiom X of **ST***, we have only to apply $e\forall$ in **ST′**, removing whatever universal quantifiers may have been added when we took the universal closure of X. Our rules of inference are preserved in the sense just explained. Conversely, a nontrivial proof in **ST′** can only make use of the proper axioms of **ST′**, and these are all direct transforms of universal closures of axioms of **ST***. Of course, we will have some tautologies and other universally valid forms which are theorems of **ST′**, and which cannot be inversely transformed into wffs (and thus theorems) of **ST***. But clearly any contradiction deducible in **ST′** is deducible in **ST***, since the use of proper axioms of **ST′** will be essential to deducing any such contradiction. Therefore, we can see that **ST*** is consistent if and only if **ST′** is.

The advantage of a first-order formulation of **ST** is that we can apply to it all of the powerful theorems of first-order logic which we have developed, such as the theorems of completeness, compactness, and the like. Any model of the system **ST′** will have to assign some domain D to **ST′** and some subset T_n of D to each predicate letter K_n of **ST′**. Each T_n will have to be nonempty. Such a model may be more general than a simple type hierarchy, since it may not be that each T_{n+1} is the power set of T_n. In the future, when we speak of a *model* for **ST**, we will mean a model for **ST′** in the strictly defined sense of a first-order model. Such a model for **ST** will have a denumerable hierarchy T_n of nonempty sets and an interpretation of the predicate letter "$\in$" which makes the axioms of **ST** true. The relation E which interprets "$\in$" will be a subset of $D \times D$ where D is the domain of the interpretation. Since every variable in any axiom of **ST′** is restricted to range only over some type T_n, it follows that we can always consider that the domain D of a model of **ST′** is simply the union of the T_n. This is so because the sets T_n are all subsets of D and the restriction of the model D to the union W of the T_n will still be a model. $E \cap (W \times W)$ will be the induced interpretation of "$\in$".

Among the models for **ST** in this more general sense will be the simplified type hierarchies generated by some infinite set T_0. The set D which is the union of the T_n will be the domain of an interpretation of **ST′** in the obvious way: T_n will be the interpretation of K_n and the interpretation of "$\in$" will simply be the membership relation between the elements of these sets. We will speak of models for **ST** which are not simplified type hierarchies as *nonstandard* models for **ST**.[†] We can prove that nonstandard models of **ST** do exist. In Chapter 7, non-

---

[†] Generally speaking, whenever we have a first-order system F with a primitive binary predicate letter "$\in$", and whose axioms are intended to describe a set theory, a standard model for F is thought of as a model $\langle D, g \rangle$ such that the elements of D are all sets and the relation $g(\in)$ is the membership relation between the sets in D. Clearly, the simplified type hierarchies which are models of **ST** are standard models in this broad sense. It is conceivable, however, that there are models of **ST** which are not simple type hierarchies and where

standard models of **ST** of a specialized kind are shown to exist if certain extensions of **ST** are consistent. Chapter 6 contains a general discussion of the question of existence of models for foundational systems.

There is a real sense in which every higher-order theory can be translated into a first-order theory by the addition of appropriate predicate letters and the use of restricted quantification. It is for this reason (as well as the fact, previously noted, that more can be proved about first-order theories) that first-order theories have been the object of such intensive study. This observation also justifies our intuitive use of model-theoretic ideas in discussing some of the higher-order theories of this chapter. Robinson [2] gives an excellent detailed discussion of a general method for representing higher-order theories in a first-order way.

---

the interpretation of "ϵ" is still membership between sets. We are thus using the designation "standard" for **ST** in a somewhat restrictive sense. However, the difference in this case seems small enough to warrant our use of the term.

   Of course, one can always argue as to whether or not a given system really does describe a set theory. Thus, a standard model of a given system essentially means some particular model we have in mind which best represents to us the genius of the system. From this point of view, our choice of the simplified type hierarchies with an infinite domain of individuals as the standard models for **ST** seems well motivated.

Chapter 5

Zermelo–Fraenkel Set Theory

In this chapter we shall consider an alternative approach to foundations. It was first proposed by Zermelo in 1908, the same year that Russell's article, "Mathematical Logic as Based on the Theory of Types", was published. The system was later ameliorated by Fraenkel and Skolem. Our presentation will incorporate their improvements.

Zermelo's answer to the paradoxes and inconsistencies in Frege's system, as contained in his axioms for set theory, is based on the assumption that the sets which mathematicians need can be built up from certain simple sets by means of given operations. The approach is in many ways inverse to the approach of Poincaré–Russell. Russell's system is concerned with logical and linguistic principles such as impredicativity and the vicious-circle arguments; Zermelo's system is more directly concerned with mathematics and the needs of mathematical structures. One might characterize the latter approach as follows: Mathematics is (we believe) consistent. Thus, if we give a precise account of the intuitive use of sets as mathematicians use them, we shall have an adequate and correct foundation. This approach recalls the observation of Chapter 3 that, although Frege's axiom of abstraction is contradictory, the particular uses he makes of it in constructions may not be contradictory.

Of course we do not really know whether mathematics is consistent. But if it is not, any foundation will fail. Therefore, we might as well assume that it is, at least for the purposes of foundational studies. Continuing in this line of thought, we observe that mathematicians do not normally use such sets as "the set of all sets" or the "set of all sets not elements of themselves". We might contend that these contradictory notions are not really valid mathematical objects at all.

It should be made clear at this point that we are not envisaging a statistical analysis of what "most mathematicians" do in an effort to get a sort of common denominator set theory. Mathematicians do indeed use the principle of abstraction, and do not generally bother to worry about what applications of this principle may lead to contradiction, so long as contradictions do not appear in their work. What we are attempting to do here is to push beyond such a superficial statistical analysis to a deeper logical analysis. Basically, it is an attempt to seize and render explicit those hidden, intuitive principles which may unconsciously govern mathematical thinking and its concurrent tendency to avoid *de facto* contradictions.

Now, one thing stands out concerning the usage of the principle of abstraction in mathematics: When sets are defined by properties, they are usually *subsets* of some given, mathematically defined set. For example, a mathematician may pick out the set of all continuous real functions or all differentiable real functions by defining the respective notions "f is continuous" and "f is differentiable" in purely logical terms. But these are both subsets of the set of all real functions, which the mathematician considers as a valid mathematical object. It would not usually occur to the mathematician to consider any such notion as "the set of all functions there are" any more than he would be moved to consider the set of all sets. Many might even feel that such a notion as the set of all possible functions is meaningless, a function being determined by a given domain, a rule of assignment, a codomain, and so on.

Thus, in Zermelo's system, the principle of abstraction becomes a principle of *separation* or *Aussonderung*. It is a process of picking out subsets of a given set by properties rather than by defining sets *ab initio* by properties. The intuitive principle of separation can be stated as follows: For any condition P (which will be expressed by some wff of our formal theory), and any given set y, the set of all elements of y which satisfy the property P exists. Using our "$\in$" notation, we can state formally: $\vdash (y)(Ex)(z)(z \in x \equiv z \in y \land A(z))$ where the variables x, y, and z are all different, z may be free in the wff $A(z)$, and x does not appear in $A(z)$. (The wff $A(z)$ expresses the property P.)

Rather than positing the existence of sets, the separation scheme posits the existence of certain subsets of a given set. The only set whose existence can be proved solely from the separation scheme is the null set. Just take any self-contradictory property for $A(z)$. For example, we have

$$\vdash (Ex_1)(x_2)(x_2 \in x_1 \equiv x_2 \in x_3 \land x_2 \in x_2 \land x_2 \notin x_2),$$

which yields $\vdash (Ex_1)(x_2)(x_2 \notin x_1)$.

Notice that a logically true formula for $A(z)$ will not give us a (nonnull) universal set.

Exercise. Prove the last statement, i.e. that a (nonnull) universal set is not deducible from the separation scheme alone. Use a first-order language with "$\in$" as the sole predicate letter and variables as the only terms.

The principle of separation, which will be an axiom of our formal system, gives us the null set outright. Otherwise, it gives us only new sets from given sets. But where are the given sets to come from in the first place? The answer to this is that they will be sets which can be constructed from the null set by iterating certain basic operations. The basic operations involved are essentially those of power set and union. We shall need considerable additional postulation to guarantee that we have these operations available in their full generality. These additional axioms, together with the principle of separation, will constitute the system of Zermelo–Fraenkel, abbreviated as **ZF**.

As we develop the system of Zermelo–Fraenkel, we shall have in mind an intuitive model that we will be trying to describe by our axioms. The elements of this model will be members of the sets of a certain hierarchy, which can be informally described as follows: Beginning with the null set, which we denote by "0", we iterate the operation $\mathcal{P}$ of power set, obtaining the sets $\{0\}$; $\{0, \{0\}\}$; $\{0, \{\{0\}\}, \{0, \{0\}\}, \{0\}\}$; and so on. Each of the sets so obtained is finite, since the power set of any finite set is finite. However, the hierarchy of sets so obtained is denumerably infinite. We now take the union of all the sets in this hierarchy and form a set D. D is an infinite set and it has as elements all those and only those sets which are elements of some set $\mathcal{P}(\cdots \mathcal{P}(0) \cdots)$, to a finite number of iterations of $\mathcal{P}$. We begin again, iterating the operation of power set, to obtain the sets $\mathcal{P}(D)$, $\mathcal{P}(\mathcal{P}(D))$, and so on. We now take the union of all the sets so obtained, the sets $\mathcal{P}(\cdots \mathcal{P}(0) \cdots)$, D, and $\mathcal{P}(\cdots \mathcal{P}(D) \cdots)$,[†] and form a new set X. We then iterate our power set operation, and proceed as before. Briefly described, the hierarchy of sets we envisage consists of all the sets we can obtain by starting with the null set and iterating, in the indicated manner, our power set and union operations any transfinite number of times. The intuitively conceived union of *all* the sets of our hierarchy is the intuitive model for our system. In other words, a set x is in our model if and only if it is an element of some set of our hierarchy.

As vague and forbidding as our endless hierarchy may seem, it does give some semblance of order and constructiveness to the system of Zermelo–Fraenkel. Every set x in our model is an element of some set y in our hierarchy, and y is obtained by iterating our operations of power set and union some (possibly transfinite) ordinal number of times starting with 0. Now the intuitively conceived class of ordinals is well ordered in that any nonempty class of ordinals has a least element. Consider, for a set x in our model, all those ordinals α such that $x \in y_\alpha$ where y_α is a set of our hierarchy obtained by exactly α iterations of power set and union. There must be at least one such ordinal if x is in our model. There is thus a smallest ordinal α such that $x \in y_\alpha$. This ordinal is called the *rank* of the set x in our model. Therefore, associated with each set x in our intuitively conceived model is a uniquely defined ordinal number that is a measure of how far up in the hierarchy we must go before we obtain a set which has x as an element.

A set of finite rank is of finite cardinality, but not conversely. The set $\{0\}$ and the set $\{D\}$ are both one-element sets. The rank of $\{0\}$ is 2, but the rank of $\{D\}$ is $\omega + 2$ where ω is the first infinite ordinal (a fuller discussion of the ordinals in intuitive set theory and within our axiomatic system will be given later on in this chapter).

Notice that our model does not allow for the possibility of distinct no-element things as was possible in a type hierarchy. The only no-element thing in our model is the null set 0. Every other entity in our model is a set with elements. Also, our model is clearly richer than the type hierarchy with a denumerable set of individuals, which we considered as a basic model for the system **ST** of simplified types. Recall that in our simplified type hierarchies we iterate only with the power set starting with some given nonempty set. In the case of our basic model for **ST**, we start with a denumerable set. In the present case, we iterate both power set and union, and we do obtain a denumerably infinite set D to which the power set operation is applied. Thus, our present intuitive model for Zermelo–Fraenkel includes a model for **ST** as a subcollection. Moreover, the present use of the union operation implies that there is no separation into types in our present model.

We have presented our intuitive model for Zermelo–Fraenkel set theory as heuristic motivation for our axioms. Such a model proves the consistency of our system only if we can actually prove, by some rigorous argument, that the model really exists and is a model for our formal axioms. In certain fairly strong versions of axiomatic set theory, which we shall discuss briefly at the end of this chapter, it is possible to define our hierarchy rigorously by transfinite induction. However, no way is known to prove that the union of the sets in this hierarchy does form a model for **ZF** without bringing in assumptions equivalent to the consistency of **ZF**. This obviously begs the question. Chapter 6 will contain a more detailed discussion of consistency

---

[†] On course the union of all of the sets $\mathcal{P}(\cdots \mathcal{P}(0) \cdots)$ is already the set D, but we mention these sets again in order to emphasize that the union operation is applied, at each stage of the construction, to *all* the sets obtained up to that point in the hierarchy.

proofs and some of these touchy model-theoretic considerations. We should also keep in mind that the system **ZF** that we shall construct has many different models if it is consistent. In Chapter 6 we will see just why this is so.

We now consider a formalized version of Zermelo's system and we proceed to prove Peano's postulates, as well as to begin the development of ordinal theory.

5.1. Formalization of ZF

Zermelo–Fraenkel set theory, which we abbreviate as **ZF**, is a first-order theory with one primitive binary predicate letter "$\in$". As terms and term operators, there are two primitive constants, three singulary function letters, one binary function letter, and one variable-binding term operator. Instead of listing these primitive notions now, we follow an expository device of Bernays [1] and mention each one initially with the axiom governing its use. Rather than stating all the axioms at once, we state a few axioms, prove some theorems, and state a few more axioms. The theorems we prove, based on the axioms given at any point, will help motivate the succeeding axioms. The set of wffs of **ZF** is thus defined as follows: (1) All variables are terms. (2) If x and y are terms, then $(x \in y)$ is a wff. (3) If A is a wff and x is a variable, then $(x)A$, $(Ex)A$, and $(\sim A)$ are wffs. (4) If A and B are wffs, then $(A \lor B)$ is a wff. (5) The two constants, and the terms formed from the function letters and variable-binding term operator that are designated as primitive notions in the ensuing exposition, are all terms. (6) The preceding are the only wffs and terms.

We suppose the usual definitions of the other sentential connectives. The wffs of **ZF** are the same as those for Frege's system except for the introduction of our primitive term operators. Since **ZF** will be a theory in which equality is definable, the function letters and constants are all eliminable in theory. The only variable-binding term operator will be the separation operator used in the axiom of separation, whose intuitive meaning has already been discussed. The separation operator is the analogue in **ZF** of the abstraction operator of Frege's system. Like the abstraction operator, it is also eliminable in principle, since there is an axiom of extensionality in **ZF**.

We now proceed with the development of the system. We shall often refer to the terms of our system as "sets", as we have already done in our treatment of other systems of set theory.

Definition 1. $x = y$ for $(z)(z \in x \equiv z \in y)$, where x and y are any terms in which the variable z does not occur.

ZF.1. $(x)(y)(x = y \supset A(x, x) \equiv A(x, y))$, where $A(x, y)$ is obtained from (Ax, x) by replacing y for x in zero, one, or more occurrences of x in $A(x, x)$, and y is free for x in the occurrences of x that it replaces.

This is the axiom of extensionality. The usual way of introducing this axiom in **ZF** is by the more restricted form: **ZF.1***

$$(x_1)(x_2)(x_1 = x_2 \supset (x_3)(x_1 \in x_3 \equiv x_2 \in x_3)).$$

This is done, however, when there are no primitive terms or term operators other than variables. In this case, our form **ZF.1** can be deduced as a metatheorem. The proof is by induction on the number of sentential connectives and quantifiers in the wff $A(x, x)$.

Exercise. Prove the last statement by using Definition 1 and only **ZF.1*** as a proper axiom, and assuming variables are the only terms.

THEOREM 1. $\vdash (x_1)(x_2)(x_1 = x_2 \supset (x_3)(x_1 \in x_3 \equiv x_2 \in x_3))$.

Proof. (1) 1. $x_1 = x_2$ H

2. $x_1 \in x_3 \equiv x_1 \in x_3$ Taut

3. $x_1 = x_2 \supset ((x_1 \in x_3 \equiv x_1 \in x_3) \equiv (x_1 \in x_3 \equiv x_2 \in x_3))$ **ZF.1**, $e\forall$

(1) 4. $(x_1 \in x_3 \equiv x_1 \in x_3) \equiv (x_1 \in x_3 \equiv x_2 \in x_3)$ 1, 3, MP

(1) 5. $(x_1 \in x_3 \equiv x_2 \in x_3)$ 2, 4, MP

(1) 6. $(x_3)[5]$ 5, $i\forall$

7. $[1 \supset 6]$ 1, 6, eH

8. $(x_1)(x_2)[7]$ 7, $i\forall$

Theorem 1 shows how to recover the weaker statement **ZF.1*** from our more general form of the axiom of extensionality.

THEOREM 2. $\vdash (x_1)(x_1 = x_1)$.

Proof. Immediate from Definition 1.

Definition 2. $x \notin y$ for $\sim (x \in y)$ where x and y are any terms.

Definition 3. $x \neq y$ for $\sim (x = y)$ where x and y are any terms.

We now state the axiom of separation and mention one of our primitive term operators in the process.

ZF.2. $(z)(x)(x \in \{y \mid y \in z \wedge A(y)\} \equiv x \in z \wedge A(x))$, where $A(y)$ is a wff, x is free for y in $A(y)$, $A(x)$ results from $A(y)$ by replacing x for all free occurrences of y, and the variables x, y, and z are all different. The variable y is bound in the term $\{y \mid y \in z \wedge A(y)\}$ which is formed by the variable-binding term operator $\{ \mid \}$. This operator applies in the indicated manner to any wff and the variable y is always bound by such an application.

We have already indicated in our introductory discussion how the principle of separation differs from the principle of abstraction. Formally, the difference amounts to the presence of the extra condition "$y \in z$" in the separation operator in the axiom **ZF.2**. We already know from our introductory discussion how to formulate **ZF.2** without the use of a term operator.

Notice that, although the *vbto* of separation can be applied to any wff $A(y)$ whatever, **ZF.2** allows us to work only with wffs $A(y)$ of the form $y \in z \wedge B(y)$. This means that certain intuitively true principles involving the separation *vbto* cannot be proved using **ZF.1** and **ZF.2**. For example, although the wff $(y)(y \neq y \equiv y \in z \wedge y \neq y)$ is provable from **ZF.1**, where z is any variable different from y, we cannot use **ZF.2** to deduce from this that the wff $\{y \mid y \neq y\} = \{y \mid y \in z \wedge y \neq y\}$ is a theorem. Indeed, in most interpretations $\langle D, g \rangle$, the value of the left side of this equality is left undetermined by **ZF.1** and **ZF.2**.

The solution to this problem is to add the two schemes V.1 and V.2 of Section 1.8 as logical axioms to our underlying logic, where the *vbto* v of that section is the separation *vbto* of **ZF**. We henceforth assume these two principles for **ZF** (as well as for all other extensions and ramifications of **ZF** where the *vbto* of separation is used, e.g. the systems **Z**, **NBG**, **MKM**, etc.). Since these axioms are logical axioms, they are not proper axioms of **ZF** and so are not given numbers in the listing of **ZF** axioms.

Exercise. Use our underlying logic (and in particular the scheme V.1) and *Df.* 1 to establish:

$$\vdash \{x_1 \mid x_1 \neq x_1\} = \{x_1 \mid x_1 \in x_2 \wedge x_1 \neq x_1\}.$$

Notice that, although absolute complements of sets are not definable using **ZF**.2 since we cannot use **ZF**.2 to evaluate the term $\{x \mid x \notin y\}$, we can define the relative complement $z - y$ of two arbitrary sets by $\{x \mid x \in z \wedge z \notin y\}$ which can be evaluated by **ZF**.2.

ZF.3. $0 = \{x_1 \mid x_1 \in 0 \wedge x_1 \neq x_1\}$, where 0 is a primitive constant.

This axiom says that 0 is the null set. As we already saw in our informal discussion preceding the formal statement of our axioms, we can deduce the existence of a no-element set from **ZF**.2. **ZF**.3 simply tells us that 0 is this no-element set. Theorem 1 and Definition 1 assure us that any two no-element sets are equal; i.e. that there is only one no-element set. These conclusions are embodied in the next two theorems.

THEOREM 3. $\vdash (x_1)(x_1 \notin 0)$.

Proof.   (1) 1. $x_1 \in 0$ H
    (1) 2. $x_1 \in 0 \wedge x_1 \neq x_1$ 1, **ZF**.3, **ZF**.2, $e\forall$, Taut, MP
    (1) 3. $x_1 \neq x_1$ 2, Taut, MP
        4. $x_1 \in 0 \supset x_1 \neq x_1$ 1, 3, eH
        5. $x_1 = x_1 \supset x_1 \notin 0$ 4, Taut, MP
        6. $x_1 = x_1$ Th. 2, $e\forall$
        7. $x_1 \notin 0$ 5, 6, MP
        8. $(x_1)(x_1 \notin 0)$ 7, $i\forall$

THEOREM 4. $\vdash (x_1)((x_2)(x_2 \notin x_1) \equiv 0 = x_1)$.

Proof. The reader will prove Theorem 4 as an exercise.

No universal set V will be forthcoming in **ZF**. Indeed, if we could obtain such a set, then we could deduce Russell's paradox with the aid of **ZF**.2 through the application of the separation scheme to the wff $y \in V \wedge y \notin y$.

Definition 4. $\vdash x \subset y$ for $(z)(z \in x \supset z \in y)$ where x and y are terms in which z does not occur.

THEOREM 5. $\vdash (x_1)(x_2)((x_1 \subset x_2 \wedge x_2 \subset x_1) \supset x_1 = x_2)$.

Proof. The reader will prove Theorem 5 as an exercise.

THEOREM 6. $\vdash (x_1)(x_1 \subset x_1)$.

Proof. Immediate from Definition 4.

THEOREM 7. $\vdash (x_1)(x_2)(x_3)((x_1 \subset x_2 \wedge x_2 \subset x_3) \supset x_1 \subset x_3)$.

Proof. The reader will prove Theorem 7 as an exercise.

THEOREM 8. $\vdash (x_1)(0 \subset x_1)$.

Proof. The reader will prove Theorem 8 as an exercise.

These standard properties of the subset relation will all be useful in future proofs.

ZF.4. $(y)(x)(x \in \mathcal{P}(y) \equiv x \subset y)$ where x and y are different. $\mathcal{P}$ is a primitive singulary function letter.

$\mathcal{P}(y)$ is the power set of y, the set of all its subsets.

ZF.5. $(y)(z)(x)(x \in \{y, z\} \equiv (x = y \vee x = z))$ where x, y, and z are different variables. $\{ , \}$ is a primitive binary function letter where we write the arguments in the indicated manner.

We call $\{y, z\}$ the *unordered pair* of y and z. It is the set whose only elements are y and z.

Definition 5. $\{x\}$ for $\{x, x\}$ where x is any term.

Definition 6. $\langle x, y \rangle$ for $\{\{x\}, \{x, y\}\}$ where x and y are any terms. This is the Wiener–Kuratowski ordered pair definition whose history we have already traced in Chapter 4.

Exercise. Prove

$$\vdash (x_1)(x_2)(x_3)(x_4)(\langle x_1, x_2 \rangle = \langle x_3, x_4 \rangle \equiv (x_1 = x_3 \wedge x_2 = x_4)).$$

THEOREM 9. $\vdash (x_1)(x_1 \in \{x_2\} \equiv x_1 = x_2)$.

Proof. The reader will prove Theorem 9 as an exercise.

$\{x_1\}$ is thus the set whose only element is x_1.

Exercise 1. Prove $\vdash (x_1)(x_1 \in \{x_1\})$.

Exercise 2. Prove the converse of Theorem 1; i.e.

$$\vdash (x_1)(x_2)((x_3)(x_1 \in x_3 \equiv x_2 \in x_3) \supset x_2 = x_1).$$

ZF.6. $(y)(x)(x \in \bigcup(y) \equiv (Ez)(z \in y \wedge x \in z))$ where the variables x, y, and z are all different. $\bigcup$ is a primitive singulary function letter.

This is the axiom of sum set. If we imagine y as a set whose elements are sets, then $\bigcup(y)$ is the union of all sets in the collection x. As a matter of fact, every set y except 0 will be a non-empty collection of other sets, since Theorem 4 asserts the equality of all no-element things with 0. In Zermelo's original formulation, provision was made for the existence of *Urelemente*, i.e. distinct atomic individuals, but as this has no mathematical application it is a complication we have thought better to avoid.

Definition 7. $\bigcap(x)$ for $\{y \mid y \in \bigcup(x) \wedge (z)(z \in x \supset y \in z)\}$ where x is any term, y, x, and z are all different, and y and z do not occur in x.

$\bigcap(x)$ is the intersection of all the sets that are elements of x. We have been able to define $\bigcap(x)$ without further postulation by using **ZF.2**, since $\bigcup(x)$ is a set which contains $\bigcup(x)$.[†]

Definition 8. $(x \cup y)$ for $\bigcup(\{x, y\})$ where x and y are any terms.

Definition 9. $(x \cap y)$ for $\bigcap(\{x, y\})$ where x and y are any terms.

The purpose of the axiom of unordered pair now becomes clear. It enables us to obtain the usual notions of intersection and union from the corresponding infinitary notions. We prove:

THEOREM 10. $\vdash (x_1)(x_2)(x_3)(x_1 \in (x_2 \cup x_3) \equiv x_1 \in x_2 \vee x_1 \in x_3)$.

Proof. (1)   1. $x_1 \in (x_2 \cup x_3)$ H

(1)   2. $x_1 \in \bigcup(\{x_2, x_3\})$ 1, Df. 7

(1)   3. $(Ex_4)(x_4 \in \{x_2, x_3\} \wedge x_1 \in x_4)$ 2, **ZF.6**, Taut, MP

(1)   4. $a_1 \in \{x_2, x_3\} \wedge x_1 \in a_1$ 3, eE

(1)   5. $a_1 \in \{x_2, x_3\}$ 4, Taut, MP

(1)   6. $a_1 = x_2 \vee a_1 = x_3$ 5, **ZF.5**, $e\forall$, Taut, MP

(7)   7. $a_1 = x_2$ H

(1)   8. $x_1 \in a_1$ 4, Taut, MP

(1, 7)   9. $x_1 \in x_2$ 7, 8, Df. 1, $e\forall$, Taut, MP

(1)   10. $a_1 = x_2 \supset x_1 \in x_2$ 7, 9, eH

(1)   11. $a_1 = x_3 \supset x_1 \in x_3$ analogous to lines 7–10

(1)   12. $a_1 = x_2 \vee a_1 = x_3 \supset x_1 \in x_2 \vee x_1 \in x_3$ 10, 11, Taut, MP

(1)   13. $x_1 \in x_2 \vee x_1 \in x_3$ 6, 12, MP

       14. $[1 \supset 13]$ 1, 13, eH

(15)   15. $x_1 \in x_2$ H

[†] This differs from intuitive set theory on one minor point. In **ZF**, $\bigcup(0) = 0$ so $\bigcap(0) = 0$, since $\bigcap(0) \subset \bigcup(0)$. In intuitive set theory $\bigcap(x)$ is sometimes defined simply as "the set of all y such that $(z)(z \in x \supset y \in z)$", from which it follows that $\bigcap(0) = V$, the universal set. However, we do not really need such logical antics and, as we have already noted, we must positively avoid having a universal set in **ZF** to avoid deducing Russell's paradox.

16. $x_2 = x_2 \lor x_2 = x_3$ Th. 2, $e\forall$, Taut, MP

17. $x_2 \in \{x_2, x_3\}$ 16, **ZF**.5, $e\forall$, Taut, MP

(15) 18. $x_1 \in x_2 \land x_2 \in \{x_2, x_3\}$ 15, 17, Taut, MP

(15) 19. $(Ex_4)(x_1 \in x_4 \land x_4 \in \{x_2, x_3\})$ 18, iE

20. $[15 \supset 19]$ 15, 19, eH

21. $x_1 \in x_3 \supset (Ex_4)(x_1 \in x_4 \land x_4 \in \{x_2, x_3\})$ analogous to lines 15–20.

22. $(x_1 \in x_2 \lor x_1 \in x_3) \supset (Ex_4)(x_1 \in x_4 \land x_4 \in \{x_2, x_3\})$ 20, 21, Taut, MP

23. $(x_1 \in x_2 \lor x_1 \in x_3) \supset x_1 \in \bigcup(\{x_2, x_3\})$ 22, **ZF**.6, $e\forall$, Taut, MP

24. $(x_1 \in x_2 \lor x_1 \in x_3) \supset x_1 \in (x_2 \cup x_3)$ 23, $Df.\ 8$

25. $[14 \land 24]$ 14, 24, Taut, MP

26. $(x_1)(x_2)(x_3)[25]$ 25, $i\forall$

27. Concl $Df. \equiv$

THEOREM 11. $\vdash (x_1)(x_2)(x_3)(x_1 \in (x_2 \cap x_3) \equiv x_1 \in x_2 \land x_1 \in x_3)$.

Proof. The reader will prove Theorem 11 as an exercise. (*Hint*: By analogy with the foregoing proof.)

With the last two theorems, we have obtained the usual defining properties of union and intersection.

Definition 10. x' for $x \cup \{x\}$ where x is any term.

Definition 11.   1 for $0'$.

Definition 12.   2 for $1'$.

The operation x' is the successor function in Zermelo's system. The version of natural number we obtain is that of von Neumann, already discussed in Chapter 3. Intuitively, the natural numbers will be the smallest set containing 0 and closed with respect to the operation of successor. It is interesting, however, that we shall develop a theory of natural numbers and prove all of Peano's postulates without any axiom of infinity. A word of explanation is in order here.

A model of a first-order theory consists of a domain D together with interpretations for the primitive predicate symbols and term operators. It is not difficult to prove that there is no finite model for the six axioms (**ZF**.1 through **ZF**.6) that we have already postulated. In fact, once we have deduced Peano's postulates on the basis of these axioms, we shall have proved just that, since there is no finite model for the Peano axioms. To say that there is no finite model means that the set D of our interpretation is infinite. It does not mean that any given object of the domain is itself an "infinite set" under the interpretation. An infinite set in this latter sense is an object d in the domain D for which there is an infinity of distinct objects y in D, such that the yEd holds where E is the interpretation of the predicate letter "$\in$" of **ZF**.

In further explanation of this point, let us observe that we could now postulate an axiom of infinity of the following form:

$$(Ew)(0 \in w \land (x)(x \in w \supset x' \in w)).$$

The set w whose existence is postulated is some set containing the natural numbers. The natural numbers N form the *smallest* set containing 0 and closed under successor. Thus, if w is the set whose existence is posited by our axiom of infinity, N would be defined as the set

$$\{y \mid y \in w \wedge (x)(0 \in x \wedge (z)(z \in x \supset z' \in x) \supset y \in x)\}$$

by using **ZF**.2. This definition recalls Frege's, but **ZF**.2, to be applicable, requires the existence of some set w containing the natural numbers. Hence we need an explicit axiom of infinity, which was not needed in Frege's system.

In other words, the infinite hierarchy consisting of 0; $\{0\}$; $\{0, \{0\}\}$, and the rest is already part of our system, but we have no way within our system of "collecting up" this hierarchy into a single, containing, infinite set. When we finally do add an explicit axiom of infinity, it will have the precise consequence of "collecting up" this hierarchy, though the form of the axiom will differ from the one just considered. Our method of developing the natural numbers without such an explicit axiom of infinity is based essentially on Bernays [1]. We now turn to this development.

To simplify our development of the natural numbers, we add a further axiom known as the axiom of *restriction or regularity*. This axiom is not an axiom of infinity at all. In fact, its purpose is really to exclude certain undesired sets from our universe. It restricts the possible models of **ZF**, hence its name. Formally stated, the axiom is:

ZF.7. $(x)(x \neq 0 \supset (Ey)(y \in x \wedge y \cap x = 0))$ where x and y are distinct variables.

This axiom says that any nonnull set has some element which has no members in common with it. This amounts to the assertion that every nonempty set has at least one $\in$-minimal element, i.e. an element y such that no member of the given set bears the $\in$ relation to y. This property will be very useful further on in developing the theory of ordinal numbers in **ZF**.

One consequence of this axiom is that infinite "descending chains" of membership such as $x_1 \ni x_2 \ni x_3 \cdots$ with no minimal element (stopping place) are excluded from our system. Certainly such infinite descending chains are excluded from our heuristic model of **ZF**. Of course if **ZF** is consistent, it will have models other than our heuristic one. (If nothing else, it will have a denumerable model. However, in Chapter 6 we show that it will in fact have many nonisomorphic models if it is consistent.) However, **ZF**.7 serves to exclude certain undesirable features from any such model. For example, we can prove the following theorem by using **ZF**.7:

THEOREM 12. $\vdash (x_1)(x_1 \notin x_1)$.

Proof. (1) 1. $x_1 \in x_1$ H
    2. $x_1 \in \{x_1\}$ Th. 2, Th. 9, $e\forall$, Taut, MP
(1) 3. $x_1 \in x_1 \wedge x_1 \in \{x_1\}$ 1, 2, Taut MP
(1) 4. $x_1 \in (x_1 \cap \{x_1\})$ 3, Th. 11, $e\forall$, Taut, MP
(1) 5. $(Ex_2)(x_2 \in (x_1 \cap \{x_1\}))$ 4, iE
    6. $(Ex_2)(x_2 \in (x_1 \cap \{x_1\})) \equiv x_1 \cap \{x_1\} \neq 0$ Th. 4, $e\forall$, Taut, MP
(1) 7. $x_1 \cap \{x_1\} \neq 0$ 5, 6, Taut, MP
    8. $(Ex_2)(x_2 \in \{x_1\})$ 2, iE

9. $\{x_1\} \neq 0$ 8, Th. 4, $e\forall$, Taut, MP

10. $(Ex_2)(x_2 \in \{x_1\} \wedge (x_2 \cap \{x_1\}) = 0)$ 9, **ZF**.7, $e\forall$, MP

11. $a_1 \in \{x_1\} \wedge (a_1 \cap \{x_1\}) = 0$ 10, eE

12. $a_1 \in \{x_1\}$ 11, Taut, MP

13. $a_1 = x_1$ 12, Th. 9, $e\forall$, Taut, MP

14. $a_1 = x_1 \supset ((a_1 \cap \{x_1\}) = 0 \equiv (x_1 \cap \{x_1\}) = 0)$ **ZF**.1, $e\forall$

15. $(x_1 \cap \{x_1\} = 0)$ 13, 14, 11, Taut, MP

(1) 16. $(x_1 \cap \{x_1\}) = 0 \wedge (x_1 \cap \{x_1\}) \neq 0$ 7, 15, Taut, MP

17. $[1 \supset 16]$ 1, 16, eH

18. $[\sim 16 \supset \sim 1]$ 17, Taut, MP

19. $x_1 \notin x_1$ 18, Taut, MP, *Df*. 2

20. $(x_1)(x_1 \notin x_1)$ 19, $i\forall$

Theorem 12 says that no sets are elements of themselves. We have used **ZF**.7 to prove this fact, and it is certainly a theorem which is intuitively true of our heuristic model. Without **ZF**.7, the theorem would still be true of our hierarchy, but it would not be provable in **ZF**. Consequently, models in which sets were elements of themselves would be admissible. We continue with:

THEOREM 13. $\vdash (x_1)(x_2)(x_1 \in x_2 \supset x_2 \notin x_1)$.

Proof.

1. $\{x_1, x_2\} \neq 0$ **ZF**.5, Th. 2, Th. 4, Taut, MP

2. $(Ex_3)(x_3 \in \{x_1, x_2\} \wedge (x_2 \cap \{x_1, x_2\}) = 0)$ 1, **ZF**.7, $e\forall$, Taut, MP

3. $a_1 \in \{x_1, x_2\} \wedge a_1 \cap \{x_1, x_2\} = 0$ 2, eE

4. $a_1 \in \{x_1, x_2\}$ 3, Taut, MP

5. $a_1 = x_1 \vee a_1 = x_2$ 4, **ZF**.5, $e\forall$, Taut, MP

(6) 6. $x_1 \in x_2$ H

7. $x_1 \in \{x_1, x_2\}$ Th. 2, **ZF**.5, $e\forall$, Taut, MP

(6) 8. $x_1 \in x_2 \wedge x_1 \in \{x_1, x_2\}$ 6, 7, Taut, MP

(6) 9. $x_1 \in (x_2 \cap \{x_1, x_2\})$ 8, Th. 11, $e\forall$, Taut, MP

(6) 10. $(Ex_3)(x_3 \in (x_2 \cap \{x_1, x_2\}))$ 9, iE

(6) 11. $x_2 \cap \{x_1, x_2\} \neq 0$ 10, Th. 4, $e\forall$, Taut, MP

(6) 12. $a_1 = x_2 \supset a_1 \cap \{x_1, x_2\} \neq 0$ 11, **ZF**.1, $e\forall$, Taut, MP

(13) 13. $x_2 \in x_1$ H

(13) 14. $a_1 = x_1 \supset (a_1 \cap \{x_1, x_2\}) \neq 0$ analogous to lines 6–12

(6, 13) 15. $a_1 = x_1 \vee a_1 = x_2 \supset (a_1 \cap \{x_1, x_2\}) \neq 0$ 12, 14, Taut, MP

(6, 13) 16. $(a_1 \cap \{x_1, x_2\}) \neq 0$ 5, 15, MP

17. $(a_1 \cap \{x_1, x_2\}) = 0$ 3, Taut, MP

(6, 13) 18. $[16 \wedge \sim 16]$ 16, 17, Taut, MP

(6) 19. $[13 \supset 18]$ 13, 18, eH

(6)   20. $[\sim 18 \supset \sim 13]$ 19, Taut, MP

21. $[\sim 18]$ Taut

(6)   22. $x_2 \notin x_1$ 20, 21, MP, *Df.* 2

23. $x_1 \in x_2 \supset x_2 \notin x_1$ 6, 22, eH

24. $(x_1)(x_2)(x_1 \in x_2 \supset x_2 \notin x_1)$ 23, i∀

Definition 13. $\{x, y, z\}$ for $\{x, y\} \cup \{z\}$ where x, y and z are any terms.

Definition 14. $\{t_1, t_2, \ldots, t_n, t_{n+1}\}$ for $\{t_1, \ldots, t_n\} \cup \{t_{n+1}\}$ where the t_i are any terms.

Using Definition 14 and the method of proof of Theorem 8 and Theorem 9, we can see how to prove a general theorem to the effect that whenever we have a finite chain of membership relations $t_1 \in t_2 \in \ldots \in t_n$, then we never have $t_n \in t_1$. The form of the general argument is as follows: $\{t_1, t_2, \ldots, t_n\}$ is not 0, since it has elements. Thus, by **ZF.**7 it has some element a in it such that a has no elements in common with it. Now, if $t_1 \in t_2 \in \ldots \in t_n$ and $t_n \in t_1$ all hold, then every one of the sets t_i will have an element of $\{t_1, t_2, \ldots, t_n\}$ in it. But a must be one of the sets of t_i by Definition 14 and **ZF.**5, and so we have a contradiction. In particular, we shall need the following case in later theorems:

THEOREM 14. $\vdash (x_1)(x_2)(x_3)(x_1 \in x_2 \wedge x_2 \in x_3 \supset x_3 \notin x_1 \wedge x_3 \neq x_1)$.

Proof. The reader will prove Theorem 14 as an exercise. (*Hint*: Use $\{x_1, x_2, x_3\}$ and **ZF.**7 as just indicated.)

Having now examined the effect of **ZF.**7 and the way it simplifies our system, we proceed with the development of the theory of natural numbers. The approach is through the general theory of ordinal numbers.

Definition 15.   Trans(x) for $(y)(y \in x \supset y \subset x)$ where x is a term in which y does not occur.

Exercise. Prove that

$$\vdash \text{Trans}(x_1) \equiv (x_2)(x_3)(x_2 \in x_1 \wedge x_3 \in x_2 \supset x_3 \in x_1).$$

Trans(x) means that every element of x is also a subset of x.

Definition 16.   Con(x) for $(y)(z)((y \in x \wedge z \in x \wedge z \neq y) \supset z \in y \vee y \in z)$ where x is a term in which the distinct variables y and z do not occur.

Intuitively, Con(x) says that x is connected under the $\in$ relation.

Definition 17.   On(x) for Trans(x) $\wedge$ Con(x).

"On(x)" means "x is an ordinal number".

Exercise. $\vdash (x_1)(x_2)(x_3)(x_4)((\text{On}(x_1) \wedge x_2 \in x_1 \wedge x_3 \in x_1 \wedge x_4 \in x_1 \wedge x_2 \in x_3 \wedge x_3 \in x_4) \supset x_2 \in x_4)$. This shows that an ordinal is transitive under the $\in$ relation.

Perhaps a brief discussion of ordinal numbers in intuitive set theory is in order here to show how Definition 17 characterizes this intuitive notion within our system. Given a set S, by a binary relation R on S, we mean as usual some subset of $S \times S$. For the purpose of this discussion, we use the special symbol "$<$" for R and we write "$x < y$" instead of $\langle x, y \rangle \in <$. We say that a set S is *partially ordered* by a relation $<$, if $<$ is a binary relation on S such that the following three conditions hold: (1) For all $x \in S$, $x \not< x$. (2) If $x < y$, then $y \not< x$. (3) For all x, y, and z in x, if $x < y$ and $y < z$, then $x < z$. We call these three conditions *irreflexivity*, *asymmetry*, and *transitivity* respectively. Actually, condition 2 follows from 1 and 3, since if $x < y$, then $y < x$ cannot hold or else, by transitivity, we can deduce $x < x$, which violates condition 1.

As we list the properties for a partial order, it is immediately clear that they are satisfied by the usual "less than" relation between real numbers or integers. Yet, there is one important way in which a relation of partial order differs from these usual "less than" relations. In a partial order, it is possible that, for two distinct elements x and y of S, neither $x < y$ nor $y < x$ holds. In this case, we say that x and y are *not comparable* under the relation $<$ in question.

Relations of partial order are often represented by a two-dimensional display consisting of vertices, representing elements of the set S, and lines connecting the vertices with the understanding that if one vertex x is connected to another vertex y, y being situated above x, then $x < y$ holds. Two vertices that are not connected are considered not comparable, regardless of their relative positions. If two vertices x and y are connected *directly* (without any intervening vertex), and if $y < x$, then this means that there is no element z such that $y < z < x$ holds. In this case, we say that x *covers* y. The diagram below gives examples of (finite) partially ordered sets represented by displays of this kind.

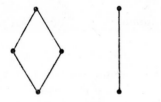  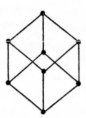

Suppose that any two distinct elements of a partially ordered set S are comparable. Then we have a *total order*, or simply an *order*. More precisely, a set S with a binary relation $<$ is said to be *totally ordered* by $<$ when S is partially ordered by $<$, and when the following condition holds: (4) For all x and y in S, either $x = y$, $x < y$ or $y < x$. We call this last condition *connectedness*. Thus, any connected partial order is a total order.

Exercise. Show that, in any total order, the three conditions of (4) are not only exhaustive but also mutually exclusive. This is called the *law of trichotomy* for total orders and is useful for proof by cases of properties of totally ordered sets.

When a relation $<$ totally orders a set S, we read "$x < y$" as "x is less than y". A total order is sometimes called a *linear order*. This terminology is partly motivated by the fact that the displays associated with total orders all look as shown on page 148.

Given a set S totally ordered by a relation $<$, we can imagine the elements of S as being "laid out" along a horizontal axis in such a way that the geometrical relation "x is to the left of y" holds if and only if $x < y$ holds.[†] This is, of course, exactly the way we represent the less than relation between integers or real numbers when we display them along the geometrical number line.

Finally, we define a *well-ordered* set S as any set S ordered by a relation $<$, and such that the following condition holds: (5) For every nonempty subset X of S, there is a (necessarily unique) smallest element of x; that is, an element $a \in x$ such that, for every other element $b \in X$, $b \neq a$, we have $a < b$.

The positive integers are well-ordered by the usual less than relation. The integers, the real numbers, and the rational numbers are not.

Given two well-ordered sets S and $S\#$, we define an *order isomorphism* of S and $S\#$ as a bijective mapping from S to $S\#$ such that order is preserved in both directions; that is, $x < y$ in S if and only if $f(x) < f(y)$ in $S\#$. In intuitive set theory, an *ordinal number* is an

equivalence class of well-ordered sets under order isomorphism. Notice that two order-isomorphic sets must have the same cardinality.

Now, our way of developing the theory of ordinals within **ZF** is to choose a representative well-ordered set from each intuitively conceived isomorphism class, rather than to work with the classes themselves. In fact, we use the $\in$ relation itself as the order relation $<$. This approach was conceived independently by Zermelo and von Neumann.

Let us see that a set that is an ordinal number as we have defined the notion is well ordered by the $\in$ relation. Irreflexivity is true generally for all sets in **ZF** by Theorem 12. Transitivity under $\in$ was established in the exercise following Definition 17, and we have already seen that irreflexivity and transitivity imply asymmetry. Con(x), which is precisely the definition of connectedness under the $\in$ relation, is also an explicit property of an ordinal number by Definition 17. This shows that any set x in **ZF** such that On(x) holds is totally ordered by $\in$. We now prove that it is actually well ordered by $\in$.

THEOREM 15. $\vdash (x_1)(x_2)(x_1 \subset x_2 \wedge x_1 \neq 0 \wedge \text{On}(x_2) \supset (Ex_3)(x_3 \in x_1$

$$\wedge (x_4)(x_4 \in x_1 \supset x_4 = x_3 \vee x_3 \in x_4))).$$

Proof. (1)   1. $x_1 \subset x_2 \wedge x_1 \neq 0 \wedge \text{On}(x_2)$ H

2. $x_1 \neq 0 \supset (Ex_3)(x_3 \in x_1 \wedge x_3 \cap x_1 = 0)$ **ZF**.7, $e\forall$

[†] This analogy is only to aid in visualizing the ordered set S. There is no intention here of building any kind of geometric theory of well-orderings or of considering how to represent geometrically those ordered sets whose cardinality is greater than $2^{\aleph_0}$, the cardinality of the continuum.

(1)   3. $a_1 \in x_1 \wedge a_1 \cap x_1 = 0$ 1, 2, Taut, MP, eE

(4)   4. $x_4 \in x_1 \wedge x_4 \neq a_1$ H

(1)   5. $a_1 \in x_2$ 1, 3, *Df*. 4, $e\forall$, Taut, MP

(1, 4)   6. $x_4 \in x_2$ 1, 4, *Df*. 4, $e\forall$, Taut, MP

(1, 4)   7. $a_1 \in x_2 \wedge x_4 \in x_2 \wedge x_4 \neq a_1$ 4, 5, 6, Taut, MP

(1)   8. $\mathrm{Con}(x_2)$ 1, Taut, MP, *Df*. 17

(1, 4)   9. $a_1 \in x_4 \vee x_4 \in a_1$ 7, 8, *Df*. 16, $e\forall$, MP

         10. $x_4 \in a_1 \wedge x_4 \in x_1 \supset x_4 \in (a_1 \cap x_1)$ Th. 11, $e\forall$, Taut, MP

         11. $x_4 \in (a_1 \cap x_1) \supset (a_1 \cap x_1) \neq 0$ Th. 4, $e\forall$, Taut, MP

         12. $x_4 \in a_1 \wedge x_4 \in x_1 \supset (a_1 \cap x_1) \neq 0$ 10, 11, Taut, MP

         13. $x_4 \in x_1 \supset (x_4 \in a_1 \supset (a_1 \cap x_1) \neq 0)$ 12, Taut, MP

(4)   14. $x_4 \in a_1 \supset (a_1 \cap x_1) \neq 0$ 4, 13, Taut, MP

(4)   15. $(a_1 \cap x_1) = 0 \supset x_4 \notin a_1$ 14, Taut, MP

(1, 4)   16. $x_4 \notin a_1$ 3, 15, Taut, MP

(1, 4)   17. $a_1 \in x_4$ 9, 16, Taut, MP

(1)   18. $(x_4 \in x_1 \wedge x_4 \neq a_1) \supset a_1 \in x_4$ 4, 17, eH

(1)   19. $x_4 \in x_1 \supset x_4 = a_1 \vee a_1 \in x_4$ 18, Taut, MP

(1)   20. $(x_4)(x_4 \in x_1 \supset x_4 = a_1 \vee a_1 \in x_4)$ 19, $i\forall$

(1)   21. $[a_1 \in x_1 \wedge 20]$ 3, 20, Taut, MP

(1)   22. $(Ex_3)(x_3 \in x_1 \wedge (x_4)(x_4 \in x_1 \supset x_4 = x_3 \vee x_3 \in x_4))$ 21, iE

         23. $[1 \supset 22]$ 1, 22, eH

         24. Concl   23, $i\forall$

The natural numbers, as we shall develop them, will be certain ordinals. The first ordinal will be 0 itself.

THEOREM 16. $\vdash \mathrm{On}(0)$.

Proof. Immediate from Theorem 3 and the relevant definitions.

Before defining the natural numbers, let us prove two more general theorems about ordinals.

THEOREM 17. $\vdash (x_1)(x_2)(\mathrm{Con}(x_1) \wedge x_2 \subset x_1 \supset \mathrm{Con}(x_2))$.

Proof.   (1)   1. $\mathrm{Con}(x_1) \wedge x_2 \subset x_1$ H

         (2)   2. $x_3 \in x_2 \wedge x_4 \in x_2 \wedge x_4 \neq x_3$ H

         (1, 2)   3. $x_3 \in x_1 \wedge x_4 \in x_1$ 1, 2, *Df*. 4, $e\forall$, Taut, MP

         (1)   4. $\mathrm{Con}(x_1)$ 1, Taut, MP

         (1, 2)   5. $x_3 \in x_4 \vee x_4 \in x_3$ 2, 3, 4, *Df*. 16, $e\forall$, Taut, MP

         (1)   6. $[2 \supset 5]$ 2, 5, eH

         (1)   7. $(x_3)(x_4)[6]$ 6, $i\forall$

                  8. $[1 \supset 7]$ 1, 7, eH

                  9. Concl   8, $i\forall$, *Df*. 16

Theorem 17 tells us that every subset of a connected set is connected. This will be useful in the following:

THEOREM 18. $\vdash (x_1)(x_2)(\mathrm{On}(x_1) \wedge x_2 \in x_1 \supset \mathrm{On}(x_2))$.

Proof.
(1)	1. $\mathrm{On}(x_1) \wedge x_2 \in x_1$ H
(1)	2. $\mathrm{Con}(x_1)$ 1, *Df.* 17, Taut, MP
(1)	3. $\mathrm{Trans}(x_1) \wedge x_2 \in x_1$ 1, *Df.* 17, Taut, MP
(1)	4. $x_2 \subset x_1$ 3, *Df.* 15, Taut, MP, $e\forall$
(1)	5. $\mathrm{Con}(x_1) \wedge x_2 \subset x_1$ 2, 4, Taut, MP
(1)	6. $\mathrm{Con}(x_2)$ 5, Th. 17, $e\forall$, MP
(7)	7. $x_3 \in x_2$ H
(8)	8. $x_4 \in x_3$ H
(1, 7)	9. $x_3 \in x_1$ 4, 7, Taut, MP, *Df.* 4, $e\forall$
(1)	10. $\mathrm{Trans}(x_1)$ 3, Taut, MP
(1, 7)	11. $x_3 \subset x_1$ 9, 10, *Df.* 15, Taut, MP, $e\forall$
(1, 7, 8)	12. $x_4 \in x_1$ 8, 11, Taut, MP, *Df.* 4, $e\forall$
	13. $x_4 \in x_3 \wedge x_3 \in x_2 \supset x_2 \notin x_4 \wedge x_2 \neq x_4$ Th. 14, $e\forall$
(7, 8)	14. $x_4 \in x_3 \wedge x_3 \in x_2$ 7, 8, Taut, MP
(7, 8)	15. $x_2 \notin x_4 \wedge x_2 \neq x_4$ 13, 14, MP
(1, 7, 8)	16. $x_2 \in x_1 \wedge x_4 \in x_1$ 1, 12, Taut, MP
(1, 7, 8)	17. $x_2 \in x_4 \vee x_4 \in x_2 \vee x_2 = x_4$ 2, 16, *Df.* 16, $e\forall$, Taut, MP
(1, 7, 8)	18. $x_4 \in x_2$ 15, 17, Taut, MP
(1, 7)	19. $x_4 \in x_3 \supset x_4 \in x_2$ 8, 18, *eH*
(1, 7)	20. $(x_4)(x_4 \in x_3 \supset x_4 \in x_2)$ 19, $i\forall$
(1, 7)	21. $x_3 \subset x_2$ 20, *Df.* 4
(1)	22. $x_3 \in x_2 \supset x_3 \subset x_2$ 7, 21, *eH*
(1)	23. $(x_3)[22]$ 22, $i\forall$
(1)	24. $\mathrm{Trans}(x_2)$ 23, *Df.* 15
(1)	25. $[6 \wedge 24]$ 6, 24, Taut, MP
	26. $[1 \supset 25]$ 1, 25, *eH*
	27. Concl 26, $i\forall$

Exercise. Prove $\vdash (x_1)(\mathrm{On}(x_1) \wedge x_1 \neq 0 \supset 0 \in x_1)$. (*Hint*: Use **ZF**.7.)

Theorem 18 tells us that every element of an ordinal is itself an ordinal. This amounts to saying that every ordinal α is precisely the set of all ordinals preceding α since the ordinals are ordered by $\in$. In terms of our intuitive model for **ZF**, the ordinals represent an important subsystem which can be defined intuitively as follows: We start with 0 (the null set), and iterate the prime operation of Definition 10. The sets obtained by any finite iteration of this operation starting with 0 are the finite ordinals (natural numbers). We collect these into a set ω, which will be the first infinite ordinal. We now continue iterating with ω', ω'', ω''', and so on. We can collect all ordinals obtained in this manner into a new set called $\omega 2$.[†] Then we continue with $(\omega 2)'$, $(\omega 2)''$, $(\omega 2)'''$, and so on. The ordinals are those sets obtained by this infinite counting process. Of course we have not yet added enough axioms to our system to obtain this hierarchy, since even ω does not yet exist without an axiom of infinity. But even after our axiom of infinity is added, further axioms will be needed to guarantee that a reasonable portion of this hierarchy is present. In no case will the ordinals themselves constitute a model of our system; however, they are the "backbone" of the system in that all sets within our system have some ordinal as a rank. (The notion of the rank of a set was explained at the beginning of the chapter.) We now turn to the promised development of the finite ordinals, the natural numbers.

Definition 18. $\mathrm{Sc}(x)$ for $(Ey)(\mathrm{On}(y) \wedge y' = x)$ where x is a term in which the variable y does not occur.

The set x is a successor if it is the successor of some ordinal. Any nonzero ordinal which is not a successor is called a limit ordinal.

Definition 19. $\mathrm{Lim}(x)$ for $\mathrm{On}(x) \wedge x \neq 0 \wedge \sim \mathrm{Sc}(x)$ where x is any term.

Definition 20. $N(x)$ for

$$\mathrm{On}(x) \wedge (x = 0 \vee \mathrm{Sc}(x)) \wedge (y)(y \in x \supset y = 0 \vee \mathrm{Sc}(y))$$

where x is some term in which y does not occur.

A natural number is an ordinal which is either 0 or a successor, and such that every element of it is also either 0 or a successor. Notice that in our intuitive hierarchy just described, infinite ordinals will not satisfy these conditions. ω, for instance, will be a limit ordinal and thus neither 0 nor the successor of another ordinal. Other infinite ordinals will have limit ordinals such as ω for elements, and so will fail to satisfy the third conjunct of Definition 20. As we shall soon see, the successor of any ordinal is an ordinal and so every ordinal will have a uniquely defined successor. But not every ordinal will have a predecessor. There are thus three kinds of ordinals: 0, successors, and limits.

From now on, we shall cease to give complete or quasicomplete formal proofs of theorems. Instead, we shall give an informal explanation that indicates the broad lines of the proof, but with sufficient detail to enable the reader versed in logic to recover the full proof if he chooses. It should be clear that this sort of informal discussion is really what constitutes a "proof" in most mathematical literature. Every such proof really constitutes an exercise of filling in the details, though the exercise is often (but not always) a routine one. The reader is invited to regard our incomplete proofs in just this light. This convention concerning proofs will apply to the remainder of this work, not just to the remainder of the present chapter.

There are several reasons for our change in procedure at this point. Chapter 3 and the initial part of the present chapter contain a number of detailed, quasicomplete formal proofs. These provide examples of most of the standard logical techniques. Since no really new techniques are used in the remaining proofs, the pedagogical value of supplying complete proofs is lessened. On the other hand, considerable space is saved. Moreover, an outline of the proof sometimes

---

[†] The reason for writing $\omega 2$ instead of 2ω is explained in a later discussion.

enables the student to understand the basic mechanism of the mathematical structure even better than does a detailed proof in which all logical devices are made explicit.

It is obviously of great importance that proofs can be formalized and that mathematics can be presented in a formal language, but it is practically unnecessary that proofs be formally presented in most cases. The main exceptions are: (1) for pedagogical purposes, (2) when the cogency of a proof is in question, and (3) when formal systems and proofs are themselves the object of mathematical study. Exceptions (3) and to a lesser extent (1) have been our basic reasons for presenting formal proofs in this book.

Let us now proceed to prove the Peano postulates for the sets satisfying the predicate $N(x)$ of Definition 20.

THEOREM 19. $\vdash N(0)$.

Proof. Immediate from Theorem 16 and the relevant definitions.

THEOREM 20. $\vdash (x_1)(0 \neq x_1')$.

Proof. $\vdash 0 = x_1' \supset x_1 \in 0$, and the conclusion follows by Theorem 3 and the rules of logic.

Theorem 20 shows that 0 is not the successor of any set. Thus, in particular it is not the successor of any natural number. The following theorem will be helpful in proving that the successor of any natural number is a natural number.

THEOREM 21. $\vdash (x_1)(\text{On}(x_1) \supset \text{On}(x_1'))$.

Proof. Assume $\text{On}(x_1)$. Since x_1' is $x_1 \cup \{x_1\}$, then $x_2 \in x_1'$ gives $x_2 \in x_1 \lor x_2 = x_1$. If $x_2 \in x_1$, then $x_2 \subset x_1$, since $\text{Trans}(x_1)$ by hypothesis. But $x_1 \subset x_1'$ trivially and so $x_2 \subset x_1'$ by Th. 7. If $x_2 = x_1$, then again $x_2 \subset x_1'$ by *Df.* 1 and *Df.* 4. Thus we have shown $x_2 \in x_1' \supset x_2 \subset x_1'$, which yields $\text{Trans}(x_1')$.

Still with the hypothesis $\text{On}(x_1)$, we assume $x_2 \in x_1'$ and $x_3 \in x_1'$ and $x_2 \neq x_3$. This gives $x_2 \in x_1 \cup \{x_1\}$ and $x_3 \in x_1 \cup \{x_1\}$. Since $x_2 \neq x_3$, both $x_3 = x_1$ and $x_2 = x_1$ cannot hold. If $x_2 \in x_1$ and $x_3 \in x_1$, then

$$x_2 \in x_3 \lor x_3 \in x_2,$$

since $\text{On}(x_1)$. If $x_2 \in x_1$ and $x_3 \in \{x_1\}$, then $x_2 \in x_3$, yielding $x_2 \in x_3 \lor x_3 \in x_2$. If $x_3 \in x_1$ and $x_2 \in \{x_1\}$, then $x_3 \in x_2$ and again $x_3 \in x_2 \lor x_2 \in x_3$. Hence, $\text{Con}(x_1')$ and our theorem is proved.

Having shown that the successor of any ordinal is an ordinal, we now prove:

THEOREM 22. $\vdash (x_1)(N(x_1) \supset N(x_1'))$.

Proof. Assume $N(x_1)$. Since $\text{On}(x_1)$ by *Df.* 20, then $\text{On}(x_1')$ by Th. 21. If $x_1 = 0$, then $x_1' \neq 0$ and so $\text{Sc}(x_1')$. If $\text{Sc}(x_1)$, then $\text{Sc}(x_1')$, since $\text{On}(x_1)$. Finally, if $x_2 \in x_1'$, then $x_2 \in x_1$ or $x_2 = x_1$. If $x_2 \in x_1$, then $x_2 = 0 \lor \text{Sc}(x_2)$, since $N(x_1)$. If $x_2 = x_1$, then $x_2 = 0 \lor \text{Sc}(x_2)$, since $x_1 = 0 \lor \text{Sc}(x_1)$, using **ZF.1**. Hence, $N(x_1')$.

The successor of any natural number is a natural number.

Proving that the successor function is injective turns out to be easy with this version of the natural numbers. The reader will recall that this was the difficult task in Frege's version, and even turned out to be impossible in the theory of types without an axiom of infinity. We have:

THEOREM 23. $\vdash (x_1)(x_2)(x_1' = x_2' \supset x_1 = x_2)$.

Proof. Assume $x_2' = x_1'$. This means that

$$(x_3)(x_3 \in x_1 \cup \{x_1\} \equiv x_3 \in x_2 \cup \{x_2\})$$

holds. Now, in particular, $\vdash x_1 \in x_1 \cup \{x_1\}$, and so we have $x_1 \in x_2 \cup \{x_2\}$ by $e\forall$ and MP. Similarly, $\vdash x_2 \in x_2 \cup \{x_2\}$ gives us $x_2 \in x_1 \cup \{x_1\}$ in the same way. If $\neq x_1 x_2$ then both $x_1 \in \{x_2\}$ and $x_2 \in \{x_1\}$ are excluded, and we thus have $x_1 \in x_2 \wedge x_2 \in x_1$. But, by Th. 13, $\vdash x_1 \in x_2 \supset x_2 \notin x_1$ which gives us $x_2 \in x_1 \wedge x_2 \notin x_1$, a contradiction. Hence $x_1 = x_2$ must hold and the theorem is established.

Theorem 23 is even stronger than what we need for it shows that the successor operator is injective on the (intuitive) domain of all sets of ZF. Hence, it is true for natural numbers in particular.

THEOREM 24. $\vdash (x_1)(x_2)(N(x_1) \wedge x_2 \in x_1 \supset N(x_2))$.

Proof. The reader will prove Theorem 24 as an exercise.

Theorem 24 will be useful in the following proof of the remaining Peano postulate, which is our metatheorem of proof by induction.

THEOREM 25. *Where $A(x)$ is any wff of ZF and x is any variable, and where $A(x')$ and $A(0)$ result from $A(x)$ by substituting x' and 0 respectively for all free occurrences of x in $A(x)$, then*

$$\vdash (A(0) \wedge (x)(N(x) \wedge A(x) \supset A(x'))) \supset (x)(N(x) \supset A(x)).$$

Proof. Assume $A(0)$ and $(x)(N(x) \wedge (A(x) \supset A(x')))$ as hypotheses.

Now suppose that $(Ex)(\sim A(x) \wedge N(x))$ holds. We show that this leads to contradiction. By eE, let a be a new constant such that

$$\sim A(a) \wedge N(a)$$

holds. Let L stand for $\{x \mid x \in a' \wedge \sim A(x)\}$. By ZF.2, $a \in L$ holds, since $a \in a' \wedge \sim A(a)$ holds. Thus $L \neq 0$ and so, by ZF.7,

$$(Ey)(y \in L \wedge y \cap L = 0)$$

holds (for some appropriately chosen variable y). Using eE, let b be a new constant such that $b \in L \wedge b \cap L = 0$ holds. Now $L \subset a'$, since $(x)(x \in L \supset x \in a')$. Thus, $b \in a'$. Also, $N(a')$ holds, since $N(a)$ (using Th. 22). Hence, $N(b)$ by Th. 24.

Now, $A(0)$ holds by hypothesis and so $b \neq 0$, since $\vdash b \in L \supset \sim A(b)$, and $b \in L$ holds. But $N(b)$ and so $Sc(b)$, which means $n = c'$ holds for some constant c for which we have $On(c)$.

11*

Again appealing to Th. 24, we have $N(c)$, since $c \in c' = b$. Now $b \cap L = 0$ holds and so $c \notin L$, since $c \in b$. By **ZF**.2, $c \in L \equiv c \in a' \wedge \sim A(c)$ holds. We have

$$c \notin L \equiv c \notin a' \vee A(c)$$

by Taut and MP. By MP we then have $c \notin a' \vee A(c)$ and further $c \in a'$, since $c \in b \in a'$ and Trans(a'). Thus $A(c)$ holds.

Applying now our second hypothesis, we obtain $A(c) \supset A(c')$ and, by MP from the preceding paragraph, $A(c')$. But $c' = b$ holds and so we arrive at $A(b)$. On the other hand, $b \in L$ and so $\sim A(b)$ by **ZF**.2. Thus, we have the contradiction $A(b) \wedge \sim A(b)$, establishing the negation of our assumption $(Ex)(\sim A(x) \wedge N(x))$, which is $(x)(N(x) \supset A(x))$. The theorem is established.

Exercise. Prove the set-theoretic form of induction; that is,

$$(x_1)((0 \in x_1 \wedge (x_2)((N(x_2) \wedge x_2 \in x_1) \supset x_2' \in x_1)) \supset (x_2)(N(x_2) \supset x_2 \in x_1)).$$

Notice that we have not obtained Theorem 25 by application of the set-theoretic form of the principle as we did in **F** and **ST**. We do not have in **ZF** a principle of abstraction strong enough to do so. The proof of Theorem 25 is quite close to the usual algebraic proof.

We have now established that the succession of constants $0, 0', 0''$, etc. represents the natural numbers in that they satisfy the Peano postulates. This shows that no model of **ZF**.1–**ZF**.7 is finite.

As an example of a proof by induction, we have:

THEOREM 26. $\vdash (x_1)(x_2)((N(x_1) \wedge x_2 \in x_1) \supset x_2' \in x_1').$

Proof. The reader will prove Theorem 26 as an exercise. (*Hint*: The wff to be proved is equivalent to $(x_1)(N(x_1) \supset (x_2)(x_2 \in x_1 \supset x_2' \in x_1'))$ which is easily proved by induction on x_1.)

Henceforth, we will denote by "**Z**" the system whose axioms are **ZF**.1–**ZF**.7 and whose nonlogical symbols are those which have been introduced up to this point in our development of **ZF**. **Z** is an important fragment of **ZF** in which a considerable amount of mathematics can be carried on. We call **Z** *Zermelo set theory*, though it is not quite the same as Zermelo's original system which included, in particular, an axiom of infinity.

5.2. The completing axioms

We could continue a development of arithmetic without use of an axiom of infinity. Bernays [1] furnishes the details of such a development (though couched in a different system, which we will consider in a later section). However, for the full development of classical mathematics, including the theory of real numbers and infinite cardinal numbers, we need the existence of an infinite set. The way in which we have developed the theory of natural numbers makes the statement of the axiom of infinity particularly easy:

ZF.8. $(x_1)(x_1 \in \omega \equiv N(x_1))$. ω is a primitive constant.

ω represents the set of all natural numbers. As we have already remarked, the axiom of infinity allows us to construct ordinals beyond the natural numbers. We prove, for example, the following theorem:

THEOREM 27. $\vdash On(\omega)$.

Proof. If $x_1 \in \omega$, then $N(x_1)$ by **ZF**.8. Thus, for any $x_2 \in x_1$ we have $N(x_2)$ by Th. 24. This gives $x_2 \in \omega$ by **ZF**.8 and so we have

$$(x_2)\,(x_2 \in x_1 \supset x_2 \in \omega).$$

Thus, $x_1 \subset \omega$. Hence $\vdash (x_1)\,(x_1 \in \omega \supset x_1 \subset \omega)$, which means that $\vdash Trans(\omega)$.

To show $\vdash Con(\omega)$, assume $x_1 \in \omega \wedge x_2 \in \omega$. This gives $N(x_1) \wedge N(x_2)$ by **ZF**.8. We will prove by induction on x_1 that

$$\vdash (N(x_1) \wedge N(x_2) \wedge x_1 \neq x_2) \supset (x_1 \in x_2 \vee x_2 \in x_1).$$

Specifically, the property $A(x_1)$ we are using is

$$(N(x_2) \wedge x_1 \neq x_2) \supset (x_1 \in x_2 \vee x_2 \in x_1).$$

$A(0)$ is thus the wff $(N(x_2) \wedge 0 \neq x_2) \supset (0 \in x_2 \vee x_2 \in 0)$. To see that $\vdash A(0)$, assume the hypothesis $N(x_2) \wedge 0 \neq x_2$. Then, since $On(x_2)$, we have $0 \in x_2$ (see the exercise following Th. 18) and thus $0 \in x_2 \vee x_2 \in 0$ holds.

We now assume $N(x_1) \wedge A(x_1)$ as an inductive hypothesis. We want to conclude from this that $(N(x_2) \wedge (x_1' \neq x_2) \supset (x_1' \in x_2 \vee x_2 \in x_1')$ holds. Suppose, then, that $N(x_2) \wedge x_1' \neq x_2$ holds. If $x_1 = x_2$, then $x_2 \in x_1'$ by **ZF**.1 since $\vdash x_1 \in x_1'$. Hence $(N(x_2) \wedge x_1' \neq x_2) \supset (x_1' \in x_2 \vee x_2 \in x_1')$ holds in this case.

On the other hand, if $x_1 \neq x_2$, then we have the conjunction of the properties $N(x_2)$ and $x_1 \neq x_2$, which gives $x_1 \in x_2 \vee x_2 \in x_1$ upon applying MP to our inductive hypothesis. If $x_2 \in x_1$, then $x_2 \in x_1' = x_1 \cup \{x_1\}$. If $x_1 \in x_2$, then $x_1' \in x_2'$ by Th. 26, since $N(x_2)$. This means that

$$x_1' \in x_2 \vee x_1' = x_2$$

holds. But $x_1' \neq x_2$ by hypothesis and so $x_1' \in x_2$ holds. Thus, we have

$$(N(x_2) \wedge x_1' \neq x_2) \supset (x_1' \in x_2 \vee x_2 \in x_1')$$

in this case as well. Since either $x_1 = x_2$ or $x_2 \neq x_2$ must hold, the conclusion follows and our induction is established.

We have now proved

$$\vdash (x_1)\,(x_2)\,(N(x_1) \supset ((N(x_2) \wedge x_1 \neq x_2) \supset (x_1 \in x_2 \vee x_2 \in x_1))).$$

This is equivalent, by **ZF**.8 and Theorem 8 of Section 1.6, to

$$\vdash (x_1)\,(x_2)\,((x_1 \in \omega \wedge x_2 \in \omega \wedge x_1 \neq x_2) \supset (x_1 \in x_2 \vee x_2 \in x_1)),$$

which is $\vdash Con(\omega)$. Our theorem is established.

THEOREM 28. $\vdash Lim(\omega)$.

Proof. By Theorem 19 $\vdash N(0)$. Thus $0 \in \omega$ by **ZF**.8, which implies $0 \neq \omega$ by Theorem 4. It remains for us to prove that $\vdash \sim Sc(\omega)$. To do this, assume that $Sc(\omega)$. This means

$(Ex)(\text{On}(x) \wedge x' = \omega)$ holds. Let a be some new constant such that $\text{On}(a) \wedge a' = \omega$ holds, by eE. Now $\vdash a \in a'$ and so $a \in \omega$ holds, which gives $N(a)$ by **ZF**.8. By Th. 22, $N(a')$ holds which gives $a' \in \omega$ by **ZF**.8 again. But $a' = \omega$ gives $\omega \in \omega$ by **ZF**.1. By Th. 12, $\omega \notin \omega$, and this contradiction establishes $\vdash \sim \text{Sc}(\omega)$ and thus that $\vdash \text{Lim}(\omega)$ by Th. 27 and *Df*. 19.

The ordinal ω is the first infinite ordinal and the smallest limit number. This is true because every ordinal less than ω (remember that it is the $\in$ relation which orders the ordinals) is a natural number by **ZF**.8, and all nonzero natural numbers are successors. By Theorem 21, the successor of any ordinal is an ordinal, so we can obtain further infinite ordinals by iterating the prime operator. The intuitive hierarchy we obtain has already been described in previous discussion.

In the infinite hierarchy of ordinals, the successors of ω such as ω', ω'', ω''', and so on will all be successor ordinals. Will there be other limit numbers, and how will we obtain them in our system? We would want the next limit number to be obtained by gathering together the hierarchy 0, 1, 2, ..., ω, ω', ω'', ... into a new set $\omega 2$, just as we gathered together the natural numbers into the set ω. But can we perform such an operation in our system?

A natural attempt to perform the desired collecting operation would be to use our separation scheme **ZF**.2. First, we could define the predicate $W(x)$ as the predicate:

$$\text{On}(x) \wedge (N(x) \vee x = \omega \vee \text{Sc}(x)) \wedge (z)(z \in x \supset N(z) \vee z = \omega \vee \text{Sc}(z)).$$

Intuitively, $x \in \omega 2 \equiv W(x)$ should hold. But to apply **ZF**.2, we need some containing set y so that the set $\{x \mid x \in y \wedge W(x)\}$ will really be $\omega 2$. Where will we get such a set?

$\mathcal{P}(\omega)$ is a likely candidate, since $\omega 2$ is, intuitively for the moment, denumerable whereas $\mathcal{P}(\omega)$ is not (Cantor's theorem is provable in our system). But the elements of $\mathcal{P}(\omega)$ are all subsets of ω, and $\omega 2$ clearly contains a host of sets which are not subsets of ω, such as ω'.

In short, we are in the same position with respect to the existence of $\omega 2$ as we were with respect to the existence of ω before our axiom of infinity. Suppose we chose to solve the problem by adding another axiom positing the existence of $\omega 2$. Then we would begin again with our prime operator and obtain $(\omega 2)'$, $(\omega 2)''$, $(\omega 2)'''$ and so on, and we would again have the problem of the existence of an $\omega 3$, which would be the next limit ordinal after $\omega 2$. In order to have anything like our indefinite ordinal hierarchy, we must surmount this problem, for it is surely unsatisfactory to think of adding a transfinite number of axioms, one for each time the counting process breaks down.

The answer to the difficulty is the *axiom of replacement* expounded by Fraenkel. It is really this axiom which is original with Fraenkel, though his work has also served to greatly influence the form of our presentation. The clue to the solution lies in the following observation: There is an obvious 1–1 correspondence between the set ω of natural numbers and the hierarchy consisting of ω, ω', ω'', and so on. Just let ω correspond to 0, ω' to 1, ω'' to 2, and generally let the nth successor of ω correspond to the nth successor of 0. Though ω is a set of our system (by **ZF**.8), the hierarchy ω, ω', ω'', ... does not form a set of our system. Yet, the cardinality of our hierarchy ω, ω', ω'', ..., is the same as the cardinality of ω. This result seems unnatural, for why should one denumerable set such as ω be in our system and another one be excluded? On what basis can we decide which denumerable collections are to be included and which are to be excluded? It will be seen that the axiom of replacement remedies this ambiguity.

Fraenkel's axiom of replacement intuitively says that if the domain of a function is a set of our system, then the range of the function is also a set of our system. Such a function must be given by a functional relation $F(x, y)$ expressible in our system. That is, $F(x, y)$ must be a wff of **ZF** and must be *functional in x*. This latter condition means that

$$\vdash (x)(y)(z)(F(x, y) \wedge F(x, z) \supset y = z),$$

where z is free for y in $F(x, y)$ and z is substituted for all free occurrences of y in $F(x, y)$ to obtain $F(x, z)$.

Once we have the axiom of replacement, the collection

$$E = \{\omega, \omega', \omega'', \ldots\}$$

will be a set of our system (more will be said about how this is so once the axiom of replacement has been formally presented). $\omega2$ is then definable as $\omega \cup E$. We can begin again with $(\omega2)$, $(\omega2)'$, $(\omega2)''$, ... We again have a correspondence between ω and this hierarchy. Applying the axiom of replacement again, we can collect this hierarchy into a set G and take the union $\omega2 \cup G$. This gives us $\omega3$. We can continue in this manner to extend our counting process without having to posit new *ad hoc* axioms at each stage, since the replacement scheme will give us the necessary tools to repeat the same "collecting up" process at each stage.

We now proceed to a formal statement of the replacement axiom.

ZF.9. Let $F(x, y)$ be any wff of **ZF**, and let $F(x, z)$ be the result of substituting z for y in all the latter's free occurrences in $F(x, y)$ where z is free for y in $F(x, y)$. Then every instance of the following is an axiom:

$$(r)((x)(y)(z)(F(x, y) \wedge F(x, z) \supset z = y) \supset (Ew)(y)(y \in w \equiv (Ex)(x \in r \wedge F(x, y)))),$$

where the variables r, x, y, z, and w are all distinct and w does not occur in $F(x, y)$.

ZF.9 says that for any set r, and for any wff functional in x, there exists a set w which is precisely the set of images of the elements of r under the functional relation expressed by the wff.

ZF.9 is a very strong axiom indeed. In fact, **ZF.2** can be obtained from it. To see this, let $A(x)$ be some wff of **ZF**. Define $F(x, y)$ as $A(x) \wedge x = y$, where y is free for x in $A(x)$. Clearly $F(x, y)$ is functional in x and so, by **ZF.9**, we have

$$\vdash (Ew)(y)(y \in w \equiv (Ex)(x \in r \wedge A(x) \wedge x = y))$$

which is equivalent to $\vdash (r)(Ew)(y)(y \in w \equiv y \in r \wedge A(y))$. But this last theorem is just the axiom of separation without the use of a special term operator.

The restriction of **ZF.9** to functional relations is not only motivated by our intended usage of extending our counting process as previously explained. It is also necessary to prevent the deduction of Russell's paradox. Without the hypothesis of functionality, we can obtain the existence of a universal set as follows: Let $F(x, y)$ be the wff $x \subset y$. $F(x, y)$ is clearly not functional in x, since $\vdash (y)(0 \subset y)$. Now if we have

$$\vdash (r)(Ew)(y)(y \in w \equiv (Ex)(x \in r \wedge F(x, y))),$$

then we have $\vdash (Ew)(y)(y \in w \equiv (Ex)(x \in \{0\} \wedge x \subset y))$ by $e\forall$. This gives $\vdash (Ew)(y)(y \in w \equiv 0 \subset y)$. Consequently, $\vdash (Ew)(y)(y \in w)$.

To deduce Russell's paradox, it suffices to take $y \notin y$ for $A(y)$ in **ZF**.2. Formally, we have:

1. $(r)(Ew)(y)(y \in w \equiv y \in r \wedge y \notin y)$ **ZF**.2
2. $(Ew)(y)(y \in w)$ from our foregoing demonstration of the existence of the universal set
3. $(y)(y \in b)$ 2, eE
4. $(Ew)(y)(y \in w \equiv y \in b \wedge y \notin y)$ 1, $e\forall$
5. $(y)(y \in a \equiv y \in b \wedge y \notin y)$ 4, eE
6. $a \in a \equiv a \in b \wedge a \notin a$ 5, $e\forall$
7. $a \in b$ 3, $e\forall$
8. $a \in a \equiv a \notin a$ 6, 7, Taut, MP

This shows that if ever a universal set is forthcoming in **ZF**, we can immediately deduce Russell's paradox. In particular, the replacement scheme without the hypothesis of functionality would allow for just such a development.

In our discussion prior to the introduction of **ZF**.9, we indicated that the axiom of replacement would allow us to collect up a denumerable hierarchy such as $\omega, \omega', \omega'', \ldots$ into a set E. However, it is not immediately clear how to do this with **ZF**.9. We shall indicate how this can be done, but a fuller discussion of the matter is best delayed until we have shown how to define functions as sets within set theory.[†] This is the subject of Section 5.3, and so we shall complete our discussion of the use of **ZF**.9 in that section. We wish to close the present section with a brief discussion of operations on ordinals in intuitive set theory.

In intuitive set theory, an ordinal number is an equivalence class of well-ordered sets under isomorphism. Thus, the operations of ordinal addition and ordinal multiplication are defined if we define these operations for well-ordered sets in an isomorphic invariant way. Given two disjoint well-ordered sets A and B, we define the ordinal sum $A+B$ of A and B as the union $A \cup B$ with the following ordering: For all x and y, if $x \in A$ and $y \in B$, then $x < y$. If x and y are both in A or both in B, then they have the same order in $A \cup B$ as they have in A or B respectively. The reader should verify that the ordinal sum of two disjoint well-ordered sets is well ordered. To add two well-ordered sets that are not disjoint, we substitute for one of the summands an order-isomorphic set disjoint from the other summand.

In previous discussion, we have seen that we can imagine the elements of an ordered set to be laid out in a line so that the "less than" relationship becomes the relationship of "to the left of". With this in mind, we can picture the ordinal sum of two well-ordered sets A and B as obtained by laying out the elements of B to the right of A, thus making all elements of A less than any element of B.

It will be seen from this observation that ordinal addition is not commutative. For example, $1 + \omega = \omega \neq \omega + 1$. To see this, imagine tacking a new element in front of 0. We obtain an order isomorphism with ω by letting the new element correspond to 0, letting 0 correspond to 1, 1 to 2, and so on. We thus obtain an order isomorphism of $1 + \omega$ onto ω. On the other hand, $\omega + 1 \neq \omega$, since $\omega + 1$ will have a maximal element whereas ω does not.

Since $1 = \{0\}$ is trivially order isomorphic to $\{\omega\}$, we see that $\omega' = \omega + 1$. In general, $\omega + n$ will be equal to ω followed by n primes. On the other hand $n + \omega = \omega$ for all finite ordinals n.

[†] We have already defined the notion of a wff functional in a variable x. But this definition of "function" applies to the wffs which are the linguistic objects of our system. We have not yet shown how to define within our system the notion of a function as a set of our system.

To define the ordinal product of two well-ordered sets A and B, we first form the cartesian product $A \times B$ of the two sets. Then we order this cartesian product by *reverse lexicographic ordering*; i.e. $\langle x, y \rangle < \langle a, b \rangle$ means $y < b$, or $y = b$ and $x < a$. We will use juxtaposition to indicate ordinal multiplication.

Again we can see that our operation is not commutative. $2\omega = \omega$, for consider the hierarchy $\langle 0, 0 \rangle, \langle 1, 0 \rangle, \langle 0, 1 \rangle, \langle 1, 1 \rangle, \langle 0, 2 \rangle, \ldots$ which is 2ω. This is order isomorphic to ω by the mapping f such that $f(\langle 0, 0 \rangle) = 0, f(\langle 1, 0 \rangle) = 1, f(\langle 0, 1 \rangle) = 2$, and so on. On the other hand, $\omega 2 \neq \omega$, for $\omega 2$ will be order isomorphic to $\omega + \omega$. $\omega 2$ is the hierarchy $\langle 0, 0 \rangle, \langle 1, 0 \rangle, \langle 2, 0 \rangle, \langle 3, 0 \rangle, \ldots,$ $\langle 0, 1 \rangle, \langle 1, 1 \rangle, \langle 2, 1 \rangle \ldots$, which is precisely what we obtain by laying off ω to the right of itself. In this hierarchy, $\langle 0, 0 \rangle$ represents 0, $\langle 1, 0 \rangle$ represents 1, and, in general, $\langle n, 0 \rangle$ represents n. $\langle 0, 1 \rangle$ represents ω (the first limit ordinal) and, in general, $\langle n, 1 \rangle$ represents $\omega + n$.

The reader can now see why we have used "$\omega 2$" as the symbol for the second limit ordinal instead of 2ω. The reason is that the second limit ordinal is the ordinal product of ω and 2 (also the ordinal sum $\omega + \omega$), but is not the ordinal product $2\omega = \omega$, as follows from the usual way of defining our ordinal operations in intuitive set theory. Of course it is only convention which keeps us from choosing direct lexicographic order in defining the ordinal product.

Within our system, these intuitive operations are formally introduced by transfinite induction. In the following discussion, we deal only with (finite) recursion up to ω. Transfinite recursion is dealt with in Section 5.7 below.

5.3. Relations, functions, and simple recursion

Definition 21. $R(x)$ for $(y)(y \in x \supset (Ew)(Ez)(y = \langle w, z \rangle))$, where z, y, w are all different and do not occur in x.

A set x is a relation if all its members are ordered pairs. This is the definition often given in intuitive expositions of set theory and the one we have used throughout this book in our informal treatment of sets.

Definition 22. $F(x)$ for

$$R(x) \wedge (y)(z)(w)(\langle y, z \rangle \in x \wedge \langle y, w \rangle \in x \supset z = w),$$

where w, y, z are all different, and do not occur in x.

A function is a relation which is functional; i.e. such that any first element of a pair has one and only one second element.

Definition 23. $D(x)$ for $\{y \mid y \in \bigcup(\bigcup(x)) \wedge (Ez)(\langle y, z \rangle \in x)\}$, where y and z are different, and do not occur in x.

Given a set x, the domain of x is the set of first elements of ordered pairs in x. If x has no pairs, then $D(x)$ is, of course, the set 0. Notice that $F(x)$ stands for a wff, whereas $D(x)$ stands for a term.

Exercise. Justify the use of $\bigcup(\bigcup(x))$ in Definition 23; i.e. show that $\vdash y \in D(x) \equiv (Ez)(\langle y, z \rangle \in x)$.

Definition 24. $I(x)$ for $\left\{y \mid y \in \bigcup\left(\bigcup(x)\right) \wedge (Ez)\left(\langle z, y \rangle \in x\right)\right\}$, where y and z are different and do not occur in x.

$I(x)$ is the range of x if x is a function. It is the set of images under x. A similar proof for the justification of $\bigcup\left(\bigcup(x)\right)$ in the foregoing exercise applies here.

Definition 25. $(x``z)$ for $\bigcup\left(\{y \mid y \in I(x) \wedge \langle z, y \rangle \in x\}\right)$, where y does not occur in either x or z.

Where x is a function, $x``z$ is the value of the function x applied to the argument z.

It should be noted that for any function x the wff $y = x``z$ is functional in z. In other words, every function gives rise to a functional wff. The axiom of replacement asserts a sort of converse to this, namely that any wff functional in z gives rise to a function on any given set as domain.

The usual notation for the application of a function f to its argument x is $f(x)$. However, there would be confusion with the universal quantifier if we followed that notation here (witness our change in notation for the universal quantifier in Chapter 4). Thus, $x``y$ is the image of y under the function x. However, we will still use the usual "$f(x)$" notation in our informal discussion.

Given sets x and y, and a function f, we say that f is a function "from x to y" if the domain of f is x and the range of f is a subset of y.

Definition 26. $t \times r$ for

$$\{y \mid y \in \mathcal{P}(\mathcal{P}(t \cup r)) \wedge (Eu)(Ev)(y = \langle u, v \rangle \wedge u \in t \wedge v \in r)\},$$

where t and r are any terms in which the distinct variables u, v, and y do not occur.

$t \times r$ is the cartesian product of the sets t and r, the set of all ordered pairs whose first elements are in t and whose second elements are in r.

We now wish to prove the theorem of simple recursion in **ZF**. This theorem says that given a set x and function f from x to x, and any selected element a of x, then there exists a unique function t from ω to x which satisfies the following conditions: (i) $t(0) = a$, and (ii) $t(n') = f(t(n))$ for all $n \in \omega$. We have already discussed the importance of this theorem for the development of mathematics within set theory (see Chapter 3). It is by means of this theorem, and certain stronger forms of it, that we can justify the usual informal practice of "definition by induction". In defining powers of an arbitrary element b in a monoid x, for example, we usually pronounce the magic words "let $b^0 = 1$ (1 being the identity element of the monoid) and let $b^{n'} = b \cdot b^n$" where "$\cdot$" is the binary operation in our monoid. What we are really appealing to in such cases is the existence of a function t from ω to x defined by simple recursion from the selected element 1 in x and the function $b \cdot (\ )$ (multiplication on the left by b), which is a function from x to x. t must satisfy the conditions (i) $t(0) = 1$ and (ii) $t(n') = b \cdot (t(n))$, which are precisely the conditions that b^n is supposed to satisfy. The uniqueness of functions defined by simple recursion tells us that $t(n) = b^n$ for all $n \in \omega$.

As we have previously noted (see the exercise in Chapter 3, p. 89), it is by means of the theorem of simple recursion that we can establish that any two sets of our system that satisfy the Peano postulates (which we have proved for ω) are isomorphic. This shows that we have uniquely characterized the natural numbers ω within our set theory. (We have certain complications when we look outside the theory as we shall see in Chapter 6.)

Let us now prove our theorem of simple recursion.

THEOREM 29.

$$\vdash (x_1)(x_2)((F(x_1) \wedge I(x_1) \subset D(x_1) \wedge x_2 \in D(x_1)) \supset (E! \, x_3)(F(x_3) \wedge \omega = D(x_3)$$
$$\wedge I(x_3) \subset D(x_1) \wedge x_3 \text{``} 0 = x_2 \wedge (x_4)(x_4 \in \omega \supset x_3 \text{``} x_4' = x_1 \text{``} (x_3 \text{``} x_4)))).$$

Proof. Assume the hypotheses $F(x_1) \wedge I(x_1) \subset D(x_1) \wedge x_2 \in D(x_1)$. Let us first prove the existence of the desired function. Let $\mathcal{E}$ be the term

$$\{x_5 \mid x_5 \in \mathcal{P}(\omega \times D(x_1)) \wedge \langle 0, x_2 \rangle \in x_5 \wedge (x_6)(x_7)(\langle x_6, x_7 \rangle \in x_5 \supset \langle x_6', x_1 \text{``} x_7 \rangle \in x_5)\}.$$

Elements of $\mathcal{E}$ are relations, for they are all subsets of $\omega \times D(x_1)$. Moreover, no set in $\mathcal{E}$ is empty, since every one contains at least $\langle 0, x_2 \rangle$ as an element by **ZF.2**. Now, let t be the term $\bigcap \mathcal{E}$. Clearly $t \neq 0$, since $\langle 0, x_2 \rangle \in t$. Also, $R(t)$, since $t \subset (\omega \times D(x_1))$. In fact $t \in \mathcal{E}$ holds, which can be easily established (we leave this as an exercise to the reader). Moreover, $(x_8)(x_8 \in \mathcal{E} \supset t \subset x_8)$ also holds. That is, t is a subset of every other set in the collection $\mathcal{E}$. This follows immediately from the fact that t is defined to be the intersection of all the sets in $\mathcal{E}$.

We shall now prove by induction that t is a function which satisfies the conditions stated in the conclusion of the theorem. Precisely, we want to prove by induction that $(x_6)(x_6 \in \omega \supset (E! \, x_9)(\langle x_6, x_9 \rangle \in t))$ holds. First, $\langle 0, x_2 \rangle \in t$ and so something is paired with 0. We wish to show that x_2 is the only thing paired with 0. Suppose $\langle 0, x_8 \rangle \in t \wedge x_8 \neq x_2$ holds, and consequently $\langle 0, x_8 \rangle \neq \langle 0, x_2 \rangle$. Let L be the term

$$\{x_9 \mid x_9 \in t \wedge x_9 \neq \langle 0, x_8 \rangle\}.$$

L is the same as t except that $\langle 0, x_8 \rangle$ is not in L. Thus $\sim (t \subset L)$, since $\langle 0, x_8 \rangle \in t$ by our supposition. We shall obtain a contradiction from this by showing that $L \in \mathcal{E}$. $\langle 0, x_2 \rangle \in L$, since $\langle 0, x_2 \rangle \in t$ and $\langle 0, x_2 \rangle \neq \langle 0, x_8 \rangle$, the latter being the only pair removed from t to get L. Furthermore, if $\langle x_6, x_7 \rangle \in L$, then $\langle x_6', x_1 \text{``} x_7 \rangle \in L$, for $0 \neq x_6'$ (by Th. 20) and so

$$\langle x_6', x_1 \text{``} x_7 \rangle \neq \langle 0, x_8 \rangle.$$

Finally, $L \subset t$ by the definition of L and so $L \subset \omega \times D(x_1)$. Thus, by **ZF.2**, $L \in \mathcal{E}$, from which it follows that $t \subset L$, thus contradicting our original supposition (recall that t is a subset of every set in the collection $\mathcal{E}$). x_2 is therefore the only set paired with 0 and our inductive argument holds for 0.

Assume now the induction hypothesis that $(E! \, x_9)(\langle x_6, x_9 \rangle \in t \wedge x_6 \in \omega)$ holds. Let a_1 be an ambiguous name of the unique set paired with x_6. $\langle x_6, a_1 \rangle \in t$ and so $\langle x_6', x_1 \text{``} a_1 \rangle \in t$, since $t \in \mathcal{E}$. Thus, there is something paired with $x_6'$, namely $x_1 \text{``} a_1$. We wish to show uniqueness again. Suppose, therefore, that $\langle x_6', x_8 \rangle \in t$ and $x_8 \neq x_1 \text{``} a_1$ so that $\langle x_6', x_8 \rangle \neq \langle x_6', x_1 \text{``} a_1 \rangle$. Consider the term $\{x_9 \mid x_9 \in t \wedge x_9 \neq \langle x_6', x_8 \rangle\}$, which we shall call K. Clearly $K \subset t$ holds by the definition of K, but $\sim (t \subset K)$ also holds, since $\langle x_6', x_8 \rangle \in t$ and $\langle x_6', x_8 \rangle \notin K$. We claim that $K \in \mathcal{E}$. $\langle 0, x_2 \rangle \neq \langle x_6', x_8 \rangle$, since $0 \neq x_6'$ and so $\langle 0, x_2 \rangle \in K$. If $\langle x_9, x_{10} \rangle \in K$, then $\langle x_9, x_{10} \rangle \in t$ (since $K \subset t$ holds) and so $\langle x_9', x_1 \text{``} x_{10} \rangle \in t$. Thus, $\langle x_9', x_1 \text{``} x_{10} \rangle \in K$ unless

$$\langle x_9', x_1 \text{``} x_{10} \rangle = \langle x_6', x_8 \rangle,$$

for this latter was the only pair removed from t to obtain K. Now, $\langle x_9', x_1 \text{``} x_{10} \rangle = \langle x_6', x_8 \rangle$ only if $x_9' = x_6' \wedge x_1 \text{``} x_{10} = x_8$ holds. But the successor function is injective and so $x_9' = x_6'$

implies $x_9 = x_6$. Thus,

$$\langle x_9, x_{10} \rangle = \langle x_6, x_{10} \rangle,$$

and $\langle x_6, x_{10} \rangle \in t$, since $\langle x_9, x_{10} \rangle \in t$. But by induction hypothesis, a_1 is the only set paired with x_6 in the pairs of t. Thus, $x_{10} = a_1$, and

$$\langle x_9, x_{10} \rangle = \langle x_6, a_1 \rangle.$$

From this, we have that $\langle x_6', x_1``a_1 \rangle = \langle x_9', x_1``x_{10} \rangle = \langle x_6', x_8 \rangle$ which implies that $x_1``a_1 = x_8$. But, by our supposition, $x_1``a_1 \neq x_8$ and so $\langle x_9', x_1``x_{10} \rangle \in K$. This establishes that $K \in \mathcal{E}$ from which it follows that $t \subset K$, contradicting $\sim (t \subset K)$ as previously stated. Thus, $\langle x_6', x_8 \rangle \in t$ only if $x_8 = x_1``a_1$ and uniqueness is established for x_6'. Our inductive proof is thus complete.

We have established that $F(t) \wedge \omega = D(t) \wedge I(t) \subset D(x_1)$ holds. By the definition of t, $t``0 = x_2$, since $\langle 0, x_2 \rangle \in t$. Moreover, for any

$$\langle x_4, t``x_4 \rangle \in t,$$

we have $\langle x_4', x_1``t``x_4 \rangle \in t$ which means that $t``x_4' = x_1``t``x_4$ (using $Df.$ 25 again). This establishes that

$$(Ex_3)(F(x_3) \wedge \omega = D(x_3) \wedge I(x_3) \subset D(x_1) \wedge x_3``0 = x_2 \wedge (x_4)(x_4 \in \omega \supset x_3``x_4' = x_1``x_3``x_4))$$

holds on the basis of our original hypotheses.

It remains for us to establish the uniqueness of the function whose existence we have just proved. It is an easy argument by induction that any two functions satisfying the established conditions must be equal. We leave this as an exercise for the reader.

It is interesting that, given a certain amount of general set theory such as we have in **ZF**, we can prove all the Peano postulates for the set ω where we assume only that ω is a set equipped with an element 0 and mapping $'$ from ω into itself for which the statement of Theorem 29 holds. We can also prove stronger forms of recursion. In a system of Chapter 8 this is actually carried out. The property of simple recursion is the sole axiom for the natural numbers, and the other Peano postulates plus a stronger version of recursion are proved.

Exercise. Use the theorem of simple recursion, Theorem 29, to prove the theorem of *primitive recursion*: Given any set A and any function f from $\omega \times A$ to A, and any selected element $a \in A$, then there exists a unique function h from ω to A such that $h(0) = a$, and $h(n') = f(n, h(n))$ for all $n \in \omega$. We can also prove in **ZF** (cf. Theorem 19, Section 8.5) a more general form of primitive recursion in which we have an extra parameter. Precisely, we prove that, when we are given the functions f from $B \times \omega \times A$ to A and g from B to A, we can define a unique function h from $B \times \omega$ to A which satisfies the conditions $h(b, 0) = g(b)$ and $h(b, n') = f(b, n, h(b, n))$ for all $b \in B$, $n \in \omega$ Generally speaking, we can add as many sets $B_1, \ldots, B_n$ as we choose as parameters. We have only to let $B = B_1 \times B_2 \times \ldots \times B_n$ in this general version of primitive recursion.

Repeated use of our simple recursion theorem obtains the arithmetic of the natural numbers. For each natural number x, let the function x_1 of Theorem 29 be the successor function itself. Then, there is a mapping $+_x$ from ω into ω such that $+_x``0 = x$ and $+_x``y' = (+_x``y)'$ for all $y \in \omega$. Instead of writing $+_x\text{``}y$, we write $(x+y)$. The usual properties of associativity, commutativity, and the like can be obtained by inductive proofs. Using the newly defined function of addition "$+$", we can again apply simple recursion to define multiplication.[†] Using multiplication and simple recursion (our more general forms of recursion are also helpful), we can define exponentiation, and so on. In this way, the usual arithmetic of the natural numbers and the arithmetic functions can be built up. Once we have the natural numbers, we can construct the integers, the rationals, and the reals in the standard ways. We sketch this development further on in Section 5.7.

The reader who is not already familiar with a certain amount of axiomatic set theory may feel some sense of discomfort at the way functions are handled by means of ordered pairs. Many mathematicians feel "intuitively" that a function should be something more than just a functional relation. A function, they feel, is a "mapping", a rule which actively associates elements of the range with elements of the domain. The set of ordered pairs, which we have called the function, might be called the *graph* of this mapping.

There is, perhaps, some legitimate basis for such feelings. Although the ordered pair is a device that allows us to reduce function theory to relation theory and set theory, it remains something of an artifice.

Bourbaki [1] considers the notion of ordered pair as primitive. This avoids certain artificial constructions, but it multiplies the number of basic notions and axiomatic concepts of the system.

In algebra and analysis it is, in many ways, the notion of a function and the application of a function to an argument which are more basic than the notion of a set and of one set existing as an element of another. For this and other reasons, we consider in Chapter 8 a relatively new approach to foundations in which the notion of function is made basic and in which membership is defined by means of this notion.[‡]

Now that we have dealt with functions a bit, let us consider once more our hierarchy ω, ω', ω'', ... and the problem of using **ZF**.9 to collect it up. First, we prove by induction that for every $n \in \omega$, $n \neq 0$, there is a function f whose domain is n and which has the property that $f(0) = \omega$, and, for all $x' \in n$, $f(x') = (f(x))'$. Formally, we establish

$$\vdash (x_1)(x_1 \in \omega \wedge x_1 \neq 0 \supset (Ex_2)(F(x_2) \wedge D(x_2) = x_1 \wedge x_2\text{``}0 = \omega$$
$$\wedge (x_3)(x_3' \in x_1 \supset x_2\text{``}x_3' = (x_2\text{``}x_3)'))).$$

This is an easy proof by induction on x_1. The reader should be careful to notice that there is no appeal to the theorem of simple recursion here. We are not proving the existence of a function defined on ω, but rather the existence of an infinity of functions, one defined on each natural number different from 0, such that if $n \in m \in \omega$, then the function defined on m is an extension of the function defined on n. We have thus associated with each $n \in \omega$, $n \neq 0$, a function f_n whose domain is n.

Now, for any given n, the range $I(f_n)$ of the function associated with n is uniquely determined by the axiom of extensionality. Thus, the wff $y = I(f_n)$ is functional in n. We can thus apply **ZF**.9 to this functional wff and to the set ω. This will yield the existence of a set S whose

---

[†] We define, for each natural number x, a function $\times_x$ from ω to ω using the recursion equations S.10 and S.11 of Section 1.9.

[‡] It is interesting to note, in this connection, that the original formulation of the system of von Neumann, to be studied in Section 5.6 of this chapter, used the notion of function as the basic one (see von Neumann [2]).

elements consist precisely of those sets which are the range of f_n for $n \in \omega$. The union $\bigcup(S)$ of this infinite collection is obviously the desired set E which has as elements precisely the sets ω, ω', ω'', and so on.

It is clear that this process can be applied to collect up $\omega 2$, $(\omega 2)'$, $(\omega 2)''$, and so forth, and to extend repeatedly our ordinal hierarchy.

Although it is not immediately apparent from the statement of the axiom, we can now see that the effect of **ZF**.9 is to extend our union operation. The axiom of sum set allows us only to take the union of a collection of sets when the collection is already a set of our system. The replacement axiom allows us to collect up a hierarchy, such as the hierarchy $I(f_n)$ just described, into a set S of our system to which the operation of sum set can then be applied. **ZF**.9 is thus essential in furnishing us the intuitive union operation used in describing our intuitive model for **ZF**. Sum set alone is obviously inadequate to describe this operation.

Before closing this section, we wish to say a word about how the theory of cardinal numbers is carried on in **ZF**, as well as something about the relation of **ST** to **ZF**.

We have already defined the notion of a function in **ZF**. We can define a 1–1 function by the following: $F^1(x)$ for

$$F(x) \wedge (u)(v)(u \neq v \supset x``u \neq x``v),$$

where u and v are distinct variables that do not occur in the term x. We can then define two sets to be *cardinally similar* (or *equipollent*) if there is a 1–1 function whose domain is one set and whose range is the other. Formally we have $x \operatorname{Sm} y$ for $(Ez)(F^1(z) \wedge x = D(z) \wedge y = I(z))$, where z does not occur in x or y.

In **ST**, and in other systems (such as **NF** of Chapter 7), cardinal numbers are defined in the Fregean manner as equivalence classes of sets under the relation of cardinal similarity; that is, the cardinal number of a set x is the class of all sets similar to x. Clearly we can entertain no such definition in **ZF**, for such a class of *all* sets equipollent to a given set will not exist in **ZF**. **ZF**.9 guarantees that any given collection equipollent to it will constitute a set, but not the class of all such collections. **ZF**.2 is of no use, for we need some containing set to apply the principle of separation.

The problem is solved in a way similar to that used in defining ordinal numbers. We pick a standard representative from each intuitively conceived class of cardinally similar sets. After adding the axiom of choice, **ZF**.10 (to be considered in the next section), we can prove that every set x is cardinally similar to some ordinal. We can also prove Cantor's theorem which states that for any set x there is no 1–1 correspondence between x and its power set $\mathcal{P}(x)$. This theorem does not use the axiom of choice. In fact its proof is fairly simple. We informally examine the details of the proof of Cantor's theorem in the discussion of Chapter 7, Section 7.2, and the interested reader can consult this section directly to obtain the proof of Cantor's theorem in **ZF**.

Exercise. Consult Chapter 7, Section 7.2, and verify that the proof of Cantor's theorem can be carried through in **ZF**.

Now, it is also true that a set x of **ZF** can be put into 1–1 correspondence with a subset of $\mathcal{P}(x)$ (namely the subset consisting of all one-element subsets of x). Thus, the intuitively conceived cardinality of $\mathcal{P}(x)$ will be strictly greater than that of x. Since every set is similar to some ordinal, the set $\mathcal{P}(x)$ will also be similar to an ordinal, and every ordinal similar to x will be less than (in the order relation of the ordinals) any ordinal similar to $\mathcal{P}(x)$. (Remember that order isomorphism involves cardinal similarity, since two order-isomorphic sets must admit a 1–1 correspondence between them.) But the "less than" relation between ordinals is the "$\in$" relation.

Thus, given any ordinal y similar to $\mathcal{P}(x)$ (there is one), all ordinals similar to x will be elements of y and so we can apply ZF.2 to form the set of all ordinals similar to x. Applying ZF.7, we obtain a smallest ordinal similar to x.[†] We then define the cardinal number of x to be the smallest ordinal similar to x. Clearly, if z is similar to x, then the set of all ordinals similar to z will be the same as that for x and consequently two similar sets will have the same cardinal number. We have determined a unique representative of each cardinal class. The usual theorems on cardinal numbers can be proved on the basis of this definition.[‡]

At this point it is quite easy for us to determine the precise relationship between ST and ZF. Our intuitive model for ST is a type hierarchy in which we start with some set T_0 of individuals and iterate the power set operation. Our intuitive model for ZF is obtained by starting with the null set and iterating the power set operation and the union operation any ordinal number of times. Clearly ZF is richer and more powerful than ST, since every operation available in ST is available in ZF but not conversely. Of course, the set T_0 of individuals of any model of ST must be infinite, but we have also an axiom of infinity in ZF.

We can easily find a model for ST within our intuitive model for ZF. It is sufficient, for the purposes of satisfying the axiom of infinity of ST, to assume that the set T_0 of individuals is denumerably infinite. We have already noted this fact in Chapter 4. We thus let T_0 be the set ω of the natural numbers of our intuitive model for ZF. T_1 is the set $\mathcal{P}(\omega)$, T_2 the set $\mathcal{P}(\mathcal{P}(\omega))$, and generally speaking, $T_{n+1} = \mathcal{P}(T_n)$ for all $n \geqslant 0$. Each of the sets T_n is in our intuitive model for ZF, since the power set of any set of ZF is a set of ZF. Thus, the simplified type hierarchy generated by the set ω of natural numbers of our model of ZF is a model for ST.

Exercise. In any simplified type hierarchy T based on a nonempty set T_0, we say that *types are cumulative* if and only if the following is true: For all x and for all n, x is of type n (i.e. x is an element of T_n) implies that x is of type $n+1$. Prove by induction that types are cumulative in the model for ST described in the preceding paragraph.

The foregoing model-theoretic discussion is not really a proof of the consistency of ST relative to ZF, since we have never *proved* that our intuitive model for ZF really is a model. However, we can prove the relative consistency of ST to ZF in a purely proof-theoretic way when guided by our model-theoretic considerations. The idea is to exhibit a set of wffs of ZF which represent wffs of ST in such a way that all of the axioms of ST are theorems of ZF. It will then follow that any proof of a contradiction deducible in ST can be translated into a proof of a contradiction in ZF. In this way, the inconsistency of ST will imply the inconsistency of ZF. Hence, the consistency of ZF implies the consistency of ST. We now examine this method.

In ZF, define $B_0(x)$ for $x \in \omega$, $B_1(x)$ for $x \in \mathcal{P}(\omega)$, and so on. Generally, we let $\mathcal{P}^n(\omega)$ stand

---

[†] We need the fact, which we have not strictly proved as yet, that any set whatever of ordinals is well ordered by the $\in$ relation. This is proved in Theorem 2 and Theorem 3 of Section 5.7.

[‡] A *cardinal* is thus an ordinal α none of whose elements $\beta \in \alpha$ are similar to α. ω is the smallest infinite cardinal.

for $\mathcal{P}(\ldots \mathcal{P}(\omega) \ldots)$ to any finite n (possibly zero) occurrences of $\mathcal{P}$. Then, $B_n(x)$ stands for $x \in \mathcal{P}^n(\omega)$, for all $n \geqslant 0$. We think of $B_n(x)$ as standing for "x is of type n". It should be clear that these are purely formal definitions as abbreviations of the usual kind.

Now, consider the first-order formulation **ST'** of **ST** which was described at the end of Chapter 4. We transform each wff X' of **ST'** into a wff X of **ZF** by formally replacing each occurrence of a wff $K_n(x)$ in X' by the wff $B_n(x)$ of **ZF**. If a wff B of **ST'** has no occurrences of $K_n(x)$, then B is already a wff of **ZF**, and we let the transform of B be B itself. The rules of inference are the same in the two systems and so we have only to check that every axiom of **ST'** is transformed into a theorem of **ZF**.

First, it is easy to see that the transform of the axiom of infinity **ST'**.3 is a theorem of **ZF**. The $\in$ relation on ω is an irreflexive, transitive, and strongly connected relation, and **ST'**.3 simply asserts the existence of such a relation between the objects of type 0. ω is the set of things of type 0 under our transformation.

Moreover, the axiom of extensionality **ZF**.1 immediately guarantees us that instances of the axiom of extensionality of **ST'**.2 are transformed into theorems of **ZF**. In fact, **ZF**.1 is much stronger than **ST'**.2.

As for **ST'**.1, we pose the following exercise:

Exercise. Show that every transform of an instance of **ST'**.1 is a theorem of **ZF**. (*Hint*: For the axioms of **ST'** which are the appropriate restrictions to the types of the universal closures of wffs of the form $(Ey)(K_{n+1}(y) \wedge (x)(K_n(x) \supset (x \in y \equiv A(x))))$, consider the term

$$\{x \mid B_n(x) \wedge A(x)\}$$

which exists in **ZF** by **ZF**.2. Such a term is a subset of $\mathcal{P}^n(\omega)$ and thus an element of $\mathcal{P}^{n+1}(\omega)$.)

From the preceding, we can see that any proof of a contradiction in **ST'** is directly translatable into a proof of a contradiction in **ZF**. It follows that a contradiction provable in **ST'** is provable in **ZF**, hence the relative consistency of **ST'** (and thus **ST**) to **ZF**.

Our way of translating wffs of **ST'** into wffs of **ZF** is a special case of a general technique. Suppose we have a first-order system F and a first-order system F^* and a mapping h from closed wffs of F to wffs of F^*. We say that h *preserves negation* if and only if, for every closed wff $\sim B$ of F, $h(\sim B) = \sim h(B)$. We say that h *preserves provability* if, for every closed wff B of F, $\vdash_F B$ implies $\vdash_{F^*} h(B)$. If h preserves both negation and provability, then h is an *interpretation* of F in F^*. If, given F and F^*, there exists an interpretation of F in F^*, we say that F *has a model* in F^* or *is interpretable* in F^*.

If F has a model in F^*, then F is consistent if F^* is. To see this, suppose F inconsistent. Then, everything is provable in F. In particular any closed wff B and its negation $\sim B$ are both provable. But then $h(B)$ and $h(\sim B) = \sim h(B)$ are both provable in F^* since h preserves negation and provability. Hence, F^* is inconsistent if F is.

The mapping h, which associates with every closed wff of **ST'** its transform in **ZF** as we defined it in the preceding discussions, clearly satisfies the requirements of an interpretation of **ST'** in **ZF**.

In Chapter 1, we defined a model of a first-order system in the terms of intuitive set theory. However, much of intuitive set theory (and certainly the portion we use in Chapter 1) is expressible in a strong axiomatic set theory such as **ZF**. Thus, a first-order system which has a model in the sense of Chapter 1 will usually admit an interpretation in **ZF** or some other strong formalized set theory. Let us take an example.

The system **S** of first-order arithmetic was discussed in Chapter 1, Section 1.9. There we observed that **S** has a model whose domain is the natural numbers and in which the interpretation of the function letter " $'$ " is successor (addition of 1). The function letters " $+$ " and " $\cdot$ " are interpreted as addition and multiplication respectively. This description of a model for **S** in the natural numbers remains on an intuitive, informal level. However, we can also describe an interpretation of **S** in **ZF** by a transformation process similar to the one used for **ST$'$** and **ZF** (and even more similar to our way of transforming closed wffs of **ST*** into closed wffs of **ST$'$**). First, we say that a closed wff A of **ZF** is *arithmetic* if its bound variables are restricted to natural numbers. This means that every instance of universal quantification in A has the form $(x)(x \in \omega \supset B)$ for some wff B, and every instance of existential quantification in A has the form $(Ex)(x \in \omega \wedge B)$ for some wff B.

Now, it is easy to see that the universal closure of wff axiom of the system **S** can be mapped to an arithmetic wff of **ZF** in a natural way. The wffs which will be images of axioms of **S** under this mapping will be theorems of **ZF**. The various axioms for equality in **S** are theorems of **ZF** by ZF.1. We have proved the Peano postulates in **ZF**, including the general metatheorem of proof by induction. The closures of these theorems are all arithmetic wffs, since we have continually used the restrictive condition $N(x)$ in these theorems. Moreover, the axiom of infinity of **ZF** tells us that $\vdash_{\mathbf{ZF}} N(x) \equiv x \in \omega$. Finally, the axioms of **S** which express the recursive properties of addition and multiplication can be transformed into theorems of **ZF**, since we have proved in **ZF** a theorem of recursion which allows us to define addition and multiplication operations satisfying the desired properties.

Generally speaking, we define a mapping h which takes every closed wff of **S** to a closed arithmetic wff of **ZF**. h clearly preserves negation. The rules of inference are the same in both systems, and we have transformed the universal closures of axioms of **S** into theorems of **ZF**. It follows that h preserves provability and so we have a model of **S** in **ZF**.

The method of using restricted quantification in F^* and of transforming the universal closures of axioms of F into theorems of F^* is a standard way of exhibiting a model of F in F^*. This method affords us a technique for comparing, in a purely proof-theoretic manner, the relative strength of two formalized foundational systems. Of course we are often guided by some model-theoretic considerations which suggest to us the correct way of interpreting the axioms of one theory as theorems of another.

5.4. The axiom of choice

There is yet another axiom which needs to be added to **ZF** if we desire to obtain certain theorems and results of classical analysis in our system. This is the famous *axiom of choice* whose necessity was explicitly recognized by Zermelo in 1904 (Fraenkel and Bar-Hillel [1] contains historical notes on the axiom of choice). Intuitively, the axiom affirms that an infinite choice is possible. One might at first suppose that infinite choices are already possible in **ZF** on the basis of the present axioms. In certain cases, such as functions defined by recursion, this would seem to be the case. Here, certain infinite choices are made because they are made in a constructive manner. But the axiom of choice asserts that infinite choices are *always* possible, not just under certain constructive conditions.

There are many different forms of the axiom of choice that are demonstrably equivalent on the basis of **ZF.1** through **ZF.9**. One form is as follows: For any collection x of sets such that $0 \notin x$, and such that $y \in x$ and $z \in x$ implies $y \cap z = 0$ (we say that x is a *disjoint* collection of nonempty sets), there exists a set y having one and only one element in common with each member of the collection x. We call this set a *choice set* for the disjoint collection of nonempty sets in question.

Another equivalent principle is the following: Given any collection x of nonempty sets, there is then a function f defined on x such that, for all $y \in x$, $f``y \in y$. This is a "choice function" which "picks out" an element from each member of x. Notice that here x is not necessarily disjoint, and so it is possible that f is not a 1–1 function.

Now, how legitimate for foundations are choice principles such as those just given? According to our intuitive model for **ZF**, the choice set for a given disjoint, nonempty collection x will be a subset of $\bigcup (x)$, hence an element of $\mathcal{P}(\bigcup (x))$. Since $\mathcal{P}(\bigcup (x))$ exists in our system, we might argue that a choice set must also exist. Yet, is there really such a subset of $\bigcup (x)$ that has *exactly* one element from each member of the collection x?

At this point in our examination, we are at the outer reaches of intuition, and there is no obvious answer. It is precisely with such questions that mathematical logic is supposed to deal, and so it has. Gödel [4] shows that if the other axioms of set theory (essentially the same ones we have given here) are consistent, then the extension obtained by adding the axiom of choice is also consistent. More recently, Paul Cohen has proved that the negation of the axiom of choice is also consistent with the other axioms of set theory (see Cohen [1] in which other references are given). The combined results of Gödel and Cohen establish the independence of the axiom of choice. We cannot, on the basis of the axioms **ZF.1** through **ZF.9**, prove that the axiom of choice holds, and we cannot prove that it does not hold, if the axioms **ZF.1** through **ZF.9** are consistent. Thus, we are free to decide whether or not we wish to add the axiom of choice.

Because many results of mathematics use the axiom of choice, most mathematicians would probably favor its addition. Gödel's relative consistency proof served to reduce the number of mathematicians objecting to the use of the axiom of choice. However, mathematicians who are constructivistic or intuitionistic in their views of mathematics reject not only the axiom of choice, but also the hierarchy of infinities that results from **ZF.8**.

We shall posit as a final axiom for **ZF** an extremely strong form, which is really a generalization of the Zermelo choice function. The principle of the choice function asserts the existence of a choice function defined on any collection of nonempty sets. Our form will assert the existence of such a choice function for the intuitively conceived domain of all nonempty sets of **ZF**. Formally, we posit:

ZF.10. $(x_1)(x_1 \neq 0 \supset \sigma(x_1) \in x_1)$, where σ is a primitive singulary function letter.

$\sigma(x)$ picks out exactly one element from each nonempty set x of **ZF**. It obviously picks out the same element from two equal sets.

THEOREM 30. $\vdash (x_1)(x_2)(x_1 = x_2 \supset \sigma(x_1) = \sigma(x_2))$.

Proof. Immediate from **ZF.1**.

Our form of the axiom of choice follows Bernays [1]. A similar form is found in Kelley [1]. The weaker form of the axiom would be formally stated as follows:

ZF.10*. $\vdash (x_1)((x_2)(x_2 \in x_1 \supset x_2 \neq 0) \supset (Ex_3)(F(x_3)$
$$\wedge D(x_3) = x_1 \wedge I(x_3) \subset \bigcup(x_1) \wedge (x_4)(x_3 \text{``} x_4 \in x_4)))).$$

This is the form using the choice function.

It is easy to see that the principle **ZF.10** implies **ZF.10***. For any given nonempty collection x_1, the range of the restriction of σ to x_1 will be a subset of $\bigcup(x_1)$. Thus, we can define the function x_3 explicitly as the set of ordered pairs of the form $\langle x_2, \sigma(x_2) \rangle$ for all $x_2 \in x_1$.

An interesting equivalent to the axiom of choice, which will find application in a later chapter, is attributable to MacLane [1]. Given a disjoint collection x of nonempty sets, the axiom of choice allows us to pick a choice set for x. Now, if we consider $\bigcup(x)$, the disjoint collection x amounts to a partitioning of $\bigcup(x)$ into equivalence classes. We know from elementary set theory that every function f with domain S partitions S into the equivalence classes, two elements of S being equal if they have the same image under f. Clearly, any partitioning of any set S can be induced by a properly chosen function. Given any set S and any function with domain S, let x be the collection of equivalence classes induced by f. By our choice principle, there exists a choice set u which will be a set consisting of one element from each equivalence class of S.

Thus, where R is the range of f, there exists a function g with domain R and range u, where u is our choice set for the partitioning of S under f. g will associate each element a of R with the unique element of u that is an inverse image of a under f (see following diagram). The two functions f and g are related by the equality $f(b) = (fgf)(b)$ for all $b \in S$. Thus, the axiom of choice implies that, for every function f, there is a function g such that $f = fgf$.

It is also clear that the converse is true; that is, the principle that for every function f there exists a function g such that $fgf = f$ implies the axiom of choice. Therefore, this principle is another form of the axiom of choice.

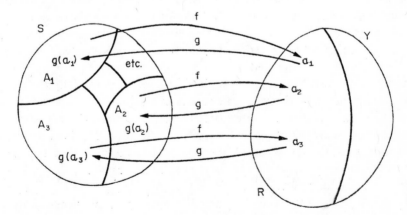

Diagram: Here A_1, A_2, and A_3 are cells in the partitioning of S by f, $f(A_i) = \{a_i\}$. g is the function from R to S, $R \subset Y$, and the choice set u is the set $\{g(a_1), g(a_2), g(a_3), \ldots\}$. $fgf = f$.

We list without explanation or proof several other equivalents of the axiom of choice: (1) An arbitrary cartesian product (possibly infinite) of nonempty sets is nonempty. (2) Every set can be well ordered. (3) Every set has a unique cardinal number. (4) Zorn's lemma: For a partially ordered set S, if every totally ordered subset of S has an upper bound, then S has a maximal element (which may not be unique).

5.5. The continuum hypothesis; descriptive set theory

The axiom of choice is not the only principle of set theory which has turned out to be independent of the axioms **ZF**.1–**ZF**.9. One of Cantor's prime motivations in developing infinite set theory was to generate tools for counting infinite sets and infinite processes with the same precision one counts finite sets and processes. The tools he developed for this purpose were, of course, the cardinal and ordinal numbers. Once developed, these tools were then applied to count well-known sets. It was established that the rational and algebraic numbers were countable (i.e. bijective with ω) and that the reals were uncountable. In fact, the reals were shown to have cardinality $2^{\aleph_0}$, the cardinality of $\mathcal{P}(\omega)$. The question naturally arose as to whether there were any sets having cardinality strictly between that of ω and $\mathcal{P}(\omega)$. Cantor hypothesized that the cardinality $2^{\aleph_0}$ of the continuum (the set of reals) was, in fact, the next highest cardinality beyond $\aleph_0$. This statement became known as the *continuum hypothesis*. A broader principle is the so-called *generalized continuum hypothesis* that for any infinite set X whatever, there is no cardinality strictly between that of X and $\mathcal{P}(X)$ (notice that, for finite sets, this is not true).

Progress was made on these questions when Sierpinski proved that the generalized continuum hypothesis implies the axiom of choice. Gödel subsequently proved the consistency not only of the axiom of choice but of the generalized continuum hypothesis with the other axioms of set theory. However, Cohen established the independence of the continuum hypothesis as well as that of the axiom of choice. In fact Cohen established the independence of the continuum hypothesis with respect to all the **ZF** axioms including the axiom of choice. In other words, there are models of the axioms **ZF**.1–**ZF**.10 in which there are cardinal numbers strictly between $\aleph_0$ and $2^{\aleph_0}$, and there are models in which this is not the case.

Most set theorists feel that there is no clear basis for deciding between the continuum hypothesis and its negation as to which should be added as a new principle of set theory. Thus, since Cohen's independence results, research into set theory has been directed toward trying to find some natural principles of set theory which would permit us to determine such questions as the continuum hypothesis. However, the results of this research has been a plethora of new independent principles, but no natural one which determines the continuum problem.

These developments have tended to undermine the view of set theory as a privileged language for mathematics. The completeness theorem of logic tells us that each independent principle of set theory determines two incompatible models in which **ZF**.1–**ZF**.10 are true. We have not one but a multiplicity of set theoretical universes, and we have very few criteria for deciding which of these universes is better for carrying on mathematics than another.

The foundational view of set theory is that mathematics is, or in some sense is reducible to, set theory. In view of the results of the recent years, we now have to decide exactly which set theory it is that mathematics may reduce to. In any case, it seems increasingly clear that the exhaustive study of set theory is not necessarily relevant to foundations. Such a realization leads to a viewpoint which considers set theory as one theory among others to be studied for its own sake as a branch of mathematics much as analysis, algebra, topology, probability theory, geometry, etc. are studied.

This new viewpoint (which, in fact, is quite close to Cantor's original viewpoint) is reflected in the recent resurgence of so-called *descriptive set theory*. Here set theory is frankly pursued for its own sake and without any real intention of trying to settle foundational questions. Instead of studying the universe of all sets, one studies selectively certain interesting sub-universes in which some of the principles undecidable in **ZF** are determined. For example, it is known that every topologically closed subset of the real numbers is either countable or bijective with the reals. In other words, there can be no closed set X of real numbers whose cardinality is strictly between $\aleph_0$ and $2^{\aleph_0}$.

We will not enter into any consideration of descriptive set theory in this work since, as we have explained above, it is not really foundational either in conception or practice. In the last section of Chapter 8, we attempt to draw some philosophical conclusions regarding foundations in the light of these recent developments in set theory.

5.6. The systems of von Neumann–Bernays–Gödel and Mostowski–Kelley–Morse

Several ways of modifying **ZF** so as to obtain a more flexible set theory have been considered. We shall consider two here that are easily described on the basis of the foregoing exposition. The first is a system which we will call **NBG** (von Neumann–Bernays–Gödel).

The primitive terms of **NBG** and the wffs of **NBG** are exactly the same as for **ZF**. We begin with the following definitions:

Definition 1. $x = y$ for $(z)(z \in x \equiv z \in y)$ where x and y are any terms in which z does not occur.

This is the same definition of equality as for **ZF**.

Definition 2. $M(x)$ for $(Ey)(x \in y)$ where x is any term in which y does not occur.

This is a new, deceptively simple definition. $M(x)$ means "x is a set" ("M" stands for *Menge*, the German word for set). Thus, an object of our system is a set if there is another object to which it belongs. A set is something which belongs to something else.

The reader may at first protest that we have been talking about sets all along and so we hardly need to define what we mean by this notion. In **NBG**, however, we make a distinction between *class* and *set*. We will think of our variables x_1, x_2, and the like as referring to classes. Sets thus appear as special kinds of classes, namely classes which belong to other classes as elements. Some classes will indeed appear which do not belong to any other class. Such classes will be called *proper classes*. We state:

Definition 3. $\Pr(x)$ for $\sim M(x)$, where x is any term.

A proper class is a class which is not a set.

Our intuitive universe consists of classes, some of which are sets and some of which are proper classes. If we think of the universe of sets as being determined, it is easy to describe the universe of classes with respect to it; classes are collections of sets. Since it is only sets which can belong to something, every class is a collection of sets. But what is our universe of sets going to be? The answer is: the sets described by **ZF**.

Our intuitive model for **NBG** can now be described as follows: Consider the intuitive model for **ZF** described on pages 136–137 of this book. Let V denote the intuitively conceived class of all the sets of **ZF**. V must be a proper class, for no universal set exists in **ZF** (if **ZF** is consistent). Now, by a class of **NBG**, we will mean any subclass of V. Some subclasses of V can be obtained from the axioms of **ZF** alone. These are the *sets* of **NBG**. The *proper* classes will be those subclasses of V which cannot be obtained as elements of some set of the hierarchy generated by starting with the null set 0 and iterating the operations of power set and union any ordinal number α of times, where α is an ordinal (thus a set) of **ZF**.[†] Finally, there *are* proper classes, since V is proper.

The next question is natural; can we give any simple criterion that distinguishes a proper class from a set? The answer is "yes". The criterion is *bigness*. Let us explain in greater detail.

By our axiom of replacement **ZF**.9, if the domain of a functional relation is a set of **ZF**, then its range will be a set of **ZF**. Thus, if a class X has the same cardinality as a set Y, then X is also a set, since it is the range of a 1–1 function whose domain is the set Y. It follows that any proper class of **NBG** is too big to be put into a 1–1 correspondence with any set. The only proper classes are those collections, such as V, which are bigger than any set.[‡]

Let us now return to formal considerations and see how we actually set up our system **NBG** in order to describe this intuitive universe of classes and sets.

Let $(x)A$ be some wff of **NBG**. We say that the initial occurrence of universal quantification in A is *restricted to sets* when A is of the form $(x)(M(x) \supset B)$, where B is some wff of **NBG**. Where the initial occurrence of universal quantification is not restricted to sets for some wff $(x)A$, the *restriction of x to sets* is defined as the wff $(x)(M(x) \supset A)$. Intuitively, we replace the wff "for all x, A is true" by "for all sets x, A is true".

Similarly, the *restriction of x to sets* for the initial occurrence of existential quantification in some wff $(Ex)A$ is the wff $(Ex)(M(x) \wedge A)$, unless A is of the form $(M(x) \wedge B)$ for some wff B. In this latter case, we say that the initial occurrence of existential quantification is *restricted to sets*. Intuitively, the restriction of existential quantification to sets replaces a statement "there is an x such that A is true" by "there is a set x such that A is true".

By a *predicative* wff of **NBG**, we mean any wff in which every occurrence of quantification (existential or universal) is restricted to sets. Given any wff A of **NBG**, we obtain the *restriction of A to sets* by restricting all occurrences of quantification in every wff that is a part of A to sets. (We begin with the smallest subformulas and work up when we restrict a wff to sets.) Thus, to

---

† Previously, in describing our intuitive model for **ZF**, we have spoken of iterating our basic operations "any transfinite number of times". But the number of times we can iterate these operations obviously depends on the ordinals available to serve as a basis for our hierarchy of sets. For **ZF**, the ordinals available are precisely those definable by means of the operations furnished to us by the axioms of **ZF**. As we shall see later, in discussing the axiom of inaccessible cardinals, it is possible to conceive of the existence of ordinals beyond those definable from the axioms of **ZF** alone. This shows that there are limitations on the richness of our model for **ZF**. The reader should keep this in mind when trying to visualize intuitively our model for **ZF**.

‡ Of course the axiom of replacement applies only to 1–1 functions that are expressible by some wff of **ZF**. It is possible that, viewed from outside the system, there exists a 1–1 correspondence between a proper class and a set of the system. What we can be sure of is that no such 1–1 correspondence can ever be talked about or expressed within our system (if the system is consistent). A more detailed discussion of such model-theoretic considerations is to be found in Chapter 6.

every wff of **NBG**, there is associated in an obvious way a predicative wff which is its restriction to sets.

With these definitions in mind, we can now state the distinctive axiom of **NBG**, the *axiom scheme of abstraction*:

NBG.0. Let $A(x)$ be any predicative wff of **NBG**. Then every instance of the following scheme is an axiom:

$$(Ey)\,(x)\,(x \in y \equiv (M(x) \land A(x))),$$

where y is different from x and y does not occur in $A(x)$.

NBG.0 says that for every predicative wff of **NBG** there exists a class y of all those *sets* x such that A is true. The class y thus defined may or may not be a set. We immediately obtain the existence of a universal class, a class of all *sets*.

THEOREM 1. $\vdash (Ex_1)\,(x_2)\,(x_2 \in x_1 \equiv M(x_2))$.

Proof. By **NBG**.0, $(Ex_1)\,(x_2)\,(x_2 \in x_1 \equiv M(x_2) \land x_2 = x_2)$ is an axiom. Since obviously $\vdash x_2 = x_2$ (as in **ZF**), the conclusion easily follows.

We could, of course, use any universally true property to obtain Theorem 1.

NBG.0 seems quite close to Frege's original axiom of abstraction, and so one might fear that Russell's paradox is deducible. All that we can prove, however, is that the class of all sets satisfying the Russell condition is a proper class. To see this, start with the axiom

$$(Ex_1)\,(x_2)\,(x_2 \in x_1 \equiv (M(x_2) \land x_2 \notin x_2)).$$

By eE, we obtain $(x_2)\,(x_2 \in a_1 \equiv (M(a_1) \land x_2 \notin a_1))$, and further, by $e\forall$, $a_1 \in a_1 \equiv M(a_1) \land a_1 \notin a_1$. By Taut this yields $\sim M(a_1)$. Thus, the Russell class is not a set.

NBG.1. Same as **ZF**.1.

This is the axiom of extensionality, which tells us that any two equal classes satisfy the same properties (notice that we have imposed no restriction to sets on the wff $A(x, x)$ of **NBG**.1).

NBG.1 allows us to conclude that the universal class of Theorem 1 is unique. We henceforth use "V" to denote the unique universal class of all sets.

Now that we have the axiom **NBG**.0, what classes can we obtain? We have already obtained a universal class, and by taking a self-contradictory property for $A(x)$ in **NBG**.0, we can obtain a null class. But **NBG**.0 alone will never allow us to obtain other sets. In fact, we can easily satisfy **NBG**.0 in a one-element model. This one element would be the universal class and it would not contain itself as an element. Consequently, it would be a proper null class. There would be no sets and so the one proper class would contain all sets (none). In other words, the predicate $M(x)$ would be false for all x.

The existence of such a trivial model for **NBG**.0 by itself shows the need for axioms giving us the existence of sets, for otherwise there will be nothing that can belong to classes. The other axioms of **NBG** will be axioms of set existence. Since our intuitive interpretation is that our class V of all sets is the class of all sets of **ZF**, it becomes clear how to obtain the desired system:

Axioms **NBG**.2 through **NBG**.10 are obtained by restricting each of the axioms **ZF**.2 through **ZF**.10 to sets, as well as understanding that the wffs of **ZF**.2 and **ZF**.9 are predicative wffs of **NBG**. The one exception is **ZF**.7 in which the initial universal quantifier "(x)" is not restricted to sets. Furthermore, we require that explicit conditions be added to the appropriate axioms so that the primitive terms 0 and ω are sets and the primitive terms $\{z \mid z \in t \wedge A(z)\}$, $\mathcal{P}(x)$, $\{x, y\}$, $\sigma(x)$, and $\bigcup(x)$ are all sets whenever the terms t, x, and y are sets. (That $\sigma(x)$ is a set whenever x is a set already follows from the axiom **ZF**.10 itself, except for $x = 0$.)

For instance, **NBG**.2 would be stated:

$$(z)(M(z) \supset (M(\{y \mid y \in z \wedge A(y)\}) \wedge (x)(M(x) \supset (x \in \{y \mid y \in z \wedge A(y)\} \equiv (x \in z \wedge A(x)))))),$$

where x, y and z are different, $A(y)$ is a predicative wff, and so on. Notice that the bound variable y is restricted to sets, since the wff "$y \in z$" is part of the term operator. We could add the explicit restriction in any case if it were necessary.

ZF.3 would become **NBG**.3 : $0 = \{x_1 \mid x_1 \in 0 \wedge x_1 \not\equiv x_1\} \wedge M(0)$.

The reason for making an exception of the initial quantifier of **ZF**.7 with respect to the restriction to sets is the following: The more general form of **NBG**.7 which results from this exception allows us to prove *within* **NBG** that the class V of all sets is precisely the class of the sets that are elements of some set of the hierarchy obtained by starting with 0 and iterating power set and union any transfinite number of times.[†] We prove this in Theorem 9, Section 5. 7, below.

This completes the formal description of **NBG**. The reader might, for an exercise, formulate explicitly each of the remaining axioms of **NBG**.

In some presentations of **NBG**, one uses two sorts of variables; capital letters representing classes generally, and small letters representing classes that are sets. When this is done, quantifiers using small letters are thought of as restricted to sets. See Bernays [1], Mendelson [1], and Wang and MacNaughton [1] for examples of such usage.

The use of **NBG**.0 can simplify the statement of axioms, like **NBG**.2 and **NBG**.9, which are axiom schemes involving predicative wffs. **NBG**.2 could be stated, for instance, in the following manner:

NBG.2*.

$$(x_1)(M(x_1) \supset (x_2)(Ex_3)(M(x_3) \wedge (x_4)(M(x_4) \supset (x_4 \in x_3 \equiv x_4 \in x_1 \wedge x_4 \in x_2)))),$$

without using a primitive term operator. In place of our predicative wff $A(x)$ of **NBG**.2, we now have the variable x_2, which is not restricted to sets. Notice that **NBG**.2* is a proper axiom, not an axiom scheme. Using **NBG**.0, we obtain the original form of **NBG**.2 (without a primitive term operator) in the following way:

Given some predicative wff $A(x)$, we have by **NBG**.0, $\vdash (Ew)(x)(x \in w \equiv M(x) \wedge A(x))$. Using eE, we obtain

$$(x)(x \in a \equiv M(x) \wedge A(x))$$

where a is some new constant letter. Now, applying $e\forall$ to **NBG**.2* in the name of some appropriate new variable z, and under the hypothesis $M(z)$, we obtain $(x_2)(Ex_3)(M(x_3) \wedge (x_4)(M(x_4) \supset (x_4 \in x_3 \equiv x_4 \in z \wedge x_4 \in x_2)))$. Applying $e\forall$ to this in the name of a we obtain

$$(Ex_3)(M(x_3) \wedge (x_4)(M(x_4) \supset (x_4 \in x_3 \equiv x_4 \in z \wedge x_4 \in a))).$$

[†] Again we remind the reader that this hierarchy depends, in turn, on the size of the ordinals definable within **ZF**.

Now using our various rules, and the general substitutivity of " $\equiv$ " (we may have to change the name of some bound variables), we obtain

$$(Eu)(M(u) \wedge (x)(M(x) \supset (x \in u \equiv x \in z \wedge M(x) \wedge A(x))))$$

for an appropriate new variable u. This, in turn, is equivalent to

$$(Eu)(M(u) \wedge (x)(M(x) \supset (x \in u \equiv x \in z \wedge A(x)))).$$

Eliminating dependence on our hypothesis $M(z)$ and applying $i\forall$, we finally obtain

$$(z)(M(z) \supset (Eu)(M(u) \wedge (x)(M(x) \supset (x \in u \equiv x \in z \wedge A(x)))))$$

where $A(x)$ is any predicative wff. But this is precisely **ZF**.2 (without the term operator) restricted to sets.

We have noted that this new form of our axiom is finite, since the wff $A(x)$ no longer appears in it. A similar technical maneuver shows that we can eliminate the occurrence of $F(x, y)$ in **ZF**.9, with the use of **NBG**.0 again being essential. This latter situation shows us that the introduction of the distinction between "set" and "class" amounts, in a certain sense, to incorporating part of the metalanguage of **ZF** into our new system.

It would seem, then, that **NBG** is clearly stronger than **ZF**. Actually, there is really not so much difference, for **NBG** is consistent relative to **ZF** (see Novak-Gál [1]). Furthermore, and somewhat surprisingly, **NBG** is finitely axiomatizable whereas **ZF** is not. If we replace **NBG**.2 and **NBG**.9 by their finite counterparts considered above, then the only remaining nonfinite axiom is **NBG**.0 (the axiom of extensionality can easily be rendered finite, essentially by taking Theorem 1 of Section 5.1 in its stead). However, it has been shown that **NBG**.0 can be replaced by a finite set of axioms concerned with classes such that **NBG**.0 is provable as a metatheorem. The reader should consult Gödel [4] or Mendelson [1] for a proof of this fact.

We have already described an intuitive model of **NBG**, and we have pointed out that because of the axiom of replacement **NBG**.9, bigness is the criterion which distinguishes between sets and classes. For this reason, **ZF** and **NBG** are often said to avoid the paradoxes by "size-limitation". This point deserves to be amplified.

The theory of cardinal numbers is performed in **NBG** in the same way as in **ZF**. Thus, the cardinal number of a set X of **NBG** is the smallest ordinal number which can be put into 1–1 correspondence with X. Such an ordinal number is itself a set of a certain kind, namely a set which is Trans and Con under the $\in$ relation. In this theory, only sets are assigned cardinal numbers. Proper classes are "too big" to have cardinal numbers in **NBG**, since any class X having a cardinal number in **NBG** must admit of a 1–1 correspondence f, expressible within **NBG**, with an ordinal α. This ordinal α is a set and the domain of a function f expressible in **NBG**, and so the axiom of replacement will require that the class X, which is the range of the function f, be a set. X therefore cannot be a proper class. We have already observed that classes like V or the Russell class are proper classes in **NBG**. It follows that the Russell class and V fail to be sets because they are too big to have a cardinal number definable in **NBG**.

All this raises the possibility of adding an axiom of infinity of a new kind to **ZF**, an axiom that posits the existence of a set bigger than any set obtainable by means of the operations furnished to us by the axioms **ZF**.1 through **ZF**.10. One such axiom is the axiom of *inaccessible cardinals*. An uncountable cardinal α is *inaccessible* if (i) for all cardinals β less than α, α is not the sum of

β cardinals less than α and (ii) if X is any set with cardinality less than α, then the cardinality of $\mathcal{P}(X)$ is also less than α. This definition of an inaccessible cardinal can be expressed in **ZF** or **NBG**. The axiom of inaccessible cardinals asserts the existence of some inaccessible cardinal α.

The axiom of inaccessible cardinals has never been shown to lead to contradiction when added to **ZF** or **NBG**. However, it does lead to a strictly stronger theory if it is added as an axiom to **ZF**. That is, we cannot prove within **ZF** that there are inaccessible cardinals. This can be seen from the following two facts: (1) If an inaccessible cardinal α exists, it is possible to obtain a model for **ZF**.1 through **ZF**.10 among the sets of cardinality less than α. (2) One of the consequences of Gödel's incompleteness theorems is that one cannot prove the consistency of a foundational system by methods expressible within the system. Since the existence of a model for **ZF**.1 through **ZF**.10 implies that **ZF** is consistent, fact (2) implies that the existence of an inaccessible cardinal cannot be proved within **ZF**, which is the system **ZF**.1 through **ZF**.10, provided that **ZF** is consistent.

In Chapter 6 we shall turn to a more detailed discussion of Gödel's incompleteness theorems and this will serve to illuminate the meaning of fact (2). In the present discussion, we wish to see briefly how fact (1) operates; that is, how we obtain a model for **ZF** if we are given the existence of an inaccessible cardinal α.

First, observe that $\aleph_0$, the cardinal number of any denumerable set, satisfies all of the criteria for an inaccessible cardinal, except that it is countable. In **ZF** or **NBG**, $\aleph_0$ is the ordinal ω, since ω is the smallest denumerable ordinal. Now our axiom of infinity **ZF**.8 is the one which posits the existence of ω. Notice that the system **ZF** minus **ZF**.8 will have a model among the sets of cardinality less than ω. In particular, the model can be described as follows: Starting with the null set, we iterate our operation $\mathcal{P}$ of power set any finite number of times. The sets of our model are then the sets X that are elements of some set $\mathcal{P}^n(0)$. $\mathcal{P}^n(0)$ is obtained by iterating the power set operator some positive integral n number of times. In other words, the elements of our model are the sets which are in the union of all the $\mathcal{P}^n(0)$. All the sets in this model are of finite cardinality, thus of cardinality less than ω.

Another way of characterizing the foregoing model for **ZF** minus the axiom of infinity is to think of truncating our original model for **ZF**. This is done by considering all sets of rank less than ω. We recall, from our discussion at the beginning of this chapter, that the rank of a set X of our model is the smallest ordinal β such that X is an element of Y_β where Y_β is obtained from the null set by exactly β iterations of power set and union. Now, the fact that ω satisfies conditions (i) and (ii) of the definition of inaccessibility means precisely that none of the operations furnished to us by the axioms of **ZF** other than **ZF**.8 enables us to generate ω from cardinals smaller than ω. Part (ii) says that the operation of power set does not give ω when applied to lower cardinals. Part (i) says that neither sum set applied to a collection X of cardinality less than ω and whose elements are of cardinality less than ω, nor pairing applied to lower cardinals, nor replacement applied to lower cardinals, i.e. taking the range of a function whose domain is a cardinal less than ω, suffices to generate ω. But a cardinal such as ω is by definition the lowest ordinal of its cardinality. Thus, every ordinal less than ω ("less than" in the ordinal sense of being an element of ω) has cardinality less than ω. Thus, we can see that parts (i) and (ii) guarantee that the sets of rank less than ω are a model for **ZF** minus **ZF**.8, since these other axioms of **ZF** never allow us to generate ω and thus obtain sets of rank ω or greater. Moreover, the sets of rank less than ω must have cardinality less than ω, for the ordinals less than ω are all cardinally smaller than ω and these are the only ordinals available as cardinals for sets of rank less than ω.

It should be noted that the converse of the last statement is not true; that is, there are sets of cardinality less than ω but of rank greater than ω. For example, the set $\{\omega\}$ is of finite cardinality 1 but of rank $\omega + 2$. This is why we were careful to say that our model for ZF without the axiom of infinity was *among* the sets of cardinality less than ω. We can say that our model consists of *all* the sets of *rank* less than ω.

We have delineated this example in such detail because the axiom asserting the existence of an inaccessible cardinal α bears a relation to ZF analogous to the relation that the axiom of infinity ZF.8 bears to the system ZF minus ZF.8. In fact, some authors define an inaccessible cardinal by requiring only that the cardinal be infinite and that it satisfy conditions (i) and (ii) of our definition. Then ω qualifies as the first inaccessible cardinal, and our present axiom of inaccessible cardinals asserts the existence of an uncountable inaccessible; that is, an inaccessible beyond ω. The axiom of infinity ZF.8 can then be viewed as asserting the existence of the first (countable) inaccessible cardinal.

Reverting now to our original terminology, we can see by strict analogy with the foregoing example how the assumed existence of an (uncountable) inaccessible cardinal α implies the existence of a model for ZF. Given α, we have again the fact that conditions (i) and (ii) insure that none of the operations available to us in ZF allows us to generate α from lower cardinals. α is uncountable and so not equal to ω. Thus, in the same manner as in the foregoing example, the sets of rank less than α form a model for ZF. These sets are all of cardinality less than α.

The existence of an inaccessible α yields a model for ZF among sets of cardinality less than α. We can thus think of the cardinality of the sets of ZF as being bounded by some inaccessible cardinal α. This gives an even more precise meaning to the notion that ZF and NBG avoid the paradoxes by size limitation.

Once we have enlarged our system by adding the inaccessible α, we might consider enlarging it again by adding a second inaccessible β beyond α. We can repeatedly add new axioms of infinity in this way and push the cardinality of the sets that we assume to exist higher and higher. We must beware, however, for we do not wish to posit a set as big as V or the Russell class, for we shall then have a contradiction. One who believes in strictly finite mathematics or is a constructivist would reject even the existence of ω. He might argue that we have as much reason to reject or accept an uncountable inaccessible α as we do to reject or accept ω, since neither of the cardinals can be obtained from lower cardinals by the other operations of ZF and so must be made the object of special existential postulation.

The size limitation method of avoiding the paradoxes is different in conception from Russell's. Russell's analysis hinges on excluding certain kinds of sets, sets that draw their elements from different types. The type distinction in turn hinges on splitting our intuitive universe into classes of objects which constitute the different type levels T_0, T_1, and so on. Thus, if x^0 is an individual, the class $\{x^0\}$ exists for Russell, but is of type 1. The cardinality of the intuitively conceived set $\{x^0, \{x^0\}\}$ is 2, yet such a set is not defined in ST (unless types are cumulative, and this cannot be proved from the axioms of ST alone).

We have observed in Chapter 4 that the theory of types does have a certain intuitive appeal, because the paradoxes of set theory, such as Russell's, do have the kind of "self-reference" or vicious circle that Russell and Poincaré observed. Yet, the criterion of bigness also has an intuitive appeal. Let us think of the Russell property $x \notin x$, "not to be an element of itself". Let us think further of those sets we intuitively conceive as legitimate. How many of these are elements of themselves? Hardly any that we can think of. The set of natural numbers is not a natural number (though it is an ordinal number in **ZF**), the set of real numbers is not a real number, and so forth. Thus, the class of all sets that are not elements of themselves must be fairly big. It must contain almost every set. Consequently, it is too big to be a set itself.

It is perhaps easier for a mathematician who is not interested in formal, logical questions to understand and apply the criterion of cardinality to escape the paradoxes than to apply the principle of vicious circle for the same purpose. This may be one reason why **ZF** and **NBG** are more favored by mathematicians than is a system of type theory like **ST**. However, we shall see in Chapter 8 that the use of cardinality as the criterion for legitimizing sets has its real drawbacks. In Chapter 7 we shall consider systems formulated by Quine. These admit big sets and avoid the strict separation into types, though a residue of the type restriction remains.

Finally, we mention here the system found in Kelley [1], which is an extension of **NBG**. This system is easily described. We obtain it from **NBG** by removing the predicativity restrictions on the unspecified wffs in the axioms **NBG**.0, **NBG**.2, and **NBG**.9. For example, the new form of **NBG**.0 will be : $(Ey)(x)(x \in y \equiv M(x) \wedge A(x))$ where x and y are different and y does not occur in $A(x)$. Here, $A(x)$ is any wff, not one that is necessarily predicative. The remaining axioms (**NBG**.1, **NBG**.3 − **NBG**.8, and **NBG**.10) are carried over unchanged. Let us call this system **MKM** for "Mostowski–Kelley–Morse". Mostowski [1] discusses the system in question, and Morse [1] contains a reworking of the system in a somewhat novel form.

Notice that **NBG** is not an extension of **ZF**, for an extension of a system is a system identical to the original except in the point of having additional non-logical symbols and/or additional axioms. **NBG** differs in form from **ZF** because of the class-set distinction and the consequent use of restricted quantification. But **MKM** is strictly an extension of **NBG**, since it admits all the axioms of **NBG** as well as others that **NBG** excludes.

MKM is clearly stronger than **NBG**. Since the property $A(x)$ of **MKM**.0 does not have to be predicative, we can, by using this axiom, define classes by a formula whose bound variables refer to the universe of classes. **MKM** is more flexible than **NBG** on this account, but is more likely to be inconsistent. A recent book by J. Rubin contains a detailed development of set theory using the system **MKM**. The system of Rubin [1] also involves a different treatment of identity to allow for "atoms", that is, distinct objects without elements (recall again that in the usual treatment, all no-element things are equal to the null class).

5.7. Number systems; ordinal recursion

Beginning with the axioms of set theory, we have constructed the natural numbers, defined the number-theoretic operation of addition, and indicated the manner of definition of other number-theoretic functions such as multiplication and exponentiation. The main tool involved in constructing these number-theoretic functions was Theorem 29 of Section 5.3 and its variants which justify within set theory the recursive definitions of addition, multiplication, and the like (cf. the system S of Section 1.9).

We can also prove that the ordinal order relation $<$ on ω (which is, in fact, just the $\in$ relation on ω) is compatible with the arithmetical operations in the precise sense that $\vdash_{ZF}(x)(y)(x \in \omega \wedge y \in \omega \supset (x \in y \equiv (Ez)(z \in \omega \wedge z \neq 0 \wedge x+z = y)))$ (cf. Definition 4, Section 1.9). We can thus consider the natural numbers to be an ordered algebraic system $\langle \omega, 0, 1, +, \times, < \rangle$. Theorem 1 below characterizes this system up to isomorphism.

Exercise. Establish that $\vdash_{\mathbf{ZF}}(x)(y)(x \in \omega \wedge y \in \omega \supset (x \in y \equiv (Ez)(z \in \omega \wedge z \neq 0 \wedge x+z = y)))$. (*Hint*: An easy proof by induction on y.)

Though we can carry through our discussion of number systems in any of the variants of set theory we have presented in the foregoing, we will nevertheless suppose from now on that we are working strictly within the class-set theory **MKM**. We suppose all of the definitions and notions previously established for the sets of **ZF** to be transferred (unchanged whenever possible) to the classes and/or the sets of **MKM**. For example, we can speak of the subclass of a class, or of a relation (a class whose members are all ordered pairs), etc. Other notions such as cartesian product, range of a function, or the Boolean operations of union, intersection, and relative complement will have to be redefined for classes in obvious ways using **MKM**.0. Still other notions (such as our definition of ordered pair) will apply only to sets and are supposed carried over in that form to **MKM**.

THEOREM 1. *Let $\langle N, 0^*, 1^*, +^*, \times^*, <^* \rangle$ be any set N with two distinct nullary operations 0^* and 1^*, binary operations $+^*$ and $\times^*$, and the binary relation $<^*$ which satisfy the following properties: $+^*$ is commutative, associative, and cancellative ($a+x = b+x$ implies $a = b$ for elements of N), and has 0^* as a neutral element. $\times^*$ is commutative, associative, and cancellative for non-zero elements, has 1^* as a neutral element and is distributive over $+^*$. Finally, $<^*$ is a well-ordering of N such that $x <^* y \equiv (Ez)(z \neq 0^* \wedge x +^* z = y)$ for x, y, and z elements of N. Then the system $\langle N, 0^*, 1^*, +^*, \times^*, <^* \rangle$ is order isomorphic to $\langle \omega, 0, 1, +, \times, < \rangle$, i.e. there is a bijection from ω to N which takes 0 to 0^*, 1 to 1^*, preserves addition and multiplication, and both preserves and counter-preserves order.*

Proof. In N, define a successor operation S by: $S(x) = x +^* 1^*$. We show that $\langle N, 0^*, S \rangle$ satisfies the Peano postulates. Since $0 +^* x = x$ for all $x \in N$, $0 <^* x$ for $x \neq 0^*$. In particular, $0^* <^* 1^*$ and hence $0^* <^* x +^* 1^*$ for all $x \in N$ by the transitivity of $<^*$. Thus, $S(x) \neq 0^*$ for all x by the law of trichotomy. The injectivity of S follows immediately from the cancellability of $+^*$. It remains to show that every element of N is in the set generated by $\{0^*\}$ and closed under arbitrary iterations of S.

Because $x <^* y \equiv (Ez)(z \neq 0^* \wedge x +^* z = y)$, the order relation is preserved for multiplication by nonzero ($\neq 0^*$) elements of N. Indeed, we have, for $m \neq 0^*$ and $x <^* y$, $xm +^* zm = ym$. Moreover, zm cannot be 0^* for $zm = 0^* = z0^*$ yields $m = 0^*$ on cancelling by $z \neq 0^*$. The identity $0^* = z0^*$ (for any z) follows from the identities $z0^* +^* z0^* = z(0^* +^* 0^*) = z0^* = z0^* +^* 0^*$ upon cancellation by $z0^*$. We conclude from all this that there is no element of N between 0^* and 1^*, for if so, let m be the smallest such ($<^*$ is a well-ordering). Then $0^* <^* m <^* 1^*$. Multiplying by m, we have $0^* <^* m \times^* m <^* m <^* 1^*$, and $m \times^* m$ contradicts the minimality of m.

Let now X be a set containing 0^* and closed under S. If $N-X$ is nonempty, then it has a smallest member m. $m \neq 0^*$ since $0^* \in X$. Thus, $0^* <^* m$ and hence $1^* = m$ or $1^* <^* m$ since there is nothing between 0^* and 1^*. In fact, $1^* \neq m$ since $1^* = S(0^*)$ is in X since 0^* is. Thus, $1^* <^* m$ which implies $m = a +^* 1^*$ for some $a \neq 0^*$. But $a <^* a +^* 1^*$ and so $a \in X$. Thus, $S(a) = a +^* 1^* = m$ is in X which contradicts $m \in N-X$. Thus, $N-X$ is empty and the last Peano postulate is established.

Using now the theorem of simple recursion (Theorem 29, Section 5.3), we define a unique function h from ω to N such that: $h(0) = 0^*$, $h(n') = S(h(n)) = h(n) +^* 1^*$ (thus, in particular

$h(1) = h(0') = 1^*$). Because $\langle N, 0^*, S \rangle$ satisfies the Peano postulates, h is a bijection whose inverse g takes 0^* to 0 and preserves successor (cf. the last exercise in Section 3.2). But h must also preserve $+$, $\times$, and order, for $+$ is defined recursively in terms of 0 and $'$, $\times$ is defined recursively in terms of $'$, $+$, and 0, and $<$ is defined in terms of $+$ and 0 (cf. the system $\mathbf{S}$ of Section 1.9). On the other hand, the operations $+^*$ and $\times^*$ satisfy the same recursion equations since these latter are just special cases of such laws as associativity and distributivity. Thus, an inductive proof establishes that h preserves $+$, $\times$ and therefore $<$. Since its inverse g will also preserve $<^*$, h counter-preserves order. This establishes the theorem.

We will use N to designate the system $\langle \omega, 0, 1, +, \times, < \rangle$ of natural numbers (but continue to use ω when speaking of the lowest denumerable ordinal). The natural numbers represent a canonical number system in which addition and multiplication are defined. The canonical nature of the system is expressed by the above theorem. If we desire to include subtraction among our operations, we obtain a minimal canonical extension of N, namely the integers Z. Z is characterized as an ordered integral domain in which the nonnegative elements are well ordered. Indeed, the nonnegative elements of Z are isomorphic to N.

A model for Z can be constructed from N in several ways. One way is to define a set N^- which is disjoint from N and in bijective correspondence with $N-\{0\}$. This is the set of *opposites* of the positive natural numbers. Z is then defined as $N \cup N^-$. Defining the extensions of addition and multiplication to Z is easy, but proving that these operations satisfy the usual properties is tedious and involves taking many different cases concerning the signs of the arguments of the operations.

We sketch in more detail another construction of Z from N. We consider the cartesian product $N \times N$ and define an equivalence relation E on this set by: $\langle a, b \rangle E \langle c, d \rangle$ for $a+d = = b+c$. The integers are the E-equivalence classes $|\langle a, b \rangle|$ of ordered pairs of natural numbers. Addition of integers is defined by defining it componentwise on couples, while multiplication is defined on couples by: $\langle a, b \rangle \times \langle c, d \rangle = \langle ac+bd, ad+bc \rangle$. Intuitively, the couple $\langle a, b \rangle$ represents the difference of a and b. Thus, each negative number will contain a unique couple of the form $\langle 0, b \rangle$ and each positive number a couple of the form $\langle a, 0 \rangle$. 0 will be represented by $\langle 0, 0 \rangle$ or by any diagonal element $\langle a, a \rangle$ and 1 by any couple of the form $\langle a+1, a \rangle$. The opposite of an integer $|\langle a, b \rangle|$ is defined by: $- |\langle a, b \rangle| = |\langle b, a \rangle|$, and subtraction is defined as the addition of opposites. Finally, a couple $\langle a, b \rangle$ is declared *positive* if $b < a$. Thus, the couples are ordered by: $\langle a, b \rangle < \langle c, d \rangle$ if and only if the difference $\langle c, d \rangle - \langle a, b \rangle$ is positive.

Exercise. Verify that the above construction gives a model for Z. Do not forget to verify that the equivalence relation is compatible with the operations and relations defined on couples. Verify also that the correspondence $n \mapsto |\langle n, 0 \rangle|$ embeds N into Z.

Z is a canonical system in which the operations of addition, multiplication, and subtraction (the inverse of addition) are performable. Its canonical character is expressed not only by its being a minimal extension of N for which subtraction is defined, but (equivalently as it turns out) by its being an initial object in the category of rings with unit (see Chapter 8 for this terminology). It is natural to seek a minimal extension Q of Z in which not only addition, multiplication, and subtraction are definable, but also division for all non-zero elements (the inverse of multiplication). This is accomplished by the well-known construction of the field of fractions over the integral domain Z. We describe it briefly.

A *fraction* is a couple $\langle a, b \rangle$ of integers where $b \neq 0$. The set of all fractions is thus the set $Z \times (Z - \{0\})$. We define an equivalence relation E on fractions by: $\langle a, b \rangle E \langle c, d \rangle$ if and only if $ad = bc$. We define operations on fractions as follows: $\langle a, b \rangle + \langle c, d \rangle = \langle ad + bc, bd \rangle$; $\langle a, b \rangle \times \langle c, d \rangle = \langle ac, bd \rangle$; $\langle a, b \rangle - \langle c, d \rangle = \langle ad - bc, bd \rangle$ (thus $-\langle a, b \rangle = \langle (-a), b \rangle$); and $\langle a, b \rangle \div \langle c, d \rangle = \langle ad, bc \rangle$ provided $c \neq 0$ (since Z is an integral domain and $b \neq 0$ by definition, $bd \neq 0$). We then show that the relation E is compatible with these operations. This allows us to induce corresponding operations on the set Q of equivalence classes of fractions. Q is the set of *rational* numbers into which Z is embedded by the association $z \mapsto |\langle z, 1 \rangle|$ ($|\langle z, 1 \rangle|$ represents the equivalence class of $\langle z, 1 \rangle$). Q is the smallest infinite field and represents a canonical number system in which all four basic operations can be performed.

The order relation on Q is handled as follows: A fraction $\langle a, b \rangle$ is *positive* if a and b are positive integers. A rational number is positive if it contains at least one positive fraction. Finally, $\langle a, b \rangle < \langle c, d \rangle$ if and only if $\langle c, d \rangle - \langle a, b \rangle$ is positive. It is easy to see that this order relation is invariant under equivalence.

In Q, the order relation is no longer discrete but is *dense* in the sense that between any two rationals there is another (and thus an infinity). As it turns out, the order relation on Q is also canonical: it is, up to order-isomorphism, the only denumerable dense linear (total) order without first or last element. Since any totally ordered field has dense order and is thus infinite, Q is also canonically characterized as the smallest totally ordered field. In particular, then, no finite field is (totally) orderable.

All these canonical characterizations of N, Z, and Q are important in that they allow us to "throw away" the particular constructions of them, just as we have disregarded each particular set-theoretical construction of the natural numbers once the Peano postulates were proved. In fact, as we have already seen (e.g. in our proof of Theorem 1 above), the property of simple recursion (Theorem 29, Section 5.3) is all we need to characterize the natural numbers and to start the process of building up our number systems. In Chapter 8, we carry out the program of proving the Peano postulates assuming only the property of simple recursion.[†]

For the full development of mathematics, one needs at least two further canonical number systems, namely the reals R and the complex numbers C. The reals are characterized as the (up to isomorphism) unique complete (totally) ordered field. Completeness means that all Cauchy sequences converge or (equivalently for totally ordered fields) that every bounded nonempty set of elements of the field has a least upper bound. The complex numbers constitute the smallest algebraically closed field containing the reals and are not ordered (and not orderable since they are Cauchy complete and, if ordered, would therefore be isomorphic to R). The reals can be constructed from Q either as equivalence classes of Cauchy sequences under an appropriately defined equivalence relation or else as certain subsets of rationals (the so-called Dedekind cuts). Since these constructions are quite widespread in the literature, we do not give details here.

The complex numbers are constructible from the reals as ordered pairs (no restriction) with appropriately defined operations. Thus, the reals contain the rationals plus irrationals like

---

[†] It should probably be mentioned that the natural numbers and their property of simple recursion constitute the essential basis of abstract algebra as well. Simple recursion characterizes the natural numbers algebraically as the free Peano algebra on one generator. It allows us to construct free monoids over arbitrary sets and free word algebras of any type. All other algebraic structures are then forthcoming as appropriate quotients of word algebras.

$\sqrt{2}$, e, and π. The complex numbers contain the reals plus the roots of all real polynomials such as x^2-1, x^2-2, etc.

Once these number systems are at hand, one can construct function spaces such as R^X and C^X where X is any set. Though the resulting structures are usually not numerical ones (division cannot generally be performed), these vector spaces have enough regular structure to carry most of the modelling that mathematical analysis needs. The important point is that, since most of these structures of analysis are built up using our canonical number systems, they constitute a "core" of mathematics which is reproducible in one way or another within almost every foundational system. More will be said of this in Chapter 8.

The canonical number systems Z, Q, R, and C extend the natural numbers N in one direction, namely by adding supplementary operations (subtraction, division by nonzero elements, taking arbitrary roots of positive numbers, taking arbitrary roots of negative numbers). However, the order relation either does not exist (as in the case of C) or else no longer has the same properties as for N.

The transfinite ordinals extend ω in another way, namely by defining well-ordering on infinite sets of arbitrarily large cardinality. The sum and product of ordinals are also defined in such a way that they extend the sum and product operations on the finite ordinals (i.e. the elements of N). In fact, not only do the ordinals extend well-ordering to arbitrary infinite sets, they extend the property of simple recursion as well! We want to take a brief look at this.

We begin by defining the class Od of all ordinals. By **MKM**.0 we know $\vdash(Ex_1)(x_2)(x_2 \in x_1 \equiv M(x_2) \wedge On(x_2))$. Since every ordinal is a set (for example, every ordinal α is an element of its successor α'), then we have, in fact, the existence of a class Od such that $(x_2)(x_2 \in$ Od $\equiv On(x_2))$ holds. By extensionality, Od is the unique class having this property.

Since every ordinal α is, in fact, the set of all ordinals preceding it, Od is itself like an ordinal, except that, as we will see, it is a proper class and not a set. Od is $\in$-transitive, for given $a \in b \in$ Od, then $a \in b$ and $On(b)$ which implies (Theorem 18, Section 5.1) $On(a)$ and thus $a \in$ Od. We now prove that Od is also $\in$-connected.

THEOREM 2. $(x_1)(x_2)(x_1 \in$ Od $\wedge x_2 \in$ Od $\supset x_1 = x_2 \vee x_1 \in x_2 \vee x_2 \in x_1)$.

Proof. Let X be the class of all ordinals $x \in$ Od for which there exists an ordinal y not $\in$-comparable with x (i.e. such that neither $x = y$, $x \in y$, nor $y \in x$ holds). We want to show that X is empty. If it is not, then it has an $\in$-minimal element a by **MKM**.7. The class Y of all ordinals noncomparable with a is nonempty so it also has an $\in$-minimal element b. Thus, $a \neq b$, $a \notin b$, and $b \notin a$ all hold. Also, since $a = a$, a is comparable with a and so $a \notin Y$. We will obtain a contradiction from all of this.

We first establish that $a \subset b$. Let $z \in a$ hold. By Theorem 18, Section 5.1, $On(z)$. Since $a \cap X = 0$, $z \notin X$ which means that z is comparable with every ordinal. Thus, in particular, either $z = b$, $z \in b$, or $b \in z$ must hold. But $z = b$ implies $b \in a$ and thus that b is comparable with a, contradicting the above. If $b \in z$, then also $b \in a$ since Trans(a), again contradicting the above. Thus we must have $z \in b$. Since z is arbitrary, we conclude $a \subset b$.

We now prove also that $b \subset a$. Let z be any element of b. Thus $z \notin Y$ since $b \cap Y = 0$. Again $On(z)$ since $On(b)$. Thus z is comparable with a. Again, if $z = a$, then $a \in b$, contradicting the noncomparability of a and b. Similarly, if $a \in z$, then $a \in b$ by Trans(b), contradicting noncomparability. Thus, $z \in a$, and hence $b \subset a$. The two inclusions allow us to affirm

$a = b$ by extensionality. But this again contradicts the noncomparability of a and b. Hence X must be empty and Od is $\in$-connected.

COROLLARY. *Od is not a set.*

Proof. If Od is a set, it is an ordinal since it is Trans and Con. Hence Od $\in$ Od, contradicting Theorem 12, Section 5.1.

THEOREM 3. *If X is any nonempty class of ordinals, then there is exactly one $\in$-minimal member of X which is the unique smallest ordinal in X.*

Proof. Since Od is Con, so is any $X \subset$ Od. **MKM**.7 says that any such nonempty $X \subset$ Od will have an $\in$-minimal element a. Since a is comparable with every $\beta \in X$, then $a = \beta$ or $a \in \beta$ for every $\beta \in X$. Thus a is smallest and unique (cf. Theorem 15, Section 5.1).

COROLLARY. *The union of any class of ordinals is either* Od *or an ordinal. In particular the union of a set of ordinals is an ordinal.*

Proof. The union X of a class of ordinals is a subclass of Od and hence is $\in$-connected by Theorem 2. Trans(X) is immediate. If Od$-X \neq 0$, then, by Theorem 3, there is a smallest $\alpha \in$ Od$-X$ such that $\alpha \cap ($Od$-X) = 0$. Thus, $\alpha \in$ Od and $\alpha \subset X$ since $\alpha \subset$ Od. But $\alpha \notin X$, and so $\beta \in \alpha$ for every $\beta \in X$, i.e. $X \subset \alpha$. Hence $X = \alpha$ and On(X). Otherwise, Od$-X = 0$ and $X =$ Od. Finally, by **MKM**.6, the union of any set is a set. Since Od is not a set, the union of any set of ordinals must be an ordinal.

THEOREM 4. *Let $X \subset V$ be any class. If, for every set $a \in V$, $a \in X$ whenever every element of a is in X, then $X = V$.*

Proof. Let $Y = V - X$. If Y is nonempty it has, by **MKM**.7, an $\in$-minimal element m such that $m \cap Y = 0$. Thus, every element of m is in X. Hence $m \in X$, contradicting $m \in Y = V - X$. Thus, Y is empty and $V = X$.

COROLLARY. *Every property of sets which is true of every set when it is true of each of its elements is true of every set in the universe.*

Proof. Immediate from Theorem 4 by **MKM**.0.

Exercise. Formalize the statement of Theorem 4 and prove it logically equivalent to the "inverted" induction scheme: For any class $X \subset V$, if, for every $a \in V$, $a \in Y = V - X$ implies there exists $b \in a$ such that $b \in Y$, then $Y = 0$.

THEOREM 5. (Complete ordinal induction.) *Let X be any class such that, for any ordinal α, $\alpha \in X$ whenever every element $\beta \in \alpha$ is in X. Then* Od $\subset X$.

13

Proof. If $Od-X$ is nonempty, it has an $\in$-minimal element k by **MKM**.7. Thus, k is an ordinal none of whose elements are in $Od-X$. But every element of k is also an ordinal and thus an element of Od. Thus, every element of k must be a member of X. Hence, by hypothesis $k \in X$, contradicting $k \in Od-X$.

We now prove iterated induction for ordinals.

THEOREM 6. *Let X be any class such that* (1) $0 \in X$, (2) *for all ordinals α, $\alpha \in X$ implies $\alpha' \in X$, and* (3) *for all limit ordinals α, $\alpha \in X$ whenever every element of α is in X. Then, $Od \subset X$.*

Proof. If $Od-X$ is nonempty, let k be its $\in$-minimal element by **MKM**.7. Since $k \in Od$ it is an ordinal. Now every ordinal is either 0 or a successor or a limit. But $k \notin X$ so k cannot be 0. If k is a successor, $k = \alpha'$, then since $\alpha \in Od$, $\alpha \in X$ which implies $\alpha' = k \in X$, contradicting $k \notin X$. Finally, if k is a limit, then every element of k is in Od and no element of k is in $Od-X$ (k is $\in$-minimal). Thus every element of k is in X which, by hypothesis, implies $k \in X$, again contradicting $k \notin X$. Thus, all logically possible cases lead to contradiction and $Od-X$ must be empty.

Uses of transfinite induction are quite similar to uses of finite induction on finite ordinals. Whenever we have some property defined for each ordinal, we can use **MKM**.0 to obtain the class X of all ordinals for which the property is true. We can then apply one of our induction theorems to the class X to establish that the given property is true for all ordinals.

Similarly, if we want to prove that some property is true for every element of some class, one technique is to count the class with the ordinals in such a way that the original property on sets can be formulated as a property of ordinals via the counting relation. We can then apply transfinite induction on ordinals to prove that the original property is true for all the sets counted by the ordinals.

We now give transfinite versions of definition by recursion. In the next two theorems, we will be dealing with arbitrary functions F (these are classes of pairs whose first elements are unique). We can always extend such a function so that its domain becomes the universe V of all sets by picking some arbitrary set, say 0, and adding to F all pairs $\langle x, 0 \rangle$ where x does not occur as a first element of a pair in F. Thus, there is no loss of generality by supposing that an arbitrary function has domain V. Also, we will be interested in the restriction of given functions. Given any function F and any set X, the restriction of F to the domain X, noted $(F \mid X)$, is the function $F \cap (X \times I(F))$.

THEOREM 7. *Given any function F with domain V, there exists a unique function G with domain Od such that, for every ordinal α, $G``\alpha = F``(G \mid \alpha)$. We say that G is defined by recursion from F.*

Proof. We sketch the proof. The idea is that we can define the values of G inductively for each ordinal α since the value of G at α only depends on the values of G for ordinals $\beta \in \alpha$. For example, $G``0 = F``0$, $G``1 = F``(\{\langle 0, G``0 \rangle\})$, etc. More formally, we define for each ordinal $\alpha$ a function $G_\alpha$ whose domain is the set $\alpha$ and such that $G_\alpha``\beta = F``(G_\beta)$ for all $\beta \in \alpha$. We prove by transfinite induction that such a G_α exists for all $\alpha \in Od$ and that, for $\beta \in \alpha$, G_α is an extension of G_β. The desired function G is then easily obtained as the

$$\bigcup_{\alpha \in Od} G_\alpha.$$

The uniqueness of G follows immediately by transfinite induction.

We also have an iterative version of transfinite recursion.

THEOREM 8. *Given any set a, and two functions F and H each with domain V, there exists a unique function G with domain* Od *such that, for all* $\alpha \in$ Od: (1) $G``(0) = a;$ (2) $G``\alpha' = F``G``\alpha;$ (3) $Lim(\alpha) \supset G``\alpha = H``(G \mid \alpha).$

Proof. Essentially modelled on the proof of Theorem 7 except that we choose different modes of valuation for 0, successor ordinals, and limit ordinals.

The uniqueness of the function G defined by transfinite recursion allows us to define, from given functions, a new function on Od which satisfies the three recursion equations just as simple recursion allowed us to define arithmetic functions like addition, multiplication, etc. Indeed, it is clear that Theorem 29, Section 5.3, is a special case of the above Theorem 8 where we exclude infinite ordinals, thus excluding all limit ordinals, and where the domain of the defined function is therefore ω. In particular, if we drop the axiom of infinity from **MKM**, then Od $= \omega$.

We conclude our brief study of ordinal induction and recursion by defining recursively one of the most interesting functions with domain Od. By Theorem 8, there exists a unique function K on Od such that $K``0 = 0$, $K``\alpha' = \mathcal{P}(K``\alpha)$, and for

$$Lim(\alpha), \quad K``\alpha = \bigcup_{\beta \in \alpha} K``\beta.$$

We let W stand for the class which is the union $\bigcup_{\alpha \in \text{Od}} K``\alpha$ of all the sets $K``\alpha$.

A moment's reflection will lead one to see that K is nothing but the formal definition of the hierarchy of sets informally presented in the introduction to this chapter. We would therefore expect that $W = V$. This is, in fact, the case.

THEOREM 9. $W = V$, *i.e. every set* $x \in K``\alpha$ *for some* $\alpha \in$ Od.

Proof. It is an easy proof by transfinite induction (iterated version) that the hierarchy of the $K``\alpha$ is cumulative, i.e. if $x \in K``\alpha$, then $x \in K``\beta$ for all $\alpha \in \beta$. We now apply $\in$ induction to the class $W$. If $x$ is some set all of whose elements $y \in W$, then, for each $y \in x$, $y \in K``\alpha$ for some α. However, since the hierarchy is cumulative, there is some ordinal β such that $y \in K``\beta$ for all $y \in x$ (use the Corollary to Theorem 3). In other words, $x \subset K``\beta$. But then, by the definition of K, $x \in \mathcal{P}(K``\beta) = K``\beta'$. Hence, by Theorem 4, $W = V$.

Theorem 9 is important in giving us a feel for the complexity of **ZF**. It tells us, in effect, that the richness of the universe V of sets depends on the ordinals which are available in Od to serve as counters for the universe V.

We can also use the hierarchy K to give a precise definition to the rank of a set of V. Given any set x, we use **MKM**.0 to define the class X of all ordinals α such that $x \in K``\alpha$. Since any class of ordinals $X$ has a unique smallest member (Theorem 3), we define $\varrho``x =$ the smallest α such that $x \in K``\alpha$. $\varrho``x$ is the rank of x.

Exercise. Prove that $\varrho``x$ is a successor ordinal for any set x.

13*

5.8. Conway's numbers

Starting with the basic number system N, we have extended in two directions each giving an elegant theory. On the one hand, we add new operations in such a way that the properties of the basic operations of N (associativity, commutativity, etc.) remain unchanged. On the other hand, we construct a class Od of ordinals, including transfinite ones, which is well ordered and which we use for inductive arguments and recursive definitions. It is natural to wonder whether it is possible to bring these two types of extensions of N together in one single structure. The answer is that one can, and the proper construction was found by the British mathematician J. H. Conway (see Conway [1]). We close this chapter with a brief consideration of Conway's numbers.

In order to get a better idea of what is needed, let us reexamine the structure of the ordered fields Q and R. Each of these fields is *Archimedian* in the precise sense that there is some natural number $n \in N$ which is greater than any given element of Q or R. Formally: $(x)(x \in R \supset (En)(n \in N \wedge n > x))$ holds. The order relation $>$ is, of course, the usual order relation of the totally ordered field R. Imagine now that we have some totally ordered field F which contains N and which also contains transfinite ordinals and such that the order relation between ordinals in F is the same order as that between general elements of the field F (not all elements of the field will be ordinals). Then any transfinite ordinal, say ω, will be greater than every natural number n (and thus greater than Q and R if $R \subset F$). Since F is a field, ω will have a multiplicative inverse $1/\omega$. Since $\omega > n$ for every $n \in N$, $0 < 1/\omega < 1/n$ for $n \in N$, $n > 0$; i.e. $1/\omega$ is an *infinitesimal number*. An infinitesimal in an ordered field will be any number whose absolute value (definable in any ordered field) is greater than 0 but less than the reciprocal (multiplicative inverse) of every positive natural number (any ordered field must contain N). Thus, every non-Archimedian field will have both infinite numbers (numbers greater than every natural number) and infinitesimals (reciprocals of infinite numbers).

Since we are interested only in ordered fields, we will forget about trying to include the complex numbers in our global extension of N. However, we will include Q and R. We are thus interested in non-Archimedian (and therefore ordered) fields containing R.

If we want our extension to contain *all* transfinite ordinals, a second problem presents itself immediately. Od is a proper class. Thus, any field $F \supset$ Od will also be a proper class. We are thus looking for a non-Archimedian field definable on a proper class of objects! The fact that Conway conceived such a thing to be possible shows almost as much imagination as the actual construction itself. We now present this construction.

We construct an ordered hierarchy of numbers such that any given number is constructed by forming appropriate sets of previously constructed numbers. A number x will be representable as an ordered pair of sets $\langle L(x), R(x) \rangle$ read "the left of x" and "the right of x" respectively. We use variables x_L, y_L, etc., for arbitrary elements of $L(x)$ and similarly, x_R, y_R, etc., for $R(x)$. A number will actually be an equivalence class of such pairs under an appropriate equivalence relation. The pairs themselves are called *games*. The recursive rule for the construction of games and the corresponding recursive definition of the order relation "$\geqslant$" (greater than or equal to) are given as follows.

Rule 1. If L and R are any two sets of games such that no member of L is $\geqslant$ any member of R (we write $L \not\geqslant R$), then $\langle L, R \rangle$ is a game. All games are constructed this way.

Rule 2. $x \geqslant y$ if and only if no $x_R \leqslant y$ and no $y_L \geqslant x$ (we write $R(x) \nleqslant y$ and $L(y) \ngeqslant x$). $x \leqslant y$ if and only if $y \geqslant x$, and $x \nleqslant y$ means $\sim (x \leqslant y)$. $x < y$ means $x \leqslant y$ and $y \nleqslant x$.

We define two games to be equivalent, and write xEy, if $x \leqslant y$ and $y \leqslant x$ both hold. A number is an equivalence class of games under E.[†] Actually, to justify this definition we need to show that $\leqslant$ is a preorder (reflexive and transitive) on games, from which it follows immediately that E is an equivalence relation. We will presently do this, but we want first to examine and discuss our two rules. We defer for the moment, the definition of addition and multiplication of numbers.

It may seem astonishing at first that such simply stated constructions could yield a field of numbers defined on a proper class and containing the reals R and the ordinals. However, it becomes more reasonable when we consider that the construction is transfinitely iterated and that our hierarchy K defines the class V of all sets in a similarly succinct manner. Since all games are pairs, and all games are constructed from other games, the sets $L(x)$ and $R(x)$ of any game x will consist of pairs of sets of pairs of sets of pairs, etc. But how far "down" can we go? Since the axiom of regularity **MKM**.7 excludes the possibility of infinitely descending $\in$-chains, we must eventually reach the empty set of pairs. Thus, the first (and simplest) game is the pair $\langle 0, 0 \rangle$. This couple will represent the (real and ordinal) number 0. Beginning with $\langle 0, 0 \rangle$, we can form other games such as $\langle \{\langle 0, 0 \rangle\}, 0 \rangle$. This is a game since, with regard to Rule 1, no element of 0 can be $\leqslant$ any element of the left side since 0 has no elements. This game will represent the (real and ordinal) number 1.

In proving general properties of the numbers, we can use a straightforward $\in$ induction (Theorem 4, Section 5.7, and its corollary): Any property which is true of every game x when it is true of every x_L and x_R is true of all games. The "inverted" $\in$ induction is also useful (see exercise following Theorem 4, Section 5.7): For any property of games, if whenever the property holds for some game x, then there exists some x_L or some x_R for which the property also holds, then that property is false for every game (i.e. its negation is true for all games). We illustrate this immediately in the proof of a few properties of games.

THEOREM 1. *For all games x, $x \leqslant x$.*

Proof. If $x \nleqslant x$ for some x, then by Rule 2 either $x_L \geqslant x$ for some x_L or $x_R \leqslant x$ for some x_R. If $x_L \geqslant x$, then, again by Rule 2, $L(x) \ngeqslant x_L$. In particular, since $x_L \in L(x)$, $x_L \nleqslant x_L$ and the property of irreflexivity holds for x_L. If, on the other hand, $x_R \leqslant x$, then $R(x) \nleqslant x_R$ and $x_R \nleqslant x_R$. Thus, by inverted induction, irreflexivity is false for every game and $x \leqslant x$ for all games.

COROLLARY. *For all games x, $x E x$.*

Proof. Immediate by definition of E and Theorem 1.

THEOREM 2. *For all games x, $x \ngeqslant R(x)$ and $x \nleqslant L(x)$.*

Proof. If $x \geqslant x_R$ for some x_R, then by Rule 2 $R(x) \nleqslant x_R$ which yields $x_R \nleqslant x_R$, contradicting Theorem 1. Similarly, $x \leqslant x_L$ leads to $x_L \ngeqslant x_L$.

[†] Conway reserves the term "game" for a more general notion which we will call "pseudo-game".

THEOREM 3. $x \leqslant y \leqslant z$ implies $x \leqslant z$ (transitivity of $\leqslant$).

Proof. Since $y \leqslant z$ we cannot have any $z_R \leqslant y$ and hence, by induction, no $z_R \leqslant x$. Similarly, we can have no $x_L \geqslant z$. Thus, $x \leqslant z$ by Rule 2.

COROLLARY. *E is an equivalence relation.*

Proof. Immediate from its definition plus Theorem 1 and Theorem 3.

Our definition of a number as an equivalence class of games is now fully justified. Notice that in the above, we have never had to use the fact that $L(x) \not\geqslant R(x)$ for games. If we remove this condition from Rule 1, we obtain a more general class of couples, the *pseudo-games* (these correspond to the "games" of Conway). Thus Theorems 1–3 hold for pseudo-games as well as for games. However, the order relationship between $L(x)$ and $R(x)$ is necessary for the following:

THEOREM 4. *For all games* x, $L(x) < x < R(x)$ *and for any two games* x *and* y, *either* $x \leqslant y$ *or* $y \leqslant x$.

Proof. Exercise (use induction, Theorems 1–3 and Rules 1 and 2).

Thus, the numbers are totally ordered by the relation $\leqslant$.

Exercise. Prove that $\langle\{\langle 0, 0\rangle\}, \{\langle 0, 0\rangle\}\rangle$ is not a game. Prove that the game $\langle 0, 0\rangle \leqslant \langle\{\langle 0, 0\rangle\}, 0\rangle$

We now state the definitions of addition, multiplication, and opposition for games:

$x+y$ for $\langle L(x)+y \cup L(y)+x, \ R(x)+y \cup R(y)+x\rangle$; $\ -x$ for $\langle -R(x), \ -L(x)\rangle$; $\ x \cdot y$ for
$\langle\{x_L \cdot y + x \cdot y_L - x_L \cdot y_L \mid x_L \in L(x), \ \ y_L \in L(y)\} \cup \{x_R \cdot y + x \cdot y_R - x_R \cdot y_R \mid x_R \in R(x), \ y_R \in R(y)\}$,
$\{x_L \cdot y + x \cdot y_R - x_L \cdot y_R \mid x_L \in L(x), \ \ y_R \in R(y)\} \cup \{x_R \cdot y + x \cdot y_L - x_R \cdot y_L \mid x_R \in R(x), \ \ y_L \in L(y)\}\rangle$.

The proof that the class of all numbers forms an ordered ring with these operations involves some complicated inductive proofs but no essential difficulties. A few of the properties, such as the commutativity of addition, do not even necessitate inductive arguments but are immediate from the definitions. Details of these proofs can be found in Conway [1] and many of them also in Knuth [1]. The definition of multiplicative inverses is quite complicated, however, and we do not give it here.

It should now be clear how both the ordinals and the real numbers are included among the numbers. Ordinals will be represented by games of the form $\langle L, 0\rangle$ (i.e. with empty right set). The real numbers will be represented by precisely those games which satisfy Rules 1 and 2 with the additional proviso that if L is nonempty but with no greatest number, then R is nonempty with no least number and vice versa. This proviso excludes all infinite numbers and thus all infinitesimals. In short, we obtain an Archimedian field (and thus one containing the rationals Q). That the field actually contains the reals follows from the usual Dedekind cut construction.

Exercise. Construct explicitly a representation for $\frac{1}{2}$; for 2. In particular, show that $\frac{1}{2}$ can be represented as the simplest (i.e. first-constructed) number to lie between zero and one.

As can be seen from the above exercise, halving is the basic operation used to construct a dense order. Thus, the representation of the reals will be as dyadic expansions of the rationals.

There remains, however, one basic problem. We have defined the numbers as equivalence classes of games. However, it is not hard to see that any given number will have a proper class of elements. Since we cannot collect up proper classes, we cannot really form within **MKM** the field of all Conway numbers. The answer, of course, is to choose a canonical representative from each equivalence class. For example, using **MKM**.7, we can choose an $\in$-minimal member of each equivalence class. These can be collected up to form the field of all Conway numbers. Let us take a quick look at a more formal approach to the construction of the Conway numbers within **MKM**.

First, we define by transfinite recursion the following function:

$$Gp\text{``}\alpha = \left(\mathcal{P}\left(\bigcup_{\beta \in \alpha} Gp\text{``}\beta\right)\right)^2.$$

Thus, for example,

$$Gp\text{``}0 = \{0\} \times \{0\} = \{\langle 0, 0\rangle\}, \; Gp\text{``}1 = \{\langle 0, 0\rangle, \langle 0, \langle 0, 0\rangle\rangle, \langle\langle 0, 0\rangle, 0\rangle, \langle\langle 0, 0\rangle, \langle 0, 0\rangle\rangle\}, \text{ etc.}$$

(it grows rather quickly). This is clearly a cumulative hierarchy whose union

$$\bigcup_{\alpha \in \mathbf{Od}} Gp\text{``}\alpha$$

is precisely the class of all pseudo-games. We then define recursively the $\geqslant$ relation on pseudo-games and prove Theorems 1–3 for pseudo-games. We can then use $\geqslant$ to define the games, the equivalence relation E, and hence the numbers.

The class of pseudo-games is just as rich in structure as the class V of all sets as can easily be seen by comparing the recursive definitions of the two hierarchies K and Gp. We could, in fact, formulate a foundational system based on the Conway numbers and recover set theory as a special case of the numbers. In short, the Conway numbers can be viewed as an independent foundational system, a sort of absolute ordered field of numbers which can serve as a vehicle for all the usual set-theoretical notions.

Chapter 6

Hilbert's Program and Gödel's Incompleteness Theorems

WE HAVE delayed until this chapter the discussion of some basic questions dealing with the relationship between proofs within a given formal language and metamathematical reasoning *about* proofs within a language. The basic problem to be considered is the following one: Suppose we are given a formal language, such as **ZF** or **ST**, which is a first-order (or higher-order) system. How can we *prove* that such a language is or is not consistent?

To prove inconsistency it suffices to exhibit the deduction of a contradiction such as was done with our language **F** of Chapter 3. Since the notion of proof and theorem are clearly defined in such languages, the exhibition of a theorem which is a logical contradiction proves the inconsistency of the theory.

But what of consistency? To *prove* consistency we have to prove that it is impossible that any contradiction can ever be forthcoming as a theorem of our system. There are basically two ways that one can envisage doing this. The first method is based on the fact that a formal system is really a mathematical structure in its own right. To say that a contradiction is deducible in the system is to say something about the mathematical properties of the system. By engaging in a rigorous analysis of the structure of the given system, it is sometimes possible to prove that the conditions necessary to deducing a contradiction are not present. By an exhaustive analysis of the proof process, viewed as a mathematical operation, one shows that a contradiction cannot be forthcoming in the system. This method is known as *proof theory*.

The second method uses the fact that any system which has a model is consistent. Notice, though, that we must *prove* that a purported model really is a model if we are to *prove* consistency. The approach that employs models is a *model-theoretic* method of proving consistency.

It will already have occurred to the reader that there seems to be something circular in all of this. For clearly, a proof of consistency must proceed by logic and presumably by the same logic used as a basis for first-order theories. Thus, the question naturally arises as to whether we prove anything at all. Perhaps we prove nothing more than the fact that if our logic is consistent, then our logic is consistent. This trivial fact we already claim to know by the tautology $\vdash X \supset X$ where X is any statement.

Of course, on the one hand we have an intuitive logic in the metalanguage and on the other hand a formal mathematical structure which represents a logic under a suitable interpretation, but which has, as a formal system, a fairly simple mathematical structure. The more serious circularity arises when we realize that, in analyzing the mathematical structure of a formal system, we use mathematics. If the system whose consistency we are trying to prove is one such as **ZF** or **ST** in which the mathematics we are using in the metalanguage can be reproduced, then there does seem to be a real dilemma. This dilemma can be summed up as follows: To prove the consistency of a formal system we must use mathematics. Thus we prove only that the formal system is as consistent as the mathematics used to prove consistency. Moreover, if we use tools of mathematics that are at least as powerful as the mathematical methods expressible within the system, then it is not clear whether we really prove anything at all.[†] Let us say, for example, that we use the axiom of choice to prove that **ZF** is consistent when it includes the axiom of choice itself. Is this proof of any value?

One way out of this dilemma is to voluntarily restrict the mathematical tools used to prove consistency. We can try to restrict our methods to such obviously constructive principles that no possible doubt as to the cogency of the consistency proof is possible.

One example of such a consistency proof is to be found in Chapter 1. It is our proof of the consistency of any first-order predicate calculus (Theorems 3, Section 1.5). We showed first that the associated propositional form of any logical axiom is a tautology. We have a purely mechanical test to determine whether a wff is a tautology, and a purely mechanical way of finding the associated propositional form of any given wff. Moreover, the set of wffs itself is clearly defined so that we can give a mechanical test to decide whether a given expression is or is not a wff. Finally, we showed that our rules of inference preserve the tautological character of the associated propositional form. Since the theorems of any predicate calculus either are logical axioms or are obtainable from them by successive application of our rules of inference, it follows by an inductive argument that any theorem will have to be so constructed that its associated propositional form is a tautology. Since a contradiction does not have a tautology as its associated propositional form, it follows that no contradiction can be obtained as a theorem. Thus, consistency is proved.

Certainly we do not use any strong, nonconstructive principles of mathematics in this proof. However, we have used certain properties of the natural numbers, and in particular the principle of mathematical induction. (Although we have used the term "set", such as "set of expressions" or "set of wffs", it should be clear that these sets are defined in such a constructive way that general set theory, in the sense that we have been studying it, has really not been used at all.) A consistency proof that involves fewer tools of arithmetic than those we have here specified is difficult to imagine. Even though our wffs have been defined in such a constructive way, it is still true that they are infinite in number for any first-order theory. Thus, even simple properties of these systems will involve an appeal to arithmetic principles capable of dealing with infinite sets, such as the principle of induction for natural numbers.

Of course we must distinguish carefully between elementary number theory and number theory based on nonconstructive principles of analysis and set theory. Moreover, we have not yet tried to make precise the intuitive notion of "mechanical test" that was previously mentioned. This we shall do in our discussion of recursive functions. For present purposes, let us recall the language **S** of first-order arithmetic described in Chapter 1.

S is a first-order theory with one primitive binary predicative letter "$=$", two binary primitive function letters "$+$" and "$\cdot$", one singulary function letter "$\prime$", and one primitive constant letter "0". Intuitively, these represent the equality relation, the operations of addition and multiplication on natural numbers, successor (addition of 1) on natural numbers, and zero respec-

---

[†] By mathematical methods being expressible in a formal (first or higher-order) theory we mean, as usual, that there is a suitable model of the theory in which the mathematical truths under consideration are provable as theorems of the formal theory.

tively. The proper axioms for S consist essentially of the axioms for equality,[†] the Peano axioms for 0 and ′, and the recursive definitions of addition and multiplication (i.e. $x+0 = x \wedge x+y' = (x+y)'$, and $x0 = 0 \wedge xy' = xy+y$). Notice that this language is a great deal weaker than our set-theoretical languages, for we have no way of speaking about sets of natural numbers. The predicate "$\in$" of membership is not in S nor is it clear how we might define it. The axiom of induction is stated in the form $\vdash A(0) \wedge (x)(A(x) \supset A(x')) \supset (x) A(x)$ where $A(x)$ is any wff of S, and where $A(x')$ and $A(0)$ are like $A(x)$, except for containing x' and 0 wherever $A(x)$ contains free occurrences of x. On the other hand, the axioms of S are obtainable as theorems within our set theories as we have already seen.

The theorems of arithmetic that can be proved within S alone might be called *elementary* (as used here, this term means "first-order" and is not synonymous with "trivial"). Now, how might we prove the consistency of S? In the light of our prior remarks, it is clear that we would have to use arithmetic itself. We might protest that it is obvious that the natural numbers form a model of this system. However, a rigorous proof that shows this will again involve mathematical tools at least as strong as arithmetic (remember, proving that a structure is a model involves dealing with infinite sequences of elements of the domain, and other related considerations). Thus, the weakest possible assumption we can make is that the system S is consistent. This assumption is at least as weak as the assumptions necessary to prove that S is consistent. (The fact that it is actually strictly weaker than these other assumptions and that the arithmetic methods of S do not suffice to prove the consistency of S will be considered in the following discussion.) Since elementary number theory is a basic part of mathematics, we must have the consistency of S if mathematics is to be possible at all.

In the light of these comments, let us now say that a consistency proof of any system is an *absolute* proof of consistency if the proof is based on no stronger assumptions than the assumption that S is consistent. We now have a certain measure of what it means to restrict our mathematical tools of inquiry. Our general considerations on formal systems force us to use some arithmetic to prove consistency. Consequently, we choose the first-order arithmetic S as an absolute standard of restriction. The reader should know that attempts have been made to consider restrictions to certain weaker systems where the axiom of induction is not included. Such attempts are legitimate efforts to refine the exact measure of "absoluteness" for consistency proofs. We shall not enter into these rather detailed considerations here.

Notice that the consistency of any predicate calculus is absolutely provable in the sense we have defined it. On the other hand, some of our theorems of logic in Chapter 1 make use of general set theory and certainly involve principles that are stronger than those available within S. (The general definition of a model depends on set theory and so most of our theorems of model theory are nonconstructive in this sense.) It follows that these particular theorems of logic cannot be used in any absolute consistency proof. Certain simpler theorems of logic, such as the consistency of any predicate calculus, can be used in an absolute proof of consistency.

---

† Although the substitutivity of equality was not an axiom of S, we noted that it could be proved from the axioms for S as presented in Chapter 1, Section 1.9.

6.1. Hilbert's program

It was Hilbert who originally formulated the discipline of metamathematics, by which he meant the study of formal systems by weak, number-theoretic methods, and with the goal of giving an absolute consistency proof of a system in which all of mathematics can be deduced. Hilbert's goal was to justify certain nonconstructive principles of mathematics, such as the axiom of choice or definition by transfinite induction. His famous *program* consisted of first formalizing mathematics by means of a system such as **ZF** or **ST**, including all of the various nonconstructive methods. The consistency of the formal system would then be proved absolutely. Since all mathematics was presumed to be expressible in a system such as **ZF**, one would thus have proved absolutely that mathematics is consistent.

Hilbert and his disciples felt that such a program was *prima facie* reasonable, for we have already seen that the wffs of a formal system such as **ZF** are definable by constructive means, even if the interpretation of some of the wffs is not constructive. Since a consistency proof depends only on the mathematical structure of the system itself and not on the interpretation, it seemed natural to suppose that an absolute proof of consistency of any system might be possible.

The completion of such a program for a formalized version of mathematics would be a solution to the paradoxes, because it would show that unfortunate surprises such as Russell's paradox could not be lurking in the background. It would prove the system mathematically incapable of producing an inconsistency. During the period of 1920 to 1930, Hilbert's school worked diligently at the task of fulfilling Hilbert's goals.

It is interesting to contrast Hilbert's approach with the intuitionists who received brief attention in Chapter 3. Intuitionism can be schematically (though in a somewhat oversimplified way) described as the position that only number-theoretic methods are valid mathematical tools. The rest, intuitionists would assert, is just a fiction. Mathematics must be restricted to number theory. While rejecting the idea that nonconstructive mathematics was a fiction, Hilbert did accept the intuitionist restriction as being reasonable for metamathematics. In other words, in Hilbert's view mathematics does not reduce to number theory, but meaningful discussion of formal systems does.

The approach to foundations represented by Hilbert's program became known as *formalism*. Our study of formal systems in this book has not been really formalistic in the original meaning of the term, for we have not restricted ourselves in any uniform way to purely constructive methods in our discussion of formal systems. Although some of our theorems of Chapter 1 are indeed constructive, we have not made any general attempt to show that they are so (except in a few cases), or to exclude theorems which are not constructive. The point here is that formalism does not mean simply the study of formal systems, for these are surely legitimate mathematical objects in their own right. Rather, formalism means the study of formal systems by restricted methods, and with the goal of proving the consistency of systems which are adequate for mathematics.

6.2. Gödel's theorems and their import

It is well known that Gödel's incompleteness theorems resulted in the destruction of some of Hilbert's basic goals, but the exact way in which this is true is sometimes less clearly understood. Let us say that a first-order system F (or a higher-order system F) in which equality is definable is *sufficiently rich* if at least first-order arithmetic can be developed within F. Precisely, this means that within F there are functional relations or function letters for which we can prove the axioms of **S** as theorems of F, perhaps by using restricted quantification in F. In Chapter 5, Section 5.3, we showed that **ZF** was sufficiently rich by indicating how to express the notions of **S** in **ZF** in such a way that the axioms of **S** were translated into theorems of **ZF**. For any sufficiently rich first-order theory F, **S** has a model in F in the sense of Chapter 5, Section 5.3.

Clearly any system in any way adequate as a foundation for mathematics must be sufficiently rich.

We now summarize the two main Gödel results in the following way: (1) If we require that the set of proper axioms of a system be a decidable set (by which we mean roughly that we can decide by purely constructive means whether or not a given wff is or is not an axiom), then there is no way of constructing a sufficiently rich, consistent first-order (or higher-order) formal system in which all true statements of mathematics expressible within the system are provable as theorems. (2) The consistency of a sufficiently rich, consistent first-order (or higher-order) system F cannot be proved by methods that can be reflected or expressed in F. In particular, if **S** is consistent, then we cannot prove the consistency of **S** within **S**. Rather, we must use methods that go beyond those expressible within **S**, provided that **S** is consistent. There is, in our terminology, no absolute consistency proof for any consistent, sufficiently rich system.

Statements of result (1) sometimes ignore the relevance of the decidability of the axiom set. Actually, we shall need to sharpen the notion of "decidable set" by means of the theory of recursive functions, and this we shall do in our more detailed consideration of Gödel's theorems.

Moreover, result (2) is sometimes formulated by saying that it is impossible to prove the consistency of mathematics. Statement (2) does seem to destroy hope for a consistency proof based uniquely on the assumption that first-order arithmetic is consistent. It does not destroy the hope that a consistency proof might be given for mathematics that involves tools stronger than first-order arithmetic, while being less than the whole of mathematics itself. Gödel's statement of the matter in Gödel [2], p. 197, is: "I wish to note expressly that Theorem XI (and the corresponding results for M and A) do not contradict Hilbert's formalistic viewpoint. For this viewpoint presupposes only the existence of a consistency proof in which nothing but finitary means of proof is used, and it is conceivable that there exist finitary proofs that *cannot* be expressed in the formalism of P (or of M or A)." (For this translation, consult van Heijenoort [1], p. 615. Here "P" refers to the system **ST**, "M" to a system essentially the same as **NBG**, and "A" to **ZF**.)

Although Gödel admits the possibility of a finitary proof of consistency which cannot be reflected in the foregoing systems of set theory, no hopeful positive results have been obtained in this direction. (As we shall see in the next section, it is the notion of "reflecting" the proof within the system that is crucial.) Gentzen proved the consistency of **S**, but his method uses transfinite induction and thus involves the methods of general set theory. His proof is therefore not finitary.

In other words, Gödel's result (2) means that we cannot prove the consistency of a given system within the system itself. We must use means which cannot be expressed or reflected within the system. This point will become clearer when we have discussed the arithmetization of metamathematics by means of Gödel's numberings.

It was not only the result (2) that was discouraging to Hilbert, for result (1) showed the impossibility of formalizing all truths of mathematics. Whatever consistency proof may be forthcoming for a system such as **ZF**, we know that there are truths of mathematics that are expressible within the system, but that cannot be proved.

We now turn to the details of Gödel's argument in order to understand more clearly the method involved in obtaining these striking results.

6.3. The method of proof of Gödel's theorems; recursive functions

The first step in the demonstration of result (1) is to establish a 1–1 mapping **g** form the signs, wffs, and sequences of wffs of a given formal system to the natural numbers. The image of a particular sign, formula, or sequence of formulas under this mapping **g** will be called its *Gödel number*. Although we can formulate a procedure for assigning Gödel numbers to formal systems of a very general kind, we shall continue to think primarily of first-order theories (or simply structured higher-order theories). We will use a specific Gödel numbering for the system **S** as our specific example for this discussion.

The basic signs of any first-order theory are denumerable. We assign a unique odd number greater than 1 to each such sign. For **S**, we define: $g(0)$ is 3, $g(')$ is 5, $g(()$ is 7, $g())$ is 9, $g(+)$ is 11, $g(\cdot)$ is 13, $g(=)$ is 15, $g(x_n)$ is $15+2n$, where x_n is any variable.

Next, we recall that an expression is any finite sequence of signs, and that a wff is an expression. We assign Gödel numbers to expressions by the following device: Where $a_0 a_1 \ldots a_n$ is a finite sequence of basic signs a_i, $g(a_0 a_1 \ldots a_n)$ is $2^{g(a_0)} \cdot 3^{g(a_1)} \ldots p_n^{g(a_n)}$ where "p_n" stands for the nth prime in order of increasing magnitude, the 0th prime being 2 by definition.

For finite sequences of expressions, we do the same thing. If $x = a_0 a_1 \ldots a_n$ is a finite sequence of expressions, then

$$\mathbf{g}(x) \quad \text{is} \quad \prod_{i=0}^{n} p_i^{g(a_i)}.$$

Notice that our mapping **g** is 1–1 on the set that is equal to the disjoint union of the signs, wffs, and sequences of wffs. Also, given a sign, wff, or sequence of wffs, we can actually calculate its Gödel number. Conversely, we can decide whether a given integer is a Gödel number of some symbolic configuration and find the configuration in question if such is the case. One should also note that the Gödel numbers of signs are odd and those of expressions are even but with odd powers of primes in their prime factorization. The Gödel numbers of sequences of expressions are even with even powers in their prime factorization.

Exercise. Calculate the Gödel number of $(x_1 = x_1)$ and

$$(x_1 + x_2) = x_3.$$

Find the expression, *if any*, associated with 100, 8, 24, 536.

Notice that we distinguish between a sign and a one-element sequence of signs. The sign "(" and the one-element sequence of signs whose only element is "(" have different Gödel numbers.

Now that we have established our correspondence between arithmetic and the given formal language, we can observe that every property of the formal system gives rise to an arithmetic property and vice versa. The property "to be a wff of **S**", for example, will determine a unique set of expressions and in turn a unique set of numbers that represents an arithmetic counterpart of our property; namely, "to be the Gödel number of a wff". Notice, moreover, that we can represent the natural numbers within **S** itself. The number zero is represented by our constant "0" and, generally speaking, the number n is represented by the term $0''^{\cdots\prime}$ to n occurrences of the prime "$'$". We call these terms of **S** *numerals*. Given a number n, we will write "$\bar{n}$" to mean the numeral associated with n.

It should be clear that this particular way of assigning Gödel numbers is not the only way. Any 1–1 assignment that allows us to recover an expression from its number or vice versa is sufficient. Such a mapping is said to "arithmetize" metamathematics in that statements about the formal system can be reduced, in every case, to statements in number theory via the Gödel mapping.

Although many formal systems have a Gödel numbering, not every formal system has such an obvious representation of the natural numbers within it as does **S**. This point will be clearer after we introduce the notions of recursive function and recursive set.

We have used the same signs "$+$", "$\cdot$" and the like as formal signs in our system and informally in our metalanguage. We shall continue to do this as long as there is no ambiguity within a given context.

We now define the notion of recursive function. The following arithmetic functions (functions from N^m to N, where N^m is the m-fold cartesian product of N with itself) are called initial: The zero function $Z(x) = 0$ for all $x \in N$; the successor function $v(x) = x+1$, for all $x \in N$; and the projection functions $U_i^n(z_1, \ldots, z_i, \ldots, z_n) = z_i$ from N^n to N which associate a given coordinate with each n-tuple.

The following operations define an arithmetic function from certain given arithmetic functions: Substitution associates with n functions $h_1, \ldots, h_n$ of m variables each, and a given function g of n variables, a function f of m variables where f is defined by the rule

$$f(z_1, \ldots, z_m) = g(h_1(z_1, \ldots, z_m), h_2(z_1, \ldots, z_m), \ldots, h_n(z_1, \ldots, z_m)).$$

Primitive recursion associates with a function h of $n+2$ variables and a function g of n variables a function f of $n+1$ variables such that

$$f(z_1, \ldots, z_n, 0) = g(z_1, \ldots, z_n)$$

and

$$f(z_1, \ldots, z_n, y+1) = h(z_1, \ldots, z_n, y, f(z_1, \ldots, z_n, y)).$$

We have already discussed the question of the existence of such a function f and the necessity for an existence proof. We also allow the case in which f has no parameters z_i. In this case, f is a one-variable function satisfying the conditions $f(0) = k$ and $f(x+1) = h(x, f(x))$ where k is some given constant. In Chapter 5, we proved the theorem of *simple* recursion, $f(0) = k$ and $f(x+1) = h(f(x))$, within **ZF**. The existence proof for primitive recursion is not appreciably more difficult. (We made it the object of an exercise in Chapter 5.) We shall give a proof of this theorem within one of the foundational systems of Chapter 8.

Finally, we have the operation of minimalization that associates a function f of n variables with a given function h of $n+1$ variables by the rule $f(z_1, \ldots, z_n) = \min y(h(z_1, \ldots, z_n, y) = 0)$, where "min y" means "the smallest y such that". Of course it may be that such a minimum is not defined for a given h and for certain values $z_1, \ldots, z_n$. *We therefore restrict our allowed use of minimalization to those cases in which there is a minimum for every set of values* of the parameters z_i. (More liberal use of minimalization leads to a larger class of functions, the so-called partial recursive ones, but we will not enter into this here.)

We now define a *recursive function* to be any function that is either initial or that can be obtained from initial functions by a finite number of applications of our three operations (where minimalization is restricted in the indicated manner). A *recursive set* of natural numbers is defined as a set of natural numbers whose characteristic function (the function that is 0 on the set and 1 elsewhere) is recursive. A relation R of degree n among natural numbers is a set of n-tuples of natural numbers, and so we define a *relation to be recursive* if its characteristic function is recursive.

A *primitive recursive function* is a recursive function that can be obtained from initial functions without any use of minimalization. We define the notions of *primitive recursive set* and *relation* by adding the modification "primitive" in the corresponding definitions for recursiveness.

The intuitive notion of a recursive function is one whose values can be calculated by a machine or "effectively computed". Intuitively, our initial functions certainly can be calculated and our operations of primitive recursion, substitution, and minimalization appear to preserve computability. We then seem justified in saying that recursive functions are computable. When we speak of "computable", we mean "computable in principle", since it is obviously possible to consider recursive functions of such complexity that it would be practically impossible to compute certain values in a reasonable amount of time.

Similarly, a recursive set of natural numbers is intuitively a "decidable" set in the sense often used in this book, a set for which we can decide in some effective way whether a given object is or is not in the set. If our recursive functions are computable, then certainly recursive sets are decidable. Given any recursive set X and any natural number n, it suffices to compute the image of n under the characteristic function of X. Since the characteristic function is computable, we can effectively calculate its image. If the image is 0, then $n \in X$. If the image is 1, then $n \notin X$.[†] In this way we can effectively decide membership in the set.

Because of the correspondence between the linguistic objects of a formal system such as S and the natural numbers afforded by Gödel numbering, we can speak of a set of wffs as *recursive* if its corresponding set of Gödel numbers is recursive. Similarly, a relation among wffs or signs of a formal system is *recursive* if the corresponding relation on natural numbers is also recursive. In particular, our requirement that the axiom set of a formal system be "mechanically decidable" can be rigorously defined by requiring that its set of Gödel numbers be recursive. A system whose set of axioms is recursive is said to be *axiomatic*. All the foundational systems we have considered (or shall consider) in this book are axiomatic. S is also axiomatic.

Generally speaking, if we want to talk about sets of objects as recursive, we first establish a 1–1 function from the class of all the objects in consideration to the set of natural numbers.[‡] Then the notion of recursiveness carries over from the natural numbers in the same way as for our Gödel numbering of signs and expressions of a formal system.

[†] In analysis, the characteristic function is usually 1 on the set and 0 off the set, but we generally reverse the procedure in number theory. The reason stems from the minimalization rule and the fact that, when we treat characteristic functions in this way, a given problem often reduces to finding the zeroes of some function. In analysis, we want the characteristic functions to be 1 on the set so that the measure of a set will be the integral of its characteristic function.

[‡] This means, of course, that the class of objects in question must be countable and that the 1 − 1 function must be effective and have a recursive set of natural numbers as its range.

The reader, if he is encountering these considerations for the first time, will have some immediate questions. Are there recursive functions that are not primitive recursive? The answer is "yes", but we shall not prove it. Since a function is a certain set of n-tuples of numbers, one might ask the questions: is a recursive function a recursive relation, and is a functional relation that is recursive a recursive function? The answer is "yes" to both of these questions, and we shall leave the task of showing this to the reader (the result follows in a fairly straightforward fashion from the definitions).

Most of the functions that possess a name, such as $+$, $\cdot$, exp, !, and the like, are recursive. In Gödel's original article, it was the primitive recursive functions which were called "recursive" and our present more general notion was developed several years later.

We feel fairly justified in believing that recursive functions are computable. But what of the converse to this? Would we be justified in thinking that all computable functions are recursive? There is no clear answer to this question, for how can we say that we shall never encounter a function which we can compute but which is not recursive? We are comparing an intuitive notion (computability) with a precisely defined notion (recursiveness). It is analogous to the "$\varepsilon - \delta$" definition of continuity in analysis, which was designed as a rigorous definition to enable us to prove theorems about continuity. There can be no "equivalence" between our intuitive conception of a continuous function and such a definition. We are always free to revise the definition if it turns out that there are functions we wish to regard as continuous but which are not continuous according to our definition.[†]

Notice that we do have an intuitive notion of computability independent of our formal definition of recursiveness. This is especially clear with respect to our Gödel numbering itself. We recognize that the function $\mathbf{g}$, which gives us our Gödel numbering of signs and expressions, is computable in the sense that we can effectively recover an expression from its Gödel number and we can effectively calculate the Gödel number of an expression. Yet, we cannot appeal to the "recursiveness" of $\mathbf{g}$, since it is only by means of $\mathbf{g}$ that the notion of recursiveness is defined for formal expressions. In any case where we would speak of the recursiveness of a set of objects by means of a 1–1 mapping to the natural numbers, it would be necessary that the mapping be intuitively effective.

The thesis that all computable functions are recursive is known as *Church's thesis*. Church's thesis is rendered more probable by the fact that many different, independently conceived ways of defining computability, such as Turing machines and Markov algorithms, have all turned out to be equivalent. This tends to show that the intuition underlying all of these conceptions was essentially the same.

Exercise 1. Prove that the union, intersection, and complement of recursive sets are recursive. The set of all recursive sets thus forms a Boolean algebra which is a subalgebra of the Boolean algebra of all subsets of the natural numbers.

Exercise 2. Prove that there are denumerably many recursive functions and denumerably many recursive sets. There are thus continuum many nonrecursive functions and nonrecursive sets.

[†] There is, in fact, a slight variance to be found in the literature concerning continuity at isolated points of the domain of a real function. Most authors regard functions as trivially continuous at such points, but the fact is that our intuition of continuity fails us here, since we intuitively consider graphs of functions defined on connected sets.

Exercise 3. A set is called *recursively enumerable* if it is the range (image) of a recursive function, or the null set. Prove that every recursive set is recursively enumerable. There are recursively enumerable sets that are not recursive, though the proof of this fact is somewhat involved.

We wish now to see what use the proof of Gödel's result (1) makes of the notion of recursiveness, and why it is desirable to introduce these considerations. Let us note here that result (1), as we have stated it, is really attributable to Gödel and Rosser and is properly called the Gödel–Rosser theorem. Gödel's original result was somewhat weaker than result (1) and involved the notion of ω-*consistency*, which we have not discussed. The reader interested in consulting the original paper by Gödel should also consult the paper Rosser [1] in conjunction with it. It is only by the incorporation of Rosser's stronger result into our discussion that we avoid the consideration of the notion of ω-consistency. We do this not because ω-consistency is a difficult notion with which to deal, but because we are interested in having before us the full strength of result (1).

Given a relation R of degree n among natural numbers, we say that R is *expressible* in S if there is a wff $A(x_1, \ldots, x_n)$ with exactly the x_i as free variables, and such that the following conditions hold: (i) If the n-tuple $\langle y_1, \ldots, y_n \rangle$ of natural numbers is in R, then $\vdash A(\bar{y}_1, \ldots, \bar{y}_n)$. Here we have substituted the numeral for y_1 for the variable x_1, the numeral for y_2 for x_2, and generally $\bar{y}_i$ for x_i. (ii) If the n-tuple $\langle y_1, \ldots, y_n \rangle$ is not in R, then $\vdash \sim A(\bar{y}_1, \ldots, \bar{y}_n)$ where again we have made the indicated substitutions. If R is expressible in S, we say that the wff $A(x_1, \ldots, x_n)$ in question *expresses* R. The crucial importance of recursiveness lies in the fact that every recursive relation is expressible in S. To briefly see how this is true, we define the notion of a function being *representable* in S.

Given a function f from N^n to N, we say that f is representable in S if there is a wff $A(x_1, \ldots, x_n, x_{n+1})$, with exactly the free variables $x_1, \ldots, x_{n+1}$ such that (i) if $f(y_1, \ldots, y_n)$ is y_{n+1}, then

$$\vdash_S A(\bar{y}_1, \ldots, \bar{y}_n, \bar{y}_{n+1}),$$

where we have substituted each numeral $\bar{y}_i$ for the corresponding free variable x_i, and (ii), for every n-tuple $\langle y_1, \ldots, y_n \rangle$ of natural numbers, we have $\vdash_S (E! x_{n+1})(A(\bar{y}_1, \ldots, \bar{y}_n, x_{n+1}))$. If f is representable we say that the wff $A(x_1, \ldots, x_n, x_{n+1})$ in question *represents* f.

Exercise 1. Prove that a number theoretic function f is representable only if the relation $f(y_1, \ldots, y_n) = y_{n+1}$ is expressible in S.

Exercise 2. Prove that a number theoretic relation R is expressible if and only if its characteristic function is representable.

The result of Exercise 2 is important, for it shows that we can prove that recursive relations are expressible in S if we can prove that recursive functions are representable. It is easy to see that our initial functions are representable. The zero function is represented by the wff $x_1 = x_1 \wedge x_2 = 0$. Similarly, the successor function is representable by the wff $x_1' = x_2$. We leave it as an exercise to show how to represent the projection functions.

Once we have shown that the initial functions are representable, we can show that our operations of primitive recursion, substitution, and minimalization can be performed within the system S. This guarantees that recursive functions are representable: since the recursive functions are functions obtainable from the initial functions by a finite number of applications of these three rules, it follows that all recursive functions are representable. The rigorous proof is obviously by induction. (Actually, we do not need minimalization. It suffices to deal with primitive recursive relations and functions as it is done in Gödel [2].) Since a relation is recursive if and only if its characteristic function is, we obtain the result that all recursive relations are expressible in S.

It is also true that a relation is expressible in S only if it is recursive, but we shall not need this fact in the proof of Gödel's theorem.

Notice that our definition of the numerals in S is purely formal and does not depend on any interpretation of S. Similarly, the notion of a number-theoretic function being representable in S, or a number-theoretic relation being expressible in S, does not depend on an interpretation of S. Of course the interpretation of the numerals as names of numbers, of $+$ as addition, and so on, clearly motivates these definitions, but the definitions themselves do not depend on this interpretation. Later on we will see that certain parts of Gödel's results do depend on interpreting S, but there are significant results which do not depend on interpretations and we must be careful to distinguish these two kinds of results. The results that do not depend on interpretations will be proved first, and the reader is invited to check, as our argument progresses, that we do not appeal to the notion of an interpretation of S in obtaining these initial results. We will explicitly introduce the question of interpretations of S only when it will be needed to extend our results.

By means of our Gödel numbering, every relation between the signs, expressions, and sequences of expressions of the system S has a uniquely associated relation on natural numbers. Since our formal definitions of such notions as "axiom", "wff", "proof", and so on are constructive for S, it turns out that the corresponding arithmetic relations are recursive. Since recursive relations are expressible in S, we can "talk about" such things as provability in S within S itself. It is this possibility of expressing statements about S within S which will be useful in obtaining our results (1) and (2).

Let us first consider the result (1). We will obtain result (1) in stages, each stage representing a result of greater subtlety. The first of these results is the following, which we call result A: There is a closed wff (sentence) W of S such that neither W nor $\sim W$ is provable as a theorem of S if S is consistent. We call such a sentence W undecidable.[†] It is important to keep in mind that no appeal whatever will be made to any interpretation or meaning of sentences in S in proving result A.

Our method of obtaining result A is that of a diagonal argument. We succeed in exhibiting a wff of S which intuitively asserts its own unprovability in S. However, as we have already stressed, no appeal will be made to any interpretations in the proof of result A. As an example of the kind of diagonal process involved, consider the set K of natural numbers defined by $K = \{g(A(x_1)) \mid \vdash \sim A(g(A(x_1)))\}$, where $A(x_1)$ is a wff with exactly x_1 as a free variable}. A natural number is in the set K if and only if it is the Gödel number of a wff $A(x_1)$ with x_1 as the sole free variable and such that the wff obtained by substituting the numeral $g(A(x_1))$ for x_1 and negating the result is provable. If K were a recursive set, then S would be inconsistent.

[†] The term "undecidable" here is predicated of individual wffs and is not the opposite of "decidable" as used in connection with decidable or recursive sets.

To show this, suppose K recursive and thus expressible in S by a wff $\overline{K(x_1)}$ with exactly x_1 as a free variable. Then, $g(K(x_1)) \in K$ implies that $\vdash \sim K(g(K(x_1)))$ by the definition of K, and that $\vdash K(g(K(x_1)))$ by the fact that $K(x_1)$ expresses K. This yields the inconsistency of S, since now a wff and its negation are both provable. On the other hand, $g(K(x_1)) \notin K$ implies $\vdash \sim K(g(K(x_1)))$ by the fact that $K(x_1)$ expresses K. But this latter fact implies $g(K(x_1)) \in K$ by the definition of K and this in turn implies that S is inconsistent as we have seen.

No way is known to prove that K is expressible in S, and it cannot be expressible in S if S is consistent. However, our proof of the result A will consist of exhibiting a diagonal relation similar to K which is recursive and thus expressible in S. This recursive diagonal relation is weaker than K and consequently we do not obtain the inconsistency of S. Rather we obtain the undecidability of a certain exhibitable wff of S if S is consistent. Let us examine this in greater detail.

After a long series of constructions, we can succeed in exhibiting a demonstrably recursive relation $\text{Bew}(x, y)$ which holds between two natural numbers x and y if and only if x is the Gödel number of a wff $A(x_1)$ containing x_1 as sole free variable, and y is the Gödel number of a proof of the wff $A(\bar{x})$ obtained by replacing the numeral $\bar{x}$ of x for the free variable x_1 in $A(x_1)$. Just as easily, we can also define the recursive relation $\text{Bew}\#(x, y)$ which holds if and only if y is the Gödel number of a proof of $\sim A(\bar{x})$. This last wff is the negation of the result of substituting $\bar{x}$ for the variable x_1 in $A(x_1)$, x being the Gödel number of $A(x_1)$.

Notice that the defining relation of the set K is precisely the relation $\text{Pr}\#(x)$, "there exists a y such that $\text{Bew}\#(x, y)$". Yet, as we have already remarked, $\text{Pr}\#$ is not recursive if S is consistent, whereas $\text{Bew}\#$ is demonstrably recursive. We can see intuitively the difference between these two relations if we think how we would go about deciding whether arbitrary natural numbers satisfy them. In the case of $\text{Bew}\#(x, y)$, suppose we are given an ordered pair $\langle x, y \rangle$ of natural numbers. We have already seen that we can effectively decide whether x is a Gödel number of an expression and that we can recover the expression that the Gödel number represents if it is. We can thus decide whether x is the Gödel number of some wff $A(x_1)$ with exactly x_1 as a free variable. If such is the case, we can exhibit $A(x_1)$. Given y, we can likewise determine whether it is the Gödel number of a sequence of wffs, and if such is the case, we can determine the sequence. We can check the exhibited sequence to see whether it is a proof, for the notion "a sequence of wffs is a proof" is constructive in S. In fact, this notion is constructive in any first-order theory whose set of axioms is recursive. Finally, we can see whether $\sim A(\bar{x})$ is the last line of the proof. We can thus decide whether or not the pair $\langle x, y \rangle$ is in the relation $\text{Bew}\#$. But for $\text{Pr}\#$, we must decide for a given $A(x_1)$ whether or not there is a proof, and this we cannot do in general. Unless we have a theory whose theorems form a decidable set, we cannot decide whether an arbitrary wff is provable or not. Let us now see how we obtain a wff that does allow for a diagonal argument.

Since $\text{Bew}(x, y)$ and $\text{Bew}\#(x, y)$ are recursive, they are expressible in the system S by two wffs. Let us call them $B(x_1, x_2)$ and $B\#(x_1, x_2)$ respectively. (Notice again that we can effectively exhibit these two wffs by reconstructing within S the definition of the recursive relations.) Let $D(x_1)$ be the wff $(x_2)(B(x_1, x_2) \supset (Ex_3)(x_3 \leqslant x_2 \wedge B\#(x_1, x_3)))$ (which can also be effectively exhibited). This says, intuitively, that if x_2 is the Gödel number of a proof of Y, then there is a number x_3 that is less than or equal to x_2 and that is the Gödel number of a proof of the negation of Y, where Y is the result of taking the wff with Gödel number x_1, etc. Again, we

14*

point out that no appeal whatsoever to this or any other interpretation of $D(x_1)$ will be made in proving result A. The reader should carefully check that this stipulation is maintained.

Let W be the closed wff $D(\bar{d})$ that results from $D(x_1)$ by substituting $\bar{d}$ (d is the Gödel number of $D(x_1)$) for x_1. W can be effectively exhibited, since $D(x_1)$ can be exhibited and since the Gödel number d of $D(x_1)$ can be effectively calculated. We now assume that S is consistent, and we wish to establish that W is undecidable. If W is provable, then $\vdash D(\bar{d})$, that is, $\vdash (x_2)(B(\bar{d}, x_2) \supset (Ex_3)(x_3 \leqslant x_2 \wedge B \# (\bar{d}, x_3)))$. The proof of W must have some Gödel number; let us call it y. By the definition of Bew, it follows that $\mathrm{Bew}(d, y)$ must hold. But $B(x_1, x_2)$ expresses Bew, and so we have $\vdash B(\bar{d}, \bar{y})$. Applying $e\forall$ to W, we obtain $\vdash B(\bar{d}, \bar{y}) \supset (Ex_3)(x_3 \leqslant \bar{y} \wedge B \# (\bar{d}, x_3))$. Applying *modus ponens* to this and $\vdash B(\bar{d}, \bar{y})$, we obtain

$$\vdash (Ex_3)(x_3 \leqslant \bar{y} \wedge B \# (\bar{d}, x_3)).$$

Since S is assumed consistent, it cannot be that $\sim W$ is provable, for we would have a contradiction with the assumption that $\vdash W$. This tells us that $\mathrm{Bew} \# (\bar{d}, z)$ is false for all z, and thus in particular for y. Hence,

$$\vdash \sim B \# (\bar{d}, \bar{z})$$

is provable for all z, since it is $B \# (x_1, x_2)$ which expresses $\mathrm{Bew} \#$. In particular, then, we can prove $\vdash \sim B \# (\bar{d}, \bar{0})$, $\vdash \sim B \# (\bar{d}, \bar{1})$, and so on up to $\vdash \sim B \# (\bar{d}, \bar{y})$ (here is where the finite limit on the existential quantifier in $D(x_1)$ is essential). We can prove, then, that $\vdash (x_3)$ $(x_3 \leqslant \bar{y} \supset \sim B \# (\bar{d}, x_3))$ holds, which contradicts the provability of $(Ex_3)(x_3 \leqslant \bar{y} \wedge B \# (\bar{d}, x_3))$ already established. Thus, our assumed provability of W leads to contradiction with the assumed consistency of S.

Suppose, on the other hand, that $\sim W$ is provable. Let v be the Gödel number of a proof of $\sim W$. By the definition of the relation $\mathrm{Bew} \#$, we have that $\mathrm{Bew} \# (d, v)$ holds. Hence, $\vdash B \# (\bar{d}, \bar{v})$. If S is consistent, then W is not provable (as we have shown in the preceding paragraph) and so for all z, $\mathrm{Bew}(d, z)$ is false. Thus, $\sim B(\bar{d}, \bar{z})$ is provable for all z. In particular, we can prove $\vdash \sim B(, \bar{d}\bar{0})$, $\vdash \sim B(\bar{d}, \bar{1})$, ..., $\vdash \sim B(\bar{p}, \bar{v})$. Putting these results together, we obtain $\vdash x_2 \leqslant \bar{v} \supset \sim B(\bar{d}, x_2)$. Since $B \# (\bar{d}, \bar{v})$ is provable, we also have $\vdash \bar{v} \leqslant x_2 \supset (Ex_3)$ $(x_3 \leqslant x_2 \wedge B \# (\bar{d}, x_3))$. Now $x_2 \leqslant \bar{v} \vee \bar{v} \leqslant x_2$ is a theorem of S,[†] and so we immediately have that $\vdash \sim B(\bar{d}, x_2) \vee (Ex_3)(x_3 \leqslant x_2 \wedge B \# (\bar{d}, x_3))$, from which it follows by $i\forall$ that $\vdash (x_2)(B(\bar{d}, x_2) \supset (Ex_3)(x_3 \leqslant x_2 \wedge B \# (\bar{d}, x_3)))$. Thus, we have shown that W is provable, contradicting the consistency of S. Hence, our assumed provability of $\sim W$ is false.

We can now state result A: if S is consistent, then it is incomplete. There is a closed wff (sentence) of S such that neither the sentence nor its negation is provable. If we agree to assume the consistency of S, we can state simply: First-order arithmetic is incomplete.

Let us assume that S is consistent for the remainder of Section 6.3.

The method of our proof clearly shows why Gödel's results hold for sufficiently rich theories. The method of proving incompleteness depends on our ability to express recursive relations in S, and this in turn requires that S have a certain internal structure. We have already noted

---

[†] In Chapter 1, Section 1.9, we discussed some simple properties of S, though we did not prove the simple properties of the relation "$\leqslant$" which we use here. However, these theorems are not difficult to obtain by carrying out in S the usual proof given in informal number theory. The reader may take it as an exercise to fill in these gaps if he desires.

that our systems of set theory are sufficiently rich and thus permit as adequate a derivation of Gödel's results as does the weaker system S. More will be said about this later on.

Essential in the proof of the expressibility of the relations Bew and Bew# is the fact that the notion of provability is constructive in S. This in turn depends essentially on the fact that our axioms form a recursive set. Furthermore, the undecidable wff W can be effectively exhibited, as we have pointed out. These observations are all relevant to the following natural question: Why not try to complete our system by adding either W or $\sim W$ as a further axiom? Theorem 7 of Chapter 1, Section 1.5 shows that if any closed wff X is unprovable in a system F, then we can consistently add $\sim X$ as an axiom. Since both W and $\sim W$ are unprovable in S, we can consistently add either one!

We can indeed consistently add either W or $\sim W$ as a new axiom, but the resulting system will still be incomplete. The reason for this is that all the requirements necessary to prove the existence of another undecidable sentence will still be present. Since we are dealing with an extension of S, we still have all the expressive power of S. Moreover, our new set of axioms will still be recursive, for we shall have added only one new axiom whose Gödel number we can calculate (since the new axiom can be effectively exhibited).

Let S^* be the system resulting from S by adding, let us say W, as an axiom. Then we can again define our diagonal wff $D^*(x_1)$ in S^*, and deduce the existence of a new undecidable sentence W^* by exactly the same argument as before. Of course, W^* will not be equivalent to W. As a matter of fact, W^* will be undecidable in S as well as in S^*. Otherwise, either W^* or $\sim W^*$ is provable in S. But S^* is an extension of S and so every statement provable in S is provable in S^*. Thus, either W^* or $\sim W^*$ is provable in S^*, contradicting the undecidability of W^* in S^*.

Since W^* was obtained in S^* by the same process that yielded W in S, we can also exhibit W^* and obtain an extension S^{**} of S by adding W^* as a new axiom. S^{**} will also be axiomatic and we obtain a new undecidable sentence W^{**}. We can iterate this process any finite number of times and thus obtain a denumerable infinity of extensions of S, each obtained from a former by adjoining an undecidable sentence as an axiom. All these undecidable sentences will be undecidable in S, and they will be pair-wise nonequivalent. We thus obtain result B: S contains a denumerable infinity of nonequivalent undecidable sentences.

We can obtain another result by introducing the notion of *recursive axiomatizability*. A first-order theory F is said to be *recursively axiomatizable* if there is a theory with the same signs, wffs, and theorems of F and whose axioms form a recursive set. We now observe that the conditions necessary to prove the existence of an undecidable sentence will be present in any consistent, recursively axiomatizable extension of S. The machinery necessary to represent our numbers by numerals and to express recursive sets is present. Moreover, the axiom set will be recursive. We can thus state result C: Any consistent, recursively axiomatizable extension of S is incomplete.

We call any system such as S, which is incomplete and recursively axiomatizable, and such that every consistent, recursively axiomatizable extension is also incomplete, *essentially incomplete*.

Let us take this opportunity to emphasize the importance of the notion of recursive axiomatizability. In Chapter 1, we proved that every consistent first-order theory has a consistent, complete extension (see Section 1.5, Theorem 8). Result C tells us that no such consistent, complete extension of S will even be recursively axiomatizable.

If we accept Church's thesis, we can formulate result C in more colloquial terms: As long as we extend S in any way that still permits us to recognize which wffs are axioms, the extension will be incomplete if it is consistent.

Our results A, B, and C have been formulated without any reference to interpretations or possible interpretations of S. Indeed, it can be shown that the whole proof that demonstrates that S has an undecidable sentence can be carried on within S itself. We shall return to this point when we discuss Gödel's result (2). For the moment, we are interested in the fact that results A, B, and C do not depend on general set theory, but only on methods available within S itself. This is like discussing the grammar of English within English, so to speak.

To obtain result (1) as we have presented it, we must now consider the general logical theorems dealing with interpretations originally discussed in Chapter 1. These theorems make use of general set theory. The reader may suppose, if he wishes, that we are now working within a given set theory such as ZF. It will be seen that this does not, in any way, alter the quality of the results we obtain.

Now one of our general theorems of logic, provable within a set theory such as ZF, is that a closed wff (sentence) of any first-order theory F must be either true or false under any interpretation of F (the reader is referred to our set-theoretic definitions of these notions in Chapter 1). Moreover, if a given wff X of F is closed, then $\sim X$ is also closed, and either X or $\sim X$ is true under any interpretation. Now result B tells us that we have an infinity of nonequivalent undecidable sentences in S. Since each of these sentences X is undecidable, neither X nor $\sim X$ is provable in S, though one of the two is true for any given interpretation. Thus, result D: Under any model whatsoever of S, S has a denumerably infinite number of nonequivalent, true, but unprovable sentences.

Now, what types of structure are models of S? The most obvious one is the set N of natural numbers itself. After all, the axioms of S are nothing but the Peano postulates plus the logic of the predicate calculus and equality, and the recursive definitions of addition and multiplication. Of course, we cannot prove within S that N really is a model, but we can prove it within a general set theory such as ZF. We call N the *standard model* or *standard interpretation* of S. In the standard interpretation, $=$ is equality, $+$ is addition on natural numbers, $\cdot$ is multiplication on natural numbers, $'$ is the successor function (addition of one), and 0 is zero. The numerals in S become names of natural numbers; i.e. $\bar{n}$ is the name of n for all natural numbers n.

Under the standard interpretation (which was the way we always intuitively thought of S anyway), every sentence of S represents an assertion which is either true or false of the natural numbers. Applying result D to the standard interpretation, we thus have result E: First-order arithmetic S contains a denumerably infinite number of nonequivalent true statements about the natural numbers which are not provable within S.

We are now approaching result (1) in its fullness. We need one more essential observation: Our proof of incompleteness can be carried through in systems other than S. To obtain our various results for a given system F, it is obviously necessary that F satisfy certain structural conditions. We must be able to represent the natural numbers by numerals in F, to express recursive relations in F, and the like. We shall not try to find the exact criteria in this discussion. However, it is worthwhile noting that there are systems somewhat weaker than S which satisfy these criteria, and for which we can deduce our incompleteness results. In particular, R. Robinson [1] has given a system with only a finite number of axioms for which Gödel's incompleteness results can be demonstrated. Robinson's system has the same signs as **S**, but the axiom scheme of induction is not assumed. Because the scheme of induction is dropped, Robinson assumes a few additional, explicit properties of the operations of addition and multiplication, as well as of the "less than" relation. Robinson's system is strictly weaker than **S**, and consequently **S** is not the absolutely minimal system for which Gödel's results are provable.

In any case, what we can observe is that all the criteria necessary to deduce the Gödel results are present in any system strong enough to be considered a foundation for mathematics. In all the systems we have treated in this book, we have either proved the Peano postulates and defined arithmetic operations or indicated that this could be done (except for a few weaker systems, which we explicitly singled out, such as **ST** without the axiom of infinity). Each of these systems is sufficiently rich in the sense defined at the beginning of Section 6.2. Moreover, they are all recursively axiomatizable, and in fact, recursively axiomatized. It therefore follows that all our results apply to any of the systems we have discussed, such as **ST**, **ZF**, **NBG** (as well as to the systems we shall consider in later chapters). In fact, Gödel used a form of **ST** in his original demonstration of his incompleteness results.

We thus obtain result (1): It is impossible to construct a consistent, sufficiently rich, recursively axiomatizable (first- or higher-order) system in which all true statements of mathematics that can be formulated within the system are provable.

The result (1), which uses some of our general logical theorems of Chapter 1 that are themselves dependent on general set theory, can be deduced only within some foundational system. We have supposed, as an example, that we were working within **ZF**, but we can obtain the same results within any foundational system strong enough for general set theory.

Within, let us say, **ZF**, we can define a structure that will represent the formal system **ZF** itself. (We can do this by means of number-theoretic relations via Gödel numberings, if not otherwise.) Result (1) then says that we can obtain within **ZF** a proof that if **ZF** is consistent, then there are true statements of **ZF** not provable within **ZF**. What happens, one may ask, if **ZF** is not consistent? Then everything is provable within it, and we can then prove both result (1) and its negation. But if **ZF** is consistent, then we can prove result (1) and not its negation. With these metamathematical observations in mind, we can state the following: If **ZF** is consistent, then result (1) holds and is provable within **ZF**. But result (1) says that if **ZF** is consistent, then it has true but unprovable statements. Thus, our new statement is just a restatement of result (1). The same situation is manifest for any system capable of reproducing general set theory. Consequently, we see that the quality of result (1) does not depend on a particular system of general set theory nor is it lessened by the fact that we use set theory in obtaining it.

Exercise. For Boolean algebraists: In an earlier exercise (see the last exercise in Section 1.6 of Chapter 1, p. 56) we defined the Lindenbaum algebra of a first-order system F to be the Boolean algebra of equivalence classes of wffs of F under the equivalence relation $\vdash \ldots \equiv$ ---. The Lindenbaum algebra of closed formulas is the subalgebra we obtain by restriction to closed wffs (show that this is a Boolean algebra). Prove that the Lindenbaum algebra of closed wffs of any essentially incomplete first-order system F is the free one on $\aleph_0$ generators. (*Hint*: Use the fact that the only denumerable, atomless Boolean algebra is the free one on $\aleph_0$ generators.)

We recall that a formal system is said to be decidable if its set of theorems is decidable (the term "decidable" was used in an intuitive sense). Now that we have the precise notion of a recursive set of wffs by means of our Gödel numbering, we can say that a first-order system is *recursively decidable* or simply *decidable* if its set of theorems is recursive. If the set of theorems of a system is not recursive, we say that the system is *recursively undecidable* (or simply *undecidable*). It can be shown that, provided S is consistent, S is recursively undecidable, and in fact that every consistent extension of S is recursively undecidable. A system with this latter property is said to be *essentially recursively undecidable*.

The method of proof involved in showing that S is recursively undecidable is as follows: We take, on natural numbers, the set defined by the property "x is the Gödel number of a wff provable in S". We show that the set T of natural numbers determined by this property is not expressible in S if S is consistent. Consequently, T is not recursive, since every recursive set is expressible in S. But T is precisely the set of Gödel numbers of the theorems of S.

Exercise. Let S_N be a system which has the same signs and wffs as S and whose theorems are precisely the closed wffs of S which are true in the standard model. Prove that S_N is not recursively axiomatizable.

We now proceed to our discussion of result (2).

We have already noted that result A is purely number theoretic, and can be reproduced within the system S itself. To begin with, we can formulate a wff of S called *Consis*, which, when interpreted as to content by means of the standard interpretation and our Gödel numbering g, says that S is consistent. *Consis* will, of course, be a statement of number theory under the standard interpretation. It will be an extremely complicated wff asserting the consistency of S by talking about the Gödel numbers of proofs of wffs and their negations and the impossibility of obtaining any wff and its negation as a theorem.

We can also formulate within S a number-theoretic statement P which asserts under the standard interpretation that neither W nor $\sim W$ is provable. To say that result A can be reproduced in S means precisely that the wff ($Consis \supset P$) can be proved in S. This is true because ($Consis \supset P$) says, under the standard interpretation, that if S is consistent, then neither W nor $\sim W$ is provable.

Now, let us think of the interpretation of W itself under the standard model for S. W asserts that if there is a number which is the Gödel number of a proof of W, then there is a number (less than or equal to the first one) which is the Gödel number of a proof of $\sim W$. This assertion is a conditional statement, and it is true only when the antecedent is false or the consequent is true. Now, if S is consistent, then W is not provable. Consequently, the antecedent is false, which makes the statement true. Thus, if S is consistent, W is true under the standard interpretation. (Of course, W is closed, and thus either W or $\sim W$ must be true under any interpretation. We have shown only that W is the one that is true under the standard interpretation if S is consistent.) In other words, the wff ($Consis \supset W$) is true under the standard interpretation.

But it is not difficult to see that the proof of ($Consis \supset W$) can also be carried through in S, just as the proof of ($Consis \supset P$) can be. In fact, the wff ($P \supset W$) means, under the standard interpretation, "If neither W nor $\sim W$ is provable, then if W is provable there exists a proof with a smaller Gödel number of $\sim W$", which is true. Moreover, $\vdash_S P \supset W$ as is clear. Thus, under the hypothesis *Consis*, we can obtain P, and from P we can obtain W. This yields $\vdash_S (Consis \supset W)$.

Now, suppose that we can prove the consistency of **S** within **S**. Suppose, in other words, that $\vdash_S$ *Consis*. Then, by *modus ponens* applied to $\vdash_S (Consis \supset W)$, we obtain $\vdash_S W$. But if **S** is consistent, W is not provable. Thus, if we can prove *Consis* within **S**, **S** is inconsistent. However, if we assume that **S** is consistent, it follows that *Consis* is not provable in **S**. In other words, we have result (2): The consistency of **S** is not provable within **S** if **S** is consistent.

That result (2) holds, as originally stated, for any consistent, sufficiently rich system is clear. All the machinery necessary for reproducing our argument for **S** as presented in the foregoing paragraphs will be present in such systems. Result (2) is often expressed by saying that we cannot prove the consistency of a system within the system itself.

Of course, we must be careful, for the interpretation of *Consis* as the wff that asserts within **S** the consistency of **S** depends on our way of reflecting within **S** metamathematical statements about **S** by means of our mapping **g**. Conceivably, there are other statements affirming the consistency of **S** which are provable within **S**. The work of Solomon Feferman in this direction has led to some positive results. However, what we can now say is that we cannot prove the consistency of **S** by any means that can be reflected in the same system, on pain of contradiction. Thus, the more careful formulation of result (2) is: We cannot prove within **S** or any other system capable of reproducing at least number theory that **S** (or the particular system under consideration) is consistent by any means that can be reflected or reproduced within the same system by our process of Gödel numbering.

Although this result does leave some hope of finding an absolute consistency proof of a foundational system such as **ZF**, the hope is slight, and no really encouraging results have been obtained that would suggest optimism. Yet, we cannot wholly exclude the possibility of finding a consistency proof of a foundational system with some type of restricted quasi-constructive tools. The mathematician G. Takeuti, and others, are working on this very problem.

6.4. Nonstandard models of S

We assume the consistency of **S** throughout Section 6.4.

Consider again the system **S** and our original undecidable sentence W. By our Theorem 7 of Chapter 1, Section 1.4, we can consistently add as an axiom to **S** either W or $\sim W$, as we have previously demonstrated. Though we can add either W or its negation, only one of the two will be true in the standard interpretation. As a matter of fact, we have already shown that it is W. Let **S**$'$ denote the system obtained from **S** by adding $\sim W$ as an axiom. If **S** is consistent, so is **S**$'$, and any consistent system has a model by our completeness theorem. Moreover, by the Löwenheim–Skolem theorem, **S**$'$ has a denumerable model, and thus a model of the same cardinality as N. But this model cannot be the natural numbers, for the wff $\sim W$ which is now an axiom of **S**$'$ must be true in any model of **S**$'$, and $\sim W$ is false in the standard model. But any model of **S**$'$ (or any extension of **S** for that matter) is *a fortiori* a model for **S**. This gives us result F: There exists for **S** a denumerable model other than the natural numbers.

We can make the "other than" of result F much more precise. Let us recall that two models of a given first-order theory are isomorphic if there is a 1–1 mapping from the domain of one model onto the domain of the other which preserves all relations and operations in both directions. Two models of a given theory are elementarily equivalent if exactly the same wffs of the theory are true in both models. Clearly, isomorphic models are elementarily equivalent (but not conversely in general). Thus, models that are not elementarily equivalent are not isomorphic. Our nonstandard model of S is therefore a structure that is not isomorphic to the natural numbers, for the wff $\sim W$ is true in the nonstandard model and false in the natural numbers.

In Exercise 2, Chapter 1, Section 1.9, we saw that S has no finite models. Now, our nonstandard model of S is denumerable, and it can be contracted to a normal model as any model can. The cardinality of the contracted model is either finite or denumerable. But it cannot be finite, since S has no finite normal models. Therefore, it must be denumerable. Our contracted model is still a model for S' (since the original model is), and no model of S' can be elementarily equivalent to the standard model. Thus, the contracted model is a nonstandard, denumerable, normal model of S.

We can restate result F: There is a denumerable, normal model of first-order arithmetic that is not isomorphic to the natural numbers. Since our nonstandard model has the same cardinality as N, the failure to be isomorphic to N implies an essential difference in structure.

As a matter of fact, we have not just one nonstandard model for S, we have at least $2^{\aleph_0}$ different, nonisomorphic models. To see this, we consider the various hierarchies of consistent extensions of S such as S^*, S^{**}, S', S'', and the like. We obtained S', for example, by adjoining $\sim W$ to S. We have already seen that no model of S' can be isomorphic to the natural numbers and this yields at least two different models of S, namely the natural numbers and our model satisfying S'. But the wff $\sim W$ is effectively exhibitable in S, that is, we can actually calculate its Gödel number. Thus, S' is a recursively axiomatizable extension of S. We thus obtain a new undecidable sentence W' in S' by deducing Gödel's theorem in S'. Again, we can consistently add either W' or $\sim W'$ as a new axiom. Since only one of these two will be true in our nonstandard model for S', we can obtain an extension S'' of S' which is consistent and which has a model different from both the natural numbers and from our nonstandard model for S'. Suppose, for example, that W' is true in our nonstandard model for S'. In this case the extension S'' obtained by adding $\sim W'$ as an axiom will have $\sim W$ as an axiom, which is false in the natural numbers, and $\sim W'$ as an axiom, which is false in our nonstandard model of S'. Thus any model of S'' will differ from both of these models. Nevertheless, S'' will have a model if S is consistent. Again, by the same reasoning as for result F, this model can be considered to be denumerable and normal.

We can iterate this process again, in fact any finite number of times, and we shall have this dual choice at each stage of the iteration. We can thus associate a different extension of S with each denumerable sequence of 0's and 1's. We start with S, which we assume to be consistent as usual. If we add W as an axiom, we put 1 as the first member of our sequence. If we add $\sim W$, we put 0 as our first member. In either case, we obtain a consistent, recursively axiomatizable extension of S, denoted respectively by $S^1 (= S^*)$ and $S^0 (= S')$. Now S^1, which is a consistent recursively axiomatizable extension of S, has an undecidable sentence $W^1 (= W^*)$, and S^0 has one which we denote by $W^0 (= W')$. Again, these undecidable sentences are effectively exhibitable in their respective systems. Continuing the process, we put 1 for the second member of our sequence if we add W^1 (respectively W^0) to obtain our new recursively axiomatizable extension S^{11} (respectively S^{01}), and we put 0 for the second member of the sequence if we add $\sim W^1$ (respectively $\sim W^0$) to obtain a recursively axiomatizable extension S^{10} (respectively S^{00}). The four possible extensions thus obtained will all be consistent (assuming the consistency of **S**), and so they will all have denumerable normal models. Furthermore, no model of any one of these extensions will be isomorphic to a model for another, since each one contains as an axiom at least one closed wff that is a negation of an axiom of any other given one.

We now iterate the process again and obtain eight different recursively axiomatizable extensions associated with the triples $\langle 0, 0, 0 \rangle$, $\langle 0, 0, 1 \rangle$, $\langle 0, 1, 0 \rangle$, $\langle 1, 0, 0 \rangle$, etc. In this way, we associate a finite recursively axiomatizable extension of **S** with each n-tuple of 0's and 1's. We call these our *finite extensions* of **S**. They are all consistent if **S** is consistent. Now we associate, with an infinite sequence of 0's and 1's, the infinite extension of **S** whose axioms are the union of all of the finite extensions obtained by dropping the tail of the sequence at any point. (Since each finite extension contains its predecessors, this union is not difficult to picture intuitively.) Any such extension must be consistent. Otherwise, there must be some finite set of axioms on which it is inconsistent, since any inconsistent set of wffs is inconsistent on a finite subset, as we have already observed in Chapter 1. But any finite set of axioms will be contained in one of our finite extensions associated with an n-tuple, and these are all consistent. Thus, our infinite extensions must be consistent, since otherwise we arrive at a contradiction.

Now each of our infinite extensions associated with an infinite sequence of 0's and 1's is consistent. Consequently, each has a denumerable model which is a model of **S** because it is a model of an extension of **S**. But if the infinite sequences of two extensions differ, the extensions cannot have any isomorphic models. To see this, let n be the smallest number such that two given, different sequences $\{a_i\}$ and $\{b_i\}$ differ in the nth place. The finite extensions represented by $a_1, a_2, \ldots, a_{n-1}$ and $b_1, b_2, \ldots, b_{n-1}$ are the same. (If $n = 1$, then $a_1, a_2, \ldots, a_{n-1}$ and $b_1, b_2, \ldots, b_{n-1}$ are vacuous, and it is **S** itself that is represented by the vacuous sequence.) Now a_n and b_n differ. Let us say that $a_n = 1$ and $b_n = 0$. What this means precisely is that the finite extension $S^{a_1, \ldots, a_n}$ was obtained from $S^{a_1, \ldots, a_{n-1}}$ by adding $W^{a_1, \ldots, a_{n-1}}$, whereas $S^{b_1, \ldots, b_n}$ was obtained from $S^{b_1, \ldots, b_{n-1}}$ by adding $\sim W^{a_1, \ldots, a_{n-1}}$. The finite extensions thus obtained are such that each contains the negation of an axiom of the other, and so no model can satisfy both at the same time. No two models of these respective systems can be elementarily equivalent as models of **S**, and so they cannot be isomorphic. Finally, our two infinite extensions are extensions of our two finite ones, and so they too cannot have isomorphic models. Any model of one of these extensions is a model for **S**.

We thus have a different denumerable model of **S** associated with each infinite extension of **S**. No two such models are elementarily equivalent, and therefore no two of them are isomorphic. Moreover, we have a 1–1 correspondence between these infinite extensions and the set of all infinite sequences of 0's and 1's. But the cardinality of this set of sequences is $2^{\aleph_0}$. Hence we see that there are at least $2^{\aleph_0}$ nonelementarily equivalent (and thus nonisomorphic) models of **S**. All of these models are denumerable. Moreover, since each of these models is a model of **S**, and **S** has no finite models, the contraction to a normal model of each of these models is denumerable. Each of the contracted models is a model of one of our infinite extensions, and so no two are isomorphic. Therefore, **S** has at least $2^{\aleph_0}$ denumerable, normal, nonisomorphic, nonelementarily equivalent models.

For the case of elementary equivalence this result is obviously best possible since any two models of **S** that are not elementarily equivalent must, by definition, differ in the truth value assigned to some wff of **S**. But the cardinality of the set of all wffs of **S** is $\aleph_0$ and so the cardinality of the set of all subsets of wffs of **S** is $2^{\aleph_0}$. Thus, there are at most $2^{\aleph_0}$ sets on which two arbitrary models of **S** can differ. Since this maximum possibility is obtained, as shown by the previous argument, the result is best possible.

The result is not necessarily best possible for the case of isomorphism, since two models can be nonisomorphic and still be elementarily equivalent.

The natural numbers N will be a model for only one of these infinite extensions of **S**. As a matter of fact, we know which one; it is $S^{111\cdots}$. It was W that was true in the standard model, and since we will obtain the undecidable sentence W^1 by proving the original incompleteness theorem in S^1, it follows that it will again be W^1 that is true in the standard model. Iterating, we see that it is the extension $S^{111\cdots}$ alone that will be satisfied by the standard model.

Of course, all of these results depend on our assumption that **S** is consistent, since **S** has no models at all if it is inconsistent.

In deducing the existence of uncountably many denumerable, normal, nonisomorphic models of **S**, we have used two facts about **S** in an essential way. The first is that **S** is essentially incomplete, and that we can effectively exhibit a new undecidable sentence in each finite extension of **S**. This enables us to obtain our hierarchy of extensions, all consistent relative to **S**. The second fact is that **S** has no finite normal models. This assures us that one can find denumerable normal models of the various extensions. Both these properties are true of any sufficiently rich system such as **ZF, ST, NBG,** and the like. We have already observed that our incompleteness results are true for such systems. It is also true that these systems possess no finite normal models, for the Peano postulates are deducible in each of these systems, and the Peano postulates exclude finite normal models. It follows that any sufficiently rich system possesses uncountably many (in fact $2^{\aleph_0}$) denumerable, normal, nonelementarily equivalent (and thus nonisomorphic) models. In particular, this is true for the various foundational systems we consider in this book, such as **ZF, ST, NBG,** and the like.

We have previously cited Tarski's result: Any system with an infinite normal model has a normal model of every infinite cardinality. From this general result, and the fact that **S** has no finite normal models, we can immediately obtain that there are nonisomorphic normal models of **S**, if **S** is consistent. Briefly, if **S** is consistent, it has a model, and thus a normal model (by contradiction). The normal model cannot be finite, and so **S** has an infinite normal model and hence normal models of every infinite cardinality. But two models are isomorphic only if they have the same cardinality. Therefore, it follows that there are nonisomorphic models of **S**. What is significant about the results of the present section is that we have shown that there are at least $2^{\aleph_0}$ nonisomorphic normal models, and that they are all of the same cardinality (denumerable). This result is much stronger than the one we obtain by applying only Tarski's result, since the nonisomorphic models of Tarski's result differ in cardinality from each other.

However, by combining Tarski's result with the method of our proof we can obtain the following strong result: For every infinite cardinal number a, any sufficiently rich first-order system has $2^{\aleph_0}$ nonisomorphic normal models of cardinality a. To see this, let us recall that the method of our proof was to obtain $2^{\aleph_0}$ different extensions of **S**, each extension consistent with respect to **S**. Thus, assuming **S** consistent, each of these extensions has a model and no model of one of these extensions can be a model of another, since they all differ by containing the negation of at least one axiom of another. But any model of an extension of **S** is a model of

S. Since **S** has no finite normal models, none of the normal (contracted) models of the extensions of **S** can be finite. Thus, each of the extensions of **S** has an infinite normal model, and thus a normal model of every infinite cardinality. Thus, for any given infinite cardinal number a, we obtain the existence of $2^{\aleph_0}$ nonisomorphic normal models of **S** of that cardinality, for we have a different normal model of that cardinality associated with each extension of the $2^{\aleph_0}$ extensions of **S**. Just as before, the result holds for any sufficiently rich system.

A result of this type can be found in Ehrenfeucht [1].

Perhaps the reader feels that there must be some contradiction involved in all this. Let us recall that the Peano postulates are among the axioms of **S** and that we have repeatedly used these postulates within set theory to characterize the natural numbers up to isomorphism. How is it that we have now proved that these same Peano postulates do not characterize the natural numbers up to isomorphism for the system **S**? Why, in short, is **S** not categorical?

The difference lies in the systems in which these Peano postulates are couched. Within a system of set theory, the Peano postulate of induction can be formulated with reference to arbitrary subsets of the natural numbers N. That is, we can quantify within the system over subsets of the natural numbers. Within **S** we do not have the notions of set and membership and we cannot quantify within **S** over sets of natural numbers. Our quantifiers in **S** refer only to the individual natural numbers themselves. The postulate of induction for **S** is formulated as an axiom scheme using wffs of **S** (again these wffs involve quantifiers that refer only to natural numbers and not to subsets of the natural numbers). Thus, the postulate of induction is weaker in **S** than in set theory. Within a set theory, we can prove that any two sets that satisfy the Peano postulates *in that system* admit of an isomorphism (again, in the system) between them. Our proof of the existence of nonisomorphic models of **S** was carried on by metamathematical reasoning outside of **S** and not within **S** at all. In fact, this metatheorem about **S** can be carried on within some axiomatic set theory such as **ZF**. This difference between the Peano postulates within **S** and within a set theory is some measure of the difference between first-order arithmetic and nonelementary arithmetic as performed within set theory.

Of course, as we have demonstrated, formalized set theories such as **ZF** will also have continuum-many nonisomorphic models of all infinite cardinalities. In fact, any consistent first-order theory F which contains the Peano postulates will not be categorical: Since the Peano postulates exclude finite normal models, F will possess a denumerable normal model and thus a normal model of every infinite cardinality. Therefore, F is not categorical. This is why we emphasize that we can prove *within* a set theory that the Peano postulates are categorical *within the theory*.

Closely related to the foregoing question is another question that is likely to occur to the reader. By means of the power set operator and Cantor's theorem, we are able to prove within a set theory such as **ZF**, which has an axiom or theorem of infinity, that there are sets of cardinality $2^{\aleph_0}$ (in fact, there is a hierarchy of sets of increasingly large uncountable cardinality). Yet, as we have already observed, there are denumerable models of **ZF**, models in which the class of all objects in the model constitutes a countable set. This seems contradictory.

The difference here is between what is going on within the given set theory and what is going on outside of it. For example, suppose we have proved within a given set theory that $\mathcal{P}(N)$, the power set of the natural numbers N, is uncountable. This means that, within the theory, there is no bijection between $\mathcal{P}(N)$ and N. (We suppose that we have proved all the Peano postulates for N and also that there is a bijection between N and a subset of $\mathcal{P}(N)$.)

Now, a bijection between $\mathcal{P}(N)$ and N is itself a set, namely a certain set of ordered pairs which, if it exists, is a subset of $\mathcal{P}(N) \times N$. To prove within a set theory that $\mathcal{P}(N)$ is uncountable is to prove that no such set exists within the theory. Thus, in any model of the theory, there is no object in the domain of the model which, under the interpretation furnished by the model, constitutes a 1–1 correspondence between the object which is the interpretation of N and the object which is the interpretation of $\mathcal{P}(N)$. This does not mean that no such bijection exists, only that it does not exist *within* the model. In particular, if we take a denumerable, normal model $\langle M, * \rangle$ (and we will have such models if the theory is consistent), then there are only a countable number of objects x available in the model to bear the $\in^*$ relation to $\mathcal{P}(N)^*$, i.e. objects x for which $x E (\mathcal{P}(N))^*$ may hold, where E is the interpretation of "$\in$" in our model. Similarly, there are only a countable number of objects available to bear the E relation to N^*. Because the Peano postulates are true under the interpretation, each of N^* and $(\mathcal{P}(N))^*$ must, however, have an infinite number of objects which bear the E relation to it. This infinite number can only be denumerable in each case, and so there is a bijection between

$$\{x \mid x E N^*\} \quad \text{and} \quad \{x \mid x E (\mathcal{P}(N))^*\}$$

where these two collections are defined in the metatheory. This bijection will be a collection (object) of the metatheory. Notice that in general none of the collections

$$\{x \mid x E N^*\}, \quad \{x \mid x E (\mathcal{P}(N))^*\},$$

or the bijection between them are objects of the model. In particular, $\{x \mid x \in \mathcal{P}(N)\}^*$, for example, is just $(\mathcal{P}(N))^*$ as follows from extensionality and the normality of the model $\langle M, * \rangle$.

The moral of all this is that cardinality is, to a certain extent, a relative notion. A given object may have a countable number of members in the context of one theory while having an uncountable number of members in the context of another.

If the reader finds this relativity repugnant, he has the alternative of appealing to the existence of an intuitively conceived, absolute realm of pre-existing sets in which all of our meta-mathematical talk about models is carried on. Some set theorists undoubtedly take some such position, and it can be defended fairly reasonably, especially if one believes in a particular set theory as being the absolute set theory. Such questions seem clearly to be more in the realm of philosophy than mathematics, for unbelievers who thoroughly reject the existence of uncountable sets are nevertheless able to learn the rules of the "set-theory" game and prove valid theorems just as well as those who believe in sets.

Finally, let us recall that considerations involving models and interpretations can always be reduced to considerations of certain syntactic translations from one theory into another. We have stressed this point in the last chapter. In this way, questions of consistency can always be viewed as questions of relative consistency between two theories.

Chapter 7

The Foundational Systems of W. V. Quine

STARTING in 1937, W. V. Quine elaborated several systems capable of reproducing a reasonably large portion of mathematics. The first system we shall study is called "New Foundations" or NF by abbreviation. Quine first introduced NF in a paper presented to the Mathematical Association in 1936, and it was the subject of an article by Quine in 1937 (see Quine [1]). The second system is that of Quine's *Mathematical Logic*, first published in 1940 and revised in 1951. We shall refer to this system as ML.

7.1. The system NF

The point of departure for NF is, to a great extent, the theory of types, particularly the system ST. Quine was strongly influenced by the work of Whitehead and Russell, and had long been seeking some way of liberalizing the type restriction to obtain a more satisfactory foundation. His answer was contained in the notion of *stratification*, which we shall presently define.

NF is a first-order theory with only one category of variables, the individual variables x_1, x_2, and so on. NF has only one primitive binary predicate letter which we shall denote by "$\in$". We shall also introduce an operator of abstraction identical with that of our system F of Chapter 3. In short, the wffs of NF are exactly the same as the wffs of Frege's system F of Chapter 3. The proper axioms of NF will, of course, be different from F. In form, NF is probably the closest of any system we study in this book to Frege's contradictory system.

We now turn to the problem of defining what is meant by the stratification of a wff. We begin by considering the wffs that do not contain any occurrence of the abstraction operator. Such wffs involve only quantifiers, variables, propositional connectives, and the primitive predicate "$\in$". We call such wffs of NF *simple*. A simple wff is *stratified* if and only if it is possible to replace each variable occurring in it by a whole number numeral in the following manner: We replace everywhere the same variable by the same numeral so that, for each occurrence of "$\in$", the numeral immediately following "$\in$" is the immediate successor of the numeral immediately preceding "$\in$".

A few examples will illustrate the idea. The wffs $(x_1 \in x_2) \supset (x_2 \in x_3)$ and $(x_1 \in x_2) \lor (x_1 \in x_3)$ are both stratified. In the first case, replace x_1 by "1", x_2 by "2", and x_3 by "3". In the second case, replace x_1 by "1", x_2 by "2", and x_3 by "2". Notice that we can replace different variables by the same numeral.

On the other hand, $(x_1 \in x_1)$, $\sim(x_1 \in x_1)$, and $(x_1 \in x_2) \land (x_2 \in x_1)$ are all unstratified.

The reader will recognize the second of these wffs as the defining condition for the class used in deducing Russell's paradox.

We can formulate an algorithm to determine whether or not a given simple wff is stratified. First, pick some arbitrary variable of the wff and replace it everywhere by 0. Then, inductively: If some $n \in y$ occurs for some numeral n and some variable y, then replace everywhere y by $n+1$. If some $y \in n$ occurs for a numeral n and a variable y, then replace everywhere y by $n-1$. If neither of these two cases occurs, then select some arbitrary new variable x and replace it everywhere by 0. Continuing in this manner, we either come to violate the criterion of stratification (and the wff is unstratified), or we continue until all variables have been replaced by numerals and stratification is exhibited.

Clearly, testing for stratification is like restoring type indices. If we considered wffs of **ST** (and for the moment exclude those containing instances of the abstraction operator) as wffs of a first-order theory formed by suppressing type superscripts (replacing with new variables where necessary to avoid confusion), then clearly every such expression would be stratified. In fact, this is precisely the stratagem of "typical ambiguity" employed by Whitehead and Russell. In their work, type indices do not appear and the reader is to restore them in any way consistent with well-formedness. Since, in most cases, it is only relative types that matter in **ST**, it is usually unimportant what type indices (numerals) are actually assigned, as long as it is possible to make an assignment consistent with well-formedness.

However, stratification will not be a criterion of well-formedness in **NF**. This is the essential difference with **ST**. We have already defined the wffs of **NF** as the same as those for **F**, and clearly there are many simple wffs that are not stratified. Stratification will be used in **NF** to define the set of axioms rather than the set of wffs. Nevertheless, because of the obvious similarity with typical ambiguity, we make the following convention: For any wff A of **NF** (simple or not) and any assignment of numerals to the variables of A, the number n that is named by the numeral assigned to a variable x of A will be called the *type* of x under the assignment. Whenever we speak of an "assignment" of numerals, we always understand, as in the foregoing, that the same numeral is assigned to every occurrence of the same variable. Thus, every variable has a uniquely defined type for any assignment of numerals to variables.

We must now extend our definition of stratification to the whole set of wffs of **NF**, those involving abstracts as well as simple ones. For a wff A of **NF** and an assignment of numerals to the variables of A, the *type* of an abstract $\{x \mid B(x)\}$ that is a part of A, under the given assignment of numerals, is $n+1$ where the type of the variable x is n under the given assignment. By an assignment of types to the *terms* of a wff A, we mean any assignment of numerals to the variables for which the type of every abstract is understood to be its type under the assignment of numerals, and the type of every variable is its type under the assignment of numerals. Finally, a wff A is said to be stratified if there is some assignment of types to the terms of A such that, for every occurrence of "$\in$" in A, the type of the term immediately following "$\in$" is the successor of (one more than) the type of the term immediately preceding "$\in$".

Notice that our way of assigning a type to an abstract is also in line with the abstraction operation in **ST**. When we circumflex a variable of type n in **ST**, we obtain an abstact of type $n+1$. Of course, we allow the use of negative numbers as types for assignments of indices in **NF**, but this is obviously nonessential.

As examples of stratified wffs involving abstracts, we have

$$x_2 \in \{x_1 \mid (x_3)(x_3 \in x_1 \equiv x_3 \in x_2)\},$$

and $\{x_1 \mid x_1 \in x_2\} \in x_3$. In the first case, let x_3 be 0 and x_1 and x_2 be 1. Then the abstract has type 2 in this assignment and so we have stratification. In the second instance, let x_1 have type 0 and x_2 type 1. The abstract has type 1 and we can assign type 2 to x_3.

As examples of unstratified wffs involving abstracts, we have $x_2 \in \{x_1 \mid x_1 \in x_2\}$ and $\{x_1 \mid x_2 \in x_1\} \in x_1$. In the first case, let n be the type of x_1 under some assignment. Then $n+1$ must be the type of the abstract and $n+1$ must be the assignment to x_2 as well. Consequently, stratification is impossible. A similar argument shows the impossibility of stratification in the second case.

If a wff is stratified, then any assignment of types to terms that conforms to the pattern of stratification is called an *adequate* assignment of types to terms. We now turn to the task of stating the axioms of **NF**.

Definition 1. $x = y$ for $(z)(z \in x \equiv z \in y)$, where z is any variable that does not occur in the terms x and y.

Depending on the terms x and y, $x = y$ may or may not be stratified. Notice that in order for $x = y$ to be stratified, there must be some adequate assignment of types such that x and y have the same type, since both x and y must have the type $n+1$ where z has type n under the assignment.

NF.1. $(x)(y)(x = y \supset (A(x, x) \equiv A(x, y)))$, where $A(x, y)$ is obtained from $A(x, x)$ by replacing y for x in zero, one, or more free occurrences of x in $A(x, x)$, and y is free for x in all of the occurrences of x that it replaces.

This is the same axiom of extensionality as that of Frege's system **F**.

NF.2. $(x)(x \in \{y \mid A(y)\} \equiv A(x))$, where $A(y)$ is stratified, x is free for y in $A(y)$, and $A(x)$ results from $A(y)$ by replacing x for y in all the latter's free occurrences in $A(y)$.

This is the axiom of abstraction in **NF**. The distinctive feature of **NF.2** is the requirement that the wff $A(y)$ be stratified. If we remove this requirement, we obtain immediately the contradictory axiom **F.2** of Chapter 3. In fact, this requirement constitutes the only difference between **NF** and **F**.

As with the system **ZF** of Chapter 5, we need to add our axiom schemes for *vbtos* if we want the underlying logic of **NF** to be complete. For example, the two wffs $x_1 = x_1$ and $(x_1 \in x_1) \vee (\sim(x_1 \in x_1))$ are logically equivalent (in fact each is a logical truth). Yet, we cannot use **NF.1** and **NF.2** to prove that

$$\{x_1 \mid x_1 = x_1\} = \{x_1 \mid (x_1 \in x_1) \vee (\sim(x_1 \in x_1))\}.$$

This is because the wff of the second term is not stratified, thus blocking appeal to **NF.2**.

To prevent this and other similar situations where variable-bound terms formed from equivalent wffs may not be provably equal, we assume the schemes V.1 and V.2 of Section 1.8 for the *vbto* of abstraction of **NF**. Since we regard these schemes as logical axioms, we do not give them numbers as proper axioms of **NF**. This completes the description of the system **NF**.

As with Frege's system **F**, as well as with **ST**, we can formulate **NF** without a term operator of abstraction and, of course, without the schemes V.1 and V.2. In addition to **NF.1**, we would have:

NF.2*. $(Ey)(x)(x \in y \equiv A(x))$ where y is not x and does not occur in the stratified wff $A(x)$.

We avoid this formulation only to lighten the technical matters while working within the system. By the axiom of extensionality, we obtain essentially the same system as with our term operator in conjunction with the schemes **V.1** and **V.2**. We have made similar observations with respect to **F**, **ST**, and **ZF**.

By the way we have defined the wffs of **NF**, we can formally apply an abstraction operator to any wff $A(x)$ whether it is stratified or not. If $A(x)$ is stratified, we say that the term $\{x \mid A(x)\}$ is *stratified*. It is only to stratified terms that we can apply **NF.2**, and this is obviously our prime method for dealing with abstracts.

In line with our practice in dealing with other systems, we refer to the terms of **NF** as *sets*. **NF.2** posits the existence of sets satisfying stratified conditions. This restriction in **NF.2** is a positive one because it directly posits the existence of certain sets while not directly excluding the existence of other sets that may not be definable by stratified wffs. Thus, Russell's paradox is not directly deducible, since the contradictory Russell condition $\sim (x \in x)$ is not stratified and thus **NF.2** cannot be applied to it. But this is not a guarantee that a set satisfying some such contradictory condition is not, by some other indirect means, definable in **NF**.

Stratification is a formal, linguistic device. We have not attempted to give an intuitive model for **NF**, nor have we tried to give an informal characterization of sets definable by stratified conditions. The reason is that, as of the present writing no model for **NF** has ever been discovered! This is in direct contrast to both **ST** and **ZF** for which we have intuitive (albeit highly nonconstructive) models.[†]

Now, one theorem of logic deducible within any set theory such as **ZF** is that a first-order theory is consistent if and only if it has a model. In the light of this theorem, our failure to find a model for **NF** after so many years might be viewed as augmenting the possibility of the inconsistency of **NF**. As we shall see later, there are other anomalies of **NF** which seem to undermine one's confidence in the consistency of the system. In particular, the axiom of choice can be disproved within the system, and Cantor's paradox is narrowly avoided.

Still, no contradiction has ever been deduced in **NF**, and several logicians have made a concerted effort to find such a contradiction. Moreover, we know from Chapter 6 that there are models of **ZF** without **ZF.10** in which the axiom of choice is false. Thus, the failure of the axiom of choice in **NF** cannot, in itself, exclude the possibility of finding a model of **NF** in **ZF** without **ZF.10**. Also, as we shall see, there is even a way of providing for most uses of the axiom of choice in **NF** by applying certain limiting conditions.

Of course, failure to deduce a contradiction in **NF** is no proof that the system is consistent. All this should leave the reader in the same position with respect to the consistency of **NF** as the author: Doubt and uncertainty. Each of the points mentioned in the above paragraph will be amplified in our later discussion, and we now proceed to the development of mathematics in **NF**.

Our method of developing mathematics in **NF** follows Frege rather than von Neumann. Since Frege's constructions were possible in type theory, they will certainly be possible in

---

[†] A model has been found by Jensen (see Jensen [1]) for a modified version of **NF**. Jensen's system will be described in a later section of this chapter.

NF, which is an obvious liberalization of **ST**. Moreover, von Neumann's natural numbers, which we used in **ZF**, will not work here. To see this, observe that the set $x \cup \{x\}$, which is the successor of x in **ZF**, is defined as $\{y \mid y = x \lor y \in x\}$. But this term is unstratified, since y and x must be of the same type (in any adequate assignment of types) in order for $y = x$ to be stratified. Thus, we cannot use **NF.2** in dealing with $x \cup \{x\}$.

Definition 2. V for $\{x_1 \mid x_1 = x_1\}$.

THEOREM 1. $\vdash (x_1)(x_1 \in V)$.

Proof. The condition $x_1 = x_1$ is stratified, and so the proof is the same as the corresponding theorem in Frege's system.

Definition 2 and Theorem 1 show that we have a universal class in **NF**. **NF** is the first system we have considered since Frege's contradictory system in which a class containing everything in our universe is definable.

COROLLARY. $\vdash V \in V$.

Proof. Immediate from Theorem 1.

The proof of Theorem 1 illustrates an important relationship between **NF** and **F**. If the defining conditions of the relevant terms are stratified, the proofs of theorems in **NF** will proceed in exactly the same manner as in **F**. This is immediately obvious, since the worry of stratification is the only difference between the two systems. We shall not attempt to maintain any sameness of numberings between theorems and definitions of this chapter and of Chapter 3, but we shall often refer to a theorem in this chapter as "the same as the corresponding theorem" in Chapter 3. The "corresponding theorem" is the one that asserts the same statement.

Definition 3. $x \subset y$ for $(z)(z \in x \supset z \in y)$ where z is a variable not appearing in the terms x and y.

Definition 4. $P(y)$ for $\{x \mid x \subset y\}$ where x is any variable not appearing in the term y.

Exercise. Prove that $P(y)$ is stratified if y is.

THEOREM 2. $\vdash (x_1)(x_1 \subset V)$.

Proof. The reader will prove Theorem 2 as an exercise.

THEOREM 3. $P(V) = V$.

Proof. Immediate from Theorems 1 and 2 and the relevant definitions.

Here we have a set in **NF** which is the same as its power set. If Cantor's theorem is provable in **NF**, we have an immediate contradiction, the Cantor paradox. Cantor's theorem (see Chapter 3, Section 3.5) says that the power set of every set has greater cardinality than the set.

15*

Since every set has the same cardinality as itself, the existence of a universal set V with the properties of the V of **NF** is inconsistent with Cantor's theorem.

However, Cantor's theorem is apparently not provable in **NF**. In any case, the usual proof of the theorem does not succeed because certain relevant conditions are not stratified. We shall discuss this in detail in the next section of this chapter.

Definition 5. $\varLambda$ for $\{x_1 \mid x_1 \neq x_1\}$.

THEOREM 4. $\vdash (x_1)(x_1 \notin \varLambda)$.

Proof. Same as the corresponding theorem in **F**, since $\varLambda$ is stratified.

In the foregoing theorem and definition, we have used the usual abbreviations "$\neq$" and "$\notin$" for applying negation to formulas involving these predicate symbols.

Definition 6. $\{x\}$ for $\{y \mid y = x\}$ where x is any term and y is any variable not occurring in x.

Definition 7. $\{x, y\}$ for $\{z \mid z = x \lor z = y\}$ where z does not occur in x or y.

$\{x\}$ is stratified if x is, and $\{x, y\}$ is stratified if and only if there is some adequate assignment of types for which x and y have the same type.

Definition 8. $\{y_1, \ldots, y_n\}$ for $\{x \mid x = y_1 \lor, \ldots, \lor x = y_n\}$ where x is some variable not occurring in any of the y_i.

Such a finite set of n elements will be stratified if and only if there is an adequate assignment f types in which each element has the same type.

Definition 9. 0 for $\{\varLambda\}$.

0 is stratified.

Definition 10. $(x \cup y)$ for $\{z \mid z \in x \lor z \in y\}$ where z does not occur in either of the terms x or y.

Definition 11. $(x \cap y)$ for $\{z \mid z \in x \land z \in y\}$ where z does not occur in either of the terms x or y.

Definition 12. $\bar{x}$ for $\{y \mid y \notin x\}$ where y does not occur in the term x.

$(x \cup y)$ and $(x \cap y)$ are stratified if and only if there is some adequate assignment of types in which x and y have the same type. $\bar{x}$ is stratified if x is and it has the same type as x under any adequate assignment of types.

The reader can begin to see one major drawback of **NF**. It is that stratification can be tested directly only when a wff or term is in an unabbreviated form. This is a tremendous practical hindrance in working within **NF**, for we depend heavily on definition to lighten an otherwise unwieldy symbolism in any set theory. We do not intend to retranslate into primitive symbolism in every instance, but we are obliged to accompany definitions with statements about stratification in order to know whether or not (or under what conditions) our axiom of abstraction is applicable to defined terms.

With these general definitions at hand, we can proceed with the Fregean definition of the natural numbers.

Definition 13. $S(x)$ for $\{z \mid (Ey)(y \in z \land z \cap \overline{\{y\}} \in x)\}$ where y and z are distinct variables not occurring in the term x.

Exercise. Prove that $S(x)$ is stratified if x is, and that $S(x)$ has the same type as x under any adequate assignment of types.

Definition 14. N for $\{x_1 \mid (x_2)(0 \in x_2 \land (x_3)(x_3 \in x_2 \supset S(x_3) \in x_2) \supset x_1 \in x_2)\}$.

N is stratified.

Definition 15. Fin for $\{x_1 \mid (Ex_2)(x_2 \in N \land x_1 \in x_2)\}$.

Definition 16. Inf for $\overline{\text{Fin}}$.

Fin is the set of all finite sets. (A set is finite if it is an element of a natural number n regarded as a class of all n-element sets.) Inf is the set of all infinite sets. Both of these terms are stratified.

THEOREM 5. $\vdash 0 \in N$.

Proof. The reader will prove Theorem 5 as an exercise. (*Hint*: Same as corresponding theorem in **F**.)

THEOREM 6. $\vdash (x_1)(x_1 \in N \supset S(x_1) \in N)$.

Proof. The reader will prove Theorem 6 as an exercise. (*Hint*: Same as corresponding theorem in **F**.)

THEOREM 7. $\vdash (x_1)(0 \neq S(x_1))$.

Proof. The reader will prove Theorem 7 as an exercise. (*Hint*: Same as corresponding theorem in **F**.)

THEOREM 8. $\vdash (x_1)(0 \in x_1 \land (x_2)(x_2 \in x_1 \land x_2 \in N \supset S(x_2) \in x_1) \supset N \subset x_1)$.

Proof. The reader will prove Theorem 8 as an exercise.

The last is the set-theoretic form of the principle of mathematical induction. Our procedure in dealing with systems in which Frege's natural numbers are used is to obtain the metatheorem of mathematical induction from the set-theoretic form by use of the abstraction operator. What we obtain in **NF** is the following:

THEOREM 9. $\vdash A(0) \land (x)(x \in N \land A(x) \supset A(S(x))) \supset (x)(x \in N \supset A(x))$

where $A(x)$ is any *stratified* wff. $A(0)$ and $A(S(x))$ result from $A(x)$ by the indicated substitutions.

Proof. Immediate from Theorem 8 by $e\forall$.

Theorem 9 is a limited form of induction, since we can induct only on stratified conditions. In **ST** we could induct on any condition, but then all conditions in **ST** are stratified. Nevertheless, we had no such limitation on mathematical induction in **ZF**; we could induct on all conditions expressible in that system. This limitation of **NF** will not be overcome, and is a permanent feature of the system.

The limitation resulting from our inability to induct on all formulas will be considerably lessened if we can deduce the remaining Peano postulate, i.e. $S(x) = S(y) \supset x = y$ where $x \in N$ and $y \in N$. We have already seen that the von Neumann natural numbers are not definable in **NF** because $x \cup \{x\}$ is unstratified for all terms x. Since we used the von Neumann natural numbers to prove the remaining Peano postulate in our system **F** of Chapter 3, it follows that we cannot use the same method to prove the theorem here. Let us review our line of reasoning in Chapter 3 in order to see the role played by the von Neumann natural numbers in proving the remaining Peano postulate.

Theorem 30 of Chapter 3,

$$\vdash (x_1)(x_2)(x_1 \in N \land x_2 \in N \land S(x_1) = S(x_2) \supset x_1 = x_2),$$

was an immediate consequence of Theorem 29,

$$\vdash (x_1)(x_2)(S(x_1) \subset S(x_2) \land x_1 \in N \land x_2 \in N \supset x_1 \subset x_2).$$

This theorem was proved by assuming the hypotheses and letting x_3 be some element of x_1. Since x_3 is an element of a natural number x_1, there is some a_1 not in x_3. (This last fact is the corollary to Theorem 28. Theorem 28 asserts that the universal set V is not finite; that is, V is not an element of a natural number.) Since a_1 is not in x_3, $(x_3 \cup \{a_1\}) \cap \overline{\{a_1\}} = x_3$. But, as follows from the very definition of the successor, $x_3 \cup \{a_1\} \in S(x_1)$. By hypothesis this yields $x_3 \cup \{a_1\} \in S(x_2)$. Finally, again applying the definition of successor and Theorem 13 of Chapter 3, we obtain that $x_3 = (x_3 \cup \{a_1\}) \cap \overline{\{a_1\}}$ is an element of x_2. Since x_3 was any element of x_1, the theorem is proved.

The whole method of the proof of Theorem 29 hinges on the existence of an a_1 not in x_3, and this fact is turn depends on the nonfiniteness of V, as we have just indicated. The nonfiniteness of V, Theorem 28, was deduced from Theorem 26, which asserts $\vdash \Lambda \notin N$, i.e. that no natural number is empty. Finally, the fact that no natural number is empty was proved (through induction) by showing that each natural number had at least one element, namely, one of the von Neumann natural numbers.

The sole use of the von Neumann natural numbers in the foregoing line of reasoning is to show that no natural number is empty. From this fact it follows that V is infinite, and from this conclusion it follows that there is something that is not in any given natural number (the corollary to Theorem 28, Chapter 3). Thus, if we can establish any of the succeeding links of this chain independently of the preceding links, we can complete the argument and prove the remaining Peano postulate without appeal to the von Neumann natural numbers. In particular, if we can prove in some independent manner that, for every element x of a natural number n, there is something that is not an element of x (corollary to Theorem 28), then our remaining postulate obviously follows by reasoning identical to that of Theorem 29 and Theorem 30 of Chapter 3. In Section 7.3 of this chapter we show how to obtain the statement of the corollary to Theorem 28 as a theorem of **NF**.

Notice, there is no obvious way of doing this. In each of our other systems such as **ST** and **ZF**, we have had to posit an axiom of infinity, though we were able, in **ZF**, to develop a theory of natural numbers without it. The question of whether or not a theorem of infinity is provable in **NF** was an open question for many years. However, in 1953, E. Specker proved that the axiom of choice does not hold in **NF**. But the axiom of choice does hold for finite sets. From this, it follows that there are infinite sets in **NF**. Thus, by this devious route the remaining Peano postulate is actually provable in **NF**. Notice that this shows the greater strength of **NF** over **ST**, since the remaining Peano postulate is not provable in **ST** without **ST.3**. In a later section we discuss the question of the axiom of choice and the theorem of infinity in more detail.

To facilitate this discussion, we need to develop the notion of cardinal number in **NF**. This we now proceed to do.

Definition 17. $\langle x, y \rangle$ for $\{\{x\}, \{x, y\}\}$ where x and y are any terms.

This is our usual definition of the notion of ordered pair.

THEOREM 10. $\langle x_1, x_2 \rangle = \langle x_3, x_4 \rangle \equiv x_1 = x_3 \wedge x_2 = x_4$.

Proof. The reader will prove Theorem 10 as an exercise.

The ordered pair $\langle x, y \rangle$ will be stratified only if there is some adequate assignment of types to x and y in which x and y have the same type.

Definition 18. R for $\{x_1 \mid (x_2)(x_2 \in x_1 \supset (Ex_3)(Ex_4)(x_2 = \langle x_3, x_4 \rangle))\}$.

R is the class of all relations. An element of R is a class of ordered pairs. R is stratified.

Definition 19. F for $\{x_1 \mid x_1 \in R \wedge (x_2)(x_3)(x_4)(\langle x_2, x_3 \rangle \in x_1 \wedge \langle x_2, x_4 \rangle \in x_1 \supset x_3 = x_4)\}$.

F is the class of all functional relations or functions. F is stratified.

In systems such as **ST** and **ZF**, the large classes such as R and F do not exist. However, such classes do exist in **NBG** or **MKM**.

Definition 20. F^1 for

$$\{x_1 \mid x_1 \in F \wedge (x_2)(x_3)(x_4)(\langle x_2, x_3 \rangle \in x_1 \wedge \langle x_4, x_3 \rangle \in x_1 \supset x_2 = x_4)\}.$$

F^1 is the class of all 1–1 functions. F^1 is stratified.

Definition 21. $D(x)$ for $\{y \mid (Ez)(\langle y, z \rangle \in x)\}$ where x is any term and the distinct variables y and z do not occur in x.

This is the domain of the relational part of x.

Definition 22. $I(x)$ for $\{y \mid (Ez)(\langle z, y \rangle \in x)\}$ where x is any term, and the distinct variables y and z do not occur in x.

This is the range of a functional relation, the set of images of a given relation. We now define the relation of cardinal similarity.

Definition 23. $x \operatorname{Sm} y$ for $(Ez)(z \in F^1 \wedge D(z) = x \wedge I(z) = y)$ where z does not occur in the terms x and y.

THEOREM 11. $\vdash (x_1)(x_1 \operatorname{Sm} x_1)$.

Proof. The set $\{x_2 \mid (Ex_3)(\langle x_3, x_3 \rangle = x_2 \wedge x_3 \in x_1)\}$ is clearly a 1–1 function whose domain and range is x_1.

THEOREM 12. $\vdash (x_1)(x_2)(x_1 \operatorname{Sm} x_2 \supset x_2 \operatorname{Sm} x_1)$.

Proof. The reader will prove Theorem 12 as an exercise.

THEOREM 13. $\vdash (x_1)(x_2)(x_3)(x_1 \operatorname{Sm} x_2 \wedge x_2 \operatorname{Sm} x_3 \supset x_1 \operatorname{Sm} x_3)$.

Proof. The reader will prove Theorem 13 as an exercise.

The last three theorems show that the relation of cardinal similarity is an equivalence relation on sets. This permits us to define the notion of the cardinal number of a set as the set of all similar sets. In Zermelo set theory, we introduced cardinals as certain ordinals. Of course we cannot construct the ordinals in **NF** as we did in **ZF** either. Instead we use the method of defining them as sets of ordinally similar sets, which is the intuitive conception of an ordinal number in intuitive set theory. In short, in **NF** we define cardinals and ordinals as the class of all classes of the given cardinal (ordinal) type, whereas in **ZF** we proceed by choosing a canonical representative from each cardinal (ordinal) class.

Definition 24. $\operatorname{Nc}(x)$ for $\{y \mid y \operatorname{Sm} x\}$ where y does not occur in x.

Definition 25. $(x+y)$ for
$$\{z \mid (Ew)(Er)(z = (w \cup r) \wedge w \in x \wedge r \in y \wedge w \cap r = \varLambda)\},$$
where the variables w, r, and z are all distinct and do not occur in the terms x and y.

Exercise. Prove that $(x+y)$ is stratified if and only if there is an adequate assignment of types to x and y in which they have the same type.

If x and y are cardinal numbers, then $x+y$ is the cardinal sum of x and y.

Definition 26. 1 for $S(0)$.

Definition 27. $\operatorname{Pu}(x)$ for $\{y \mid (Ez)(y = \{z\} \wedge z \in x)\}$, where y and z are distinct variables that do not occur in the term x.

$\operatorname{Pu}(x)$ is the set of all unit (one-element) subsets of x. $\operatorname{Pu}(x)$ is stratified if x is. We use this notion to connect cardinal addition with our operations as already defined for the natural numbers.

THEOREM 14. $\vdash 1 = \mathrm{Pu}(V)$.

Proof. The reader will prove Theorem 14 as an exercise.

THEOREM 15. $\vdash (x_1)(S(x_1) = (x_1+1))$.

Proof. The reader will prove Theorem 15 as an exercise.

With these general notions now defined in **NF**, we are ready to proceed to a closer discussion of Cantor's theorem in **NF**.

7.2. Cantor's theorem in NF

In intuitive set theory, Cantor's theorem asserts that no set is cardinally similar to its power set. In our language, this is expressed by: $(x_1)(\sim x_1 \mathrm{Sm} P(x_1))$. Now, by Theorem 3, we have $\vdash V = P(V)$ and by Theorem 11 we have that $\vdash (x_1)(x_1 \mathrm{Sm} x_1)$. Applying $e\forall$ to Theorem 11 we obtain $\vdash V \mathrm{Sm} V$. Using this together with Theorem 3 and **NF**.1, we immediately obtain $\vdash V \mathrm{Sm} P(V)$. If Cantor's theorem is provable, we can apply $e\forall$ and obtain $\sim V \mathrm{Sm} P(V)$, and we have a formal contradiction in **NF**. However, as we have indicated in previous discussion, the usual derivation of Cantor's theorem fails in **NF**. Let us see why.

In intuitive set theory, the argument for Cantor's theorem runs as follows: Let x be a set and assume that x is similar to its power set $\mathcal{P}(x)$. Let f be a function that gives the 1–1 correspondence between x and $\mathcal{P}(x)$. Now, for all $y \in x$, $\langle y, z \rangle \in f$, either $y \in z$ or $y \notin z$ holds. Let

$$w = \{y \mid y \in x \wedge (z)(\langle y, z \rangle \in f \supset y \notin z)\}.$$

Clearly, $w \subset x$, since every element of w is an element of x. Thus $w \in \mathcal{P}(x)$. Since f is a 1–1 function whose domain is x and whose range is $\mathcal{P}(x)$, there is some one element $a \in x$ such that $\langle a, w \rangle \in f$. Now, if $a \in w$, then, by the principle of abstraction, a satisfies the defining condition of w and $(z)(\langle a, z \rangle \in f \supset a \notin z)$. Applying $e\forall$ we obtain $\langle a, w \rangle \in f \supset a \notin w$. But $\langle a, w \rangle \in f$ and so, by MP, we obtain $a \notin w$. We have deduced that $a \in w$ implies $a \notin w$ which is equivalent to $a \notin w$. On the other hand, if $a \notin w$, then, since $\langle a, z \rangle \in f \supset z = w$ (f is 1–1) we have that

$$(z)(\langle a, z \rangle \in f \supset a \notin z),$$

and a satisfies the defining condition of w. By the abstraction principle, we then obtain $a \in w$. This contradicts $a \notin w$ and establishes the falsity of the assumption that x and $\mathcal{P}(x)$ are similar (that is, that there is a 1–1 correspondence between them).

The only thing preventing this line of reasoning in **NF** is the fact that the term w is not stratified. Remember that an ordered pair is stratified only if there is some adequate assignment of types in which both members of the pair are of the same type. Thus, any adequate assignment of types to w will require that y and z have the same type. But the wff $y \notin z$ is also part of w and this means that the type of z must be one higher than the type of y in any adequate assignment of types to w. Since y and z cannot have both the same type and different types, no adequate assignment of types to w is possible. This prevents the application of the principle of abstraction and thus avoids the direct deduction of the Cantor theorem (and thus the Cantor paradox) in **NF**.

Notice, however, that if we could define w as

$$\{y \mid y \in x \wedge (z)(\langle \{y\}, z \rangle \in f \wedge y \notin z)\},$$

then we would obtain a stratified condition. Taking the unit set $\{y\}$ as the first member of the pair means that $\{y\}$ and z must have the same type, which will happen precisely when the type of z is one higher than that of y. We obtain a stratified condition, but, in this case, the domain of f is no longer x but rather $\mathrm{Pu}(x)$, the set of all unit (one-element) subsets of x. We can thus prove:

THEOREM 16. $\vdash (x_1)(\sim \mathrm{Pu}(x_1)\, \mathrm{Sm}\, P(x_1))$.

Proof. Assume $\mathrm{Pu}(x_1)\, \mathrm{Sm}\, P(x_1)$. This means that there is some 1–1 function, let us call it a_1, $a_1 \in F^1$, such that $D(a_1) = \mathrm{Pu}(x_1)$ and

$$I(a_1) = P(x_1).$$

We let w stand for $\{x_2 \mid x_2 \in x_1 \wedge (x_3)(\langle \{x_2\}, x_3 \rangle \in a_1 \supset x_2 \notin x_3)\}$. The term w is stratified, and so by **NF.2** we have

$$\vdash x_2 \in w \equiv x_2 \in x_1 \wedge (x_3)(\langle \{x_2\}, x_3 \rangle \in a_1 \supset x_2 \notin x_3).$$

We easily obtain $(x_2)(x_2 \in w \supset x_2 \in x_1)$ and so $w \subset x_1$. Thus

$$w \in P(x_1) = I(a_1).$$

There is an element of $D(a_1)$, let us call it a_2, such that $\langle a_2, w \rangle \in a_1$. But $a_2 \in \mathrm{Pu}(x_1)$, and so there is an element of x_1, let us call it a_3, such that $a_2 = \{a_3\}$. Thus, $\langle \{a_3\}, w \rangle \in a_1$.

If we assume $a_3 \in w$, then, by applying **NF.2** and MP, we obtain $(x_3)(\langle \{a_3\}, x_3 \rangle \in a_1 \supset a_3 \notin x_3)$. Applying $e\forall$ we obtain

$$\langle \{a_3\}, w \rangle \in a_1 \supset a_3 \notin w.$$

Since $\langle \{a_3\}, w \rangle \in a_1$ holds from our foregoing hypotheses, we have $a_3 \notin w$ by MP. Thus, $a_3 \in w$ implies $a_3 \notin w$ which yields $a_3 \notin w$.

On the other hand, $a_1 \in F^1$ means that $(x_3)(\langle \{a_3\}, x_3 \rangle \in a_1 \supset x_3 = w)$ holds. Thus, $(x_3)(\langle \{a_3\}, x_3 \rangle \in a_1 \supset a_3 \notin x_3)$ since $a_3 \notin w$ holds. Since $a_3 \in x_1$ also holds, a_3 satisfies the defining condition of w and so $a_3 \in w$ by **NF.2**. This contradiction establishes the falsity of the assumption that

$$\mathrm{Pu}(x_1)\, \mathrm{Sm}\, P(x_1),$$

and so our theorem is established.

Theorem 16 establishes that there is no 1–1 correspondence between the one-element subsets of a given set and the set of all of its subsets. Since $\mathrm{Pu}(x)$ is a subset of $P(x)$, it easily follows that $\mathrm{Nc}(P(x))$ is greater than $\mathrm{Nc}(\mathrm{Pu}(x))$ for any set x; that is, it easily follows when the usual way of defining the "greater than" relation between cardinal numbers is introduced in the system. ($\mathrm{Nc}(x)$ is greater than $\mathrm{Nc}(y)$ will mean that y is similar to a subset of x but not conversely.)

However, Theorem 16 would seem to give rise to the possibility of a paradox in the following manner: We know that $\mathrm{Pu}(x)$ and $P(x)$ are not similar. But there is an obvious similarity between any set x and the set $\mathrm{Pu}(x)$. Just let each element $y \in x$ correspond to the set $\{y\}$. Since

Pu(x) is simply the set of all such unit subsets of x, we have a 1–1 correspondence f between any two sets x and Pu(x). We obtain Pu(V) Sm V and V Sm $P(V)$, yielding Pu(V) Sm $P(V)$ which contradicts Theorem 16. This intuitively obvious reasoning, valid in intuitive set theory, is blocked by the fact that our "obvious" 1–1 correspondence f between x and Pu(x) is not stratified. Formally, f would be defined as $\{z \mid (Ey)(y \in x \wedge \langle y, \{y\}\rangle = z)\}$. But f is not stratified, since the ordered pair $\langle y, \{y\}\rangle$ is not stratified. In any adequate assignment of types, y and $\{y\}$, the two components of the pair, must have the same type; a clear impossibility. It is thus impossible to prove in this manner that a set x is similar to its set of unit subsets Pu(x) in **NF**.

Of course, the impossibility of proving in this manner the similarity of x and Pu(x) does not, in itself, yield that there are actually sets that are not similar to their set of unit subsets. Yet, we can establish just this in **NF**. We first define:

Definition 28. Can(x) for x Sm Pu(x) where x is any term.

A set x is "Cantorian" if it is similar to its set of unit subsets. We now prove that there are non-Cantorian sets in **NF**, in particular V.

THEOREM 17. $\vdash \sim$ Can(V).

Proof. Suppose Can(V), that is, V Sm Pu(V). By Theorem 3, $V = P(V)$, and so $\vdash V$ Sm $P(V)$ by **NF**.1. By the transitivity of Sm, we deduce Pu(V) Sm $P(V)$ which contradicts Theorem 16. Thus, our assumption is false and $\vdash \sim$ Can(V).

We now have before us the paradoxical conclusion that V is a set with the same cardinality as its power set (the set of *all* its subsets). Yet V does not have the same cardinality as the set of all of its one-element subsets Pu(V). The cardinality of Pu(V) is less than that of V, since the cardinality of Pu(V) is less than that of $P(V) = V$. It is only the fact that stratification fails for ordered pairs of the form $\langle y, \{y\}\rangle$ that prevents a direct proof of contradiction in **NF**.

Of course, we would expect some type of paradoxical result to be forthcoming with the existence of a universal set V, as well as the possibility of proving Theorem 16. We finally obtain such a paradox in the form of the existence of non-Cantorian sets. We call such sets "non-Cantorian", since each intuitively conceived Cantorian set can certainly be put into 1–1 correspondence with its set of unit subsets, and Cantor would have surely preferred to dispense with the universal set rather than discard the classic form of the theorem bearing his name. This, of course, is what **ZF** does.

The existence of non-Cantorian sets in **NF** shows in a new way why finding a model for **NF** is so difficult. A model would have to be some set-theoretic structure, and one that contained non-Cantorian sets. Of course, such a model might be "nonstandard". It might be some abstract structure for which the interpretation of "$\in$" is not the usual set membership relation. Some results in set theory, such as those of Dana Scott, in which nonstandard models for **ZF** in which the continuum hypothesis fails are exhibited, show that such a model of **NF** might exist. One might ultimately reject **NF** as a foundation for purely practical reasons, e.g. because the system is too cumbersome. Nevertheless, the resolution of the question of a model for a system with such unusual properties is of genuine interest.

We now turn to the problem of the axiom of choice in **NF**.

7.3. The axiom of choice in NF and the theorem of infinity

In Chapter 5, we discussed the fact that the axiom of choice is independent with respect to the other axioms of ZF. In particular, then, the axiom of choice can be consistently added as a hypothesis to ZF (as we added it in Chapter 5). Quine (see Quine [3], p. 164), Rosser (see Rosser [2], p. 512 and p. 517), and other logicians had generally supposed that the axiom of choice was probably independent in NF (or at least "as independent" as it is in ZF, since the independence with respect to ZF had not yet been proved). However, in 1953, E. P. Specker published a proof that the axiom of choice is contradictory if added to NF. Specker's proof served to heighten concern for the possibility of inconsistency in NF, especially since Gödel had proved in 1940 that the axiom of choice is consistent with respect to the other axioms of ZF.

If we reflect on the meaning of the axiom of choice and the structure of NF as we have explored it, we can begin to see why indeed the axiom of choice might fail in NF. For a disjoint collection of nonempty sets, the axiom of choice allows us to pick a choice set having exactly one element from each set in the collection. Now clearly, in intuitive set theory, such a choice set is in 1–1 correspondence with the disjoint collection, since we have only to allow each element of the choice set to correspond to the set of the collection from which it is chosen. However, for any nonempty set x, $\mathrm{Pu}(x)$ is a disjoint collection of nonempty sets, and clearly the choice set of $\mathrm{Pu}(x)$ is just x itself. Yet, there are sets in NF for which x and $\mathrm{Pu}(x)$ are not similar, as we have just shown. Again, it is the non-Cantorian sets that deviate from our intuitive conception.

Of course, our argument in the foregoing paragraph is only heuristic, and does not itself constitute a disproof of the axiom of choice. This is so because once again our construction of the "obvious" 1–1 correspondence between the choice set and the original disjoint collection involves an unstratified definition. Specker's proof involves the axiom of choice in the form which states that the cardinal numbers of NF are well ordered by the relation $<$ of cardinal dominance (see *Df.* 30 below). Let us get a closer look at what is involved.

We define the relation $x \leqslant y$ between sets by "there is a 1–1 correspondence between x and a subset of y". Formally, we have:

Definition 29. $x \leqslant y$ for $(Ez)(z \subset y \wedge z \mathrm{\,Sm\,} x)$ where z does not occur in x or y.

We have already proved that $x \leqslant x$, since we know that $\vdash x \mathrm{\,Sm\,} x$. We obviously have transitivity as well. Another property of the relation $\leqslant$ is antisymmetry: $x \leqslant y \wedge y \leqslant x \supset x \mathrm{\,Sm\,} y$. This is the famous theorem of Schröder–Bernstein, which states that if x is similar to a subset of y and y is similar to a subset of x, then $x \mathrm{\,Sm\,} y$. This theorem can be proved without aid of the axiom of choice (see Rosser [2], p. 353).

We can now define the relation of cardinal dominance between sets:

Definition 30. $x < y$ for $x \leqslant y \wedge \sim x \mathrm{\,Sm\,} y$ where x and y are any terms.

From the foregoing properties of $\leqslant$, we have immediately that $<$ is a partial order (see our discussion in Chapter 5, Section 5.1). That is, $\vdash (x_1)(\sim(x_1 < x_1))$ and $\vdash (x_1)(x_2)(x_1 < x_2 \supset \sim(x_2 < x_1))$ hold. Furthermore, the transitivity of $<$ holds. Given any such partial order, we obtain a total order if we also have the property of connectivity, that is

$$(x_1)(x_2)(x_1 < x_2 \lor x_2 < x_1 \lor x_1 \, \text{Sm} \, x_2).$$

This says that, for any two sets, one must be similar to a subset of the other. This can be proved only with the aid of the axiom of choice. In conjunction with other facts of the ordinal numbers, it is equivalent to the axiom of choice.

The addition of the property of connectivity guarantees that the relation " $<$ " is also a total ordering on sets. A cardinal number is a set of all the sets of a given similarity type, and so cardinal numbers are ordered by comparing their representative elements. That is, given two cardinal numbers $\mathfrak{a}$ and $\mathfrak{b}$, we take $x \in \mathfrak{a}$, $y \in \mathfrak{b}$. The cardinals are ordered as are the sets x and y. This is obviously independent of the choice of the representatives x and y.

To see that this total ordering on cardinals is, in fact, a well ordering, recall that the axiom of choice is equivalent to the assertion that any set can be well ordered. This means that, for any given cardinal $\mathfrak{a}$, every element of $\mathfrak{a}$ can be put into 1–1 correspondence with some representative of an ordinal number α. An ordinal number in **NF**, we recall, is the set of all well-ordered sets of a given order type. This means that, for every cardinal number $\mathfrak{a}$, there is at least one ordinal α whose elements are similar to the elements of $\mathfrak{a}$. The class of all ordinals whose elements are similar to the elements of a given cardinal is thus nonempty, and therefore it has a least element in the ordering of the ordinals. In this way, we associate with each cardinal $\mathfrak{a}$ the smallest ordinal α (smallest in the ordering of the ordinals) whose elements are similar to the elements of $\mathfrak{a}$. Of course, the elements of any ordinal β smaller than α will have cardinality less than $\mathfrak{a}$, since α is the smallest ordinal whose elements have the cardinal type $\mathfrak{a}$. Finally, given a nonempty set of cardinal numbers, we associate with each cardinal in the set its unique smallest ordinal as previously defined. The corresponding set of ordinals is well ordered and has a smallest element, and the cardinal associated with this ordinal will be smallest in the ordering of the cardinals.

Thus, the axiom of choice yields a well ordering of the cardinal numbers. If we can show that the cardinals are not well ordered in **NF**, it will follow that the axiom of choice fails in **NF**. Specker succeeds in proving that the cardinals are not well ordered in **NF**. We shall not include the details of Specker's proof. The interested reader can consult Specker [1] and also the discussion in Quine [4], pp. 294–295.

Now, once we have established that the cardinals are not well ordered, we can observe that the set of all finite cardinals is well ordered. A cardinal in **NF** is finite or infinite according to whether the sets which make up the cardinal are finite or infinite. By Definition 15, a set is finite precisely if it belongs to some natural number. No cardinal is empty, since $x \in \text{Nc}(x)$ for all x. A finite cardinal is thus a nonempty natural number. It is not too difficult to prove that the finite cardinals are well ordered. Since the set of all cardinals is not well ordered, it follows that there is an infinite cardinal. The elements of such an infinite cardinal cannot be finite sets.

Thus, let x be any infinite cardinal and y any element of x. Since y is infinite and does not belong to any natural number, we obtain the following theorem: For any natural number n, there is a set (namely y) that is not an element of n. Formally stated, we have

$$\vdash (x_1)(x_2)(x_1 \in x_2 \land x_2 \in N \supset (Ex_3)(x_3 \notin x_1)).$$

But this theorem is precisely the corollary to Theorem 28 of Chapter 3, and we have already seen in our previous discussion that this is all that is needed to prove the remaining Peano postulate in **NF**. We thus obtain the last remaining Peano postulate in **NF** as a corollary of the failure of the axiom of choice![†]

Any of the foregoing derivative results of the disproof of the axiom of choice can be considered an axiom of infinity in **NF**. The existence of an infinite cardinal, or just the proof of the remaining Peano postulate, will imply that V is infinite; that is, V belongs to no natural number.

In Rosser [2], which is based on Quine's **NF**, the remaining Peano postulate was assumed as an additional axiom, since the foregoing devious proof of a theorem of infinity in **NF** was not known at the time.[‡] The foregoing proof of the theorem of infinity clearly shows how failure of stratification of certain terms in **NF** does not exclude the possibility of deducing the desired result by other means. The von Neumann natural numbers, used in **F** to prove the last Peano postulate, were not stratified in **NF**, thus blocking our direct method of proof in **NF**. Nevertheless, the remaining postulate is provable in **NF**. In the same way, we cannot be certain that Cantor's paradox is not deducible in **NF** simply because the usual proof of Cantor's theorem is blocked by an unstratified term.

In **ST** we encountered essentially the same difficulty as that in **NF**, i.e. that the von Neumann natural numbers are not definable in type theory. But in **ST** we knew that we had to posit an axiom of infinity since we could exhibit a model of **ST** in a finite simple type hierarchy. By a model-theoretic analysis of **ST** we could see that our axioms did not necessitate an infinite model. Since no model for **NF** is known, the possibility that a theorem of infinity could be proved was, until its proof was discovered, an open question.

Returning for a moment to the axiom of choice in **NF**, we observe that the generalized continuum hypothesis also fails in **NF**, since the axiom of choice is a consequence of the generalized continuum hypothesis. The generalized continuum hypothesis is consistent with **ZF**, and so we have another deviation of **NF** from **ZF**. Since mathematicians sometimes like to assume the continuum hypothesis in order to simplify certain questions of cardinality, and since they very often assume the axiom of choice, it would seem that we could exclude **NF** as a foundation simply by virtue of the fact that these two useful assumptions, consistent with **ZF** and other set theories, are not consistent with **NF**.

Again there is a possibility of responding to this criticism of **NF**. Clearly the non-Cantorian sets are the cause of the failure of the axiom of choice. Thus, we can still apply the axiom of choice to Cantorian sets in **NF**. Since Cantorian sets are the only ones that mathematicians ever work with anyway, the non-Cantorian ones being an oddity of **NF**, no known or useful application of the axiom of choice is excluded by restricting it to Cantorian sets. Of course, we have no proof that such a restricted use of the axiom of choice is consistent with **NF**. But there is no reason to suppose that this restricted use of the axiom leads to contradiction when added to **NF** if **NF** is consistent to begin with.

These considerations seem to show that we cannot reject **NF** as a foundation simply because the axiom of choice generally fails, for we can save the usual applications of the axiom by restricting the sets involved to Cantorian sets. Of course, the constant necessity of introducing

---

[†] The author would like to thank Nicholas Goodman for making available the details of his formal proof of the axiom of infinity in **NF** based on Specker's disproof of the axiom of choice. It should be noted that Specker [1] notes that the axiom of infinity is provable in **NF** as a corollary of the failure of the axiom of choice.

[‡] An appendix contained in a recently reprinted edition of Rosser [2] gives a simplified version of Specker's proof of the theorem of infinity and points out that it is no longer necessary to assume the fifth Peano postulate.

this supplementary condition is irritating. Clearly, though, the single most important question for **NF** is consistency. In another important paper, Specker has cast some light on this question. This paper furnishes the basis for much of our discussion in the following section.

7.4. NF and ST; typical ambiguity

We have already noted, while defining the notion of stratification, that stratification and typical ambiguity are related. Indeed, stratification is one way to deal with typical ambiguity. If we restrict the wffs of **NF** to stratified formulas, we obtain a system much like **ST**. Yet, it is not immediately clear just how strong such a system is, since typical ambiguity with respect to **ST** is more of a relaxation of formal rigor than a precise generalization of the system. In this section, we show how a rigorous and careful analysis of the notion of typical ambiguity leads to a clarification of the precise relationship between **NF** and **ST**. Our ideas follow closely the paper Specker [2].

We consider now the system **ST**, but without the axiom of infinity **ST**.3. Our first meaning of typical ambiguity is contained in the following proof-theoretic observation about **ST**, without infinity: given any wff A of **ST** without **ST**.3, let A^+ be the wff obtained from A by raising the type superscript on every term of A by exactly one. We call A^+ the *type lift* of A. Then a wff of **ST**, without axiom of infinity, is provable only if its type lift is provable. That is, if $\vdash A$ then $\vdash A^+$. To see that this is true, we observe that the type lift of any axiom **ST**.1 or **ST**.2 is an axiom of the same kind. Furthermore, the operation of type lifting preserves our rules of inference. Thus, a proof of A can be immediately translated into a proof of A^+.

Exercise. Prove rigorously, by mathematical induction on the length of the proof of A, the foregoing metatheorem (if $\vdash A$, then $\vdash A^+$) for **ST** without **ST**.3.

For the remainder of this section, **ST** *will always mean* **ST** *without* **ST**.3.

We now have at least one precise notion of typical ambiguity in **ST**. It is that provability is preserved by type lift. It is only natural to ask whether or not provability is also preserved by a lowering of types. That is, is it true that $\vdash A^+$ only if $\vdash A$? The answer is no. Without the axiom of infinity, any finite simplified type hierarchy is a model for **ST**. Thus, the domain of individuals T_0 may have only one element. In that case, $T_1 = \mathcal{P}(T_0)$ has two elements, $T_2 = \mathcal{P}(T_1)$ has four elements, and, generally speaking, T_n has 2^n elements. Moreover, we can prove within **ST** that there are at least 2^n different sets of type n. Let us see this briefly.

We define V^1 as $\{x_1^0 \mid x_1^0 = x_1^0\}$, and Λ^1 as $\{x_1^0 \mid x_1^0 \neq x_1^0\}$. We have, as usual, the theorems $\vdash (\forall x_1^0)(x_1^0 \in V^1)$ and $\vdash (\forall x_1^0)(x_1^0 \notin \Lambda^1)$. From the first of these we have $\vdash (Ex_1^0)(x_1^0 \in V^1)$ and from the second we obtain

$$\vdash \sim (Ex_1^0)(x_1^0 \in \Lambda^1).$$

Thus, by the axiom of extensionality, we easily prove $\vdash V^1 \neq \Lambda^1$, and thereby $\vdash (Ex_1^1)(Ex_2^1)(x_1^1 \neq x_2^1)$. This proves that there are at least $2^1 = 2$ things of type 1. We can now iterate this process, using our usual definitions of union, intersection, and the like. We can form the terms $\{V^1\}$, $\{V^1, \Lambda^1\}$, $\{\Lambda^1\}$, Λ^2. These sets will all be different terms of type 2, and so we can prove that there are at least $2^2 = 4$ different sets of type 2. We leave as an exercise to the reader the completion of our inductive argument.

Now, the wff $(Ex_1^1)(Ex_2^1)(x_1^1 \neq x_2^1)$ is the type lift of $(Ex_1^0)(Ex_2^0)(x_1^0 \neq x_2^0)$. If our converse rule is to hold, then this latter formula must be a theorem since the first one is. But it is easy to see that this latter formula cannot be a theorem, since we have a model for **ST** in which there is only one individual in T_0. There is thus a model in which $(Ex_1^0)(Ex_2^0)(x_1^0 \neq x_2^0)$ is false, and so it cannot be a theorem.

Suppose we now consider the theory obtained by adding to **ST** the general rule $\vdash A^+$ only if $\vdash A$.† Our argument from the preceding paragraph shows that we obtain a strictly stronger theory. In fact, finite type hierarchies will be excluded as models with the addition of such a rule. To see this, suppose that a simple type hierarchy with T_0 of finite cardinality n is a model of **ST** with our added rule. Now, as we have seen, we can prove in **ST** that there are 2^n different sets of type n. More particularly, we can prove

$$(Ex_1^n)(Ex_2^n) \ldots (Ex_{2^n}^n)(x_1^n \neq x_2^n \wedge \ldots x_1^n \neq x_{2^n}^n \wedge x_2^n \neq x_3^n \wedge \ldots \wedge \ldots \wedge x_{2^n-1}^n \neq x_{2^n}^n).$$

Call this formula A^n. A^n involves only variables of type n, and it asserts the existence of 2^n different sets of type n. With our added rule, we can therefore prove $\vdash A^{n-1}$, obtained by lowering the types in A^n by one, and this asserts the existence of 2^n sets of type $n-1$. Proceeding inductively, that is, by iterated application of our new rule, we can prove $\vdash A^0$, which is obtained by n iterations of our rule starting with $\vdash A^n$, and which asserts the existence of $2^n > n$ objects of type 0. Since A^0 is provable if we use our added rule, it is true in every type hierarchy that is a model of **ST** with our added rule. Thus, the assumption that T_0 has only n elements is false. Consequently, no finite hierarchy is a model for **ST** with the added rule.

We speak of **ST** with our added rule as **ST** *with typical ambiguity*. Our rule of typical ambiguity is not a genuine axiom of infinity since it does not enable us to prove within **ST** that some set of **ST** is infinite. Of course, the technique of the preceding paragraph can be applied to prove that V^1, say, has any given finite number of elements. But to conclude from this that V^1 is infinite involves the meta-induction that we have presented above, and this cannot be carried through within **ST** itself, even with typical ambiguity. Also, a model-theoretic proof of the consistency of **ST** with typical ambiguity relative to **ST** can be given. We sketch the details of this proof:

Let X be the set of all wffs that are theorems of **ST** with typical ambiguity according to *one application* of our new rule. Precisely, a wff A is in X if and only if A^+ is a theorem of **ST**. We wish to show that every finite subset of X is consistent relative to **ST**. Since only a finite number of wffs can be used in any proof, only a finite number of the new theorems added by one application of our rule can appear in any proof. It follows that the whole set X will be consistent relative to **ST**. Let $A_1, \ldots, A_m$ be m wffs of X. This finite set of wffs has a model if and only if the conjunction $(A_1 \wedge A_2 \wedge \ldots \wedge A_m)$ has a model. Now, $(A_1 \wedge A_2 \wedge \ldots \wedge A_m)^+$ is the same as $(A_1^+ \wedge A_2^+ \wedge \ldots \wedge A_m^+)$. Since each A_i in our finite set is in X, the wff A_i^+ is a theorem of **ST** for each i. That is, $\vdash_{ST} A_i^+$ for $1 \leqslant i \leqslant m$. Thus, $\vdash_{ST} A_1^+ \wedge A_2^+ \wedge \ldots \wedge A_m^+)$, and thus $\vdash_{ST}(A_1 \wedge A_2 \wedge \ldots \wedge A_m)^+$. The type lift of the conjunction of our wffs is a theorem of **ST**, and is thus true in every model of **ST**.

Since we have excluded **ST**.3 from **ST** in this discussion, any simple type hierarchy $T = \{T_0, T_1, \ldots\}$ will be a model for **ST**. Given T, let $U = \{U_0, U_1, \ldots\}$ be the denumerable

---

† For any wff of **ST**, let A^n represent the wff $A^{++\cdots+}$ (to n occurrences of $+$) obtained from A by exactly n applications of type lift. Then our new rule can be stated: If $\vdash_{ST} A^n$, then $\vdash A$.

collection defined by the condition that $U_i = T_{i+1}$ for all natural numbers i. Thus, $U_0 = T_1$, $U_1 = T_2$, and so on. U is also a simple type hierarchy, for the relationship between the sets T_i is given by $T_{i+1} = \mathcal{P}(T_i)$, and this relationship obviously carries over to U. Thus, U is also a model for **ST**. Now, the wff

$$(A_1 \wedge A_2 \wedge \ldots \wedge A_m),$$

because of its formal relation to its type lift $(A_1 \wedge A_2 \wedge \ldots \wedge A_m)^+$, has exactly the same truth set in U as its type lift has in T. Since the type lift is true in T (because it is a theorem and thus true in every model), the wff $(A_1 \wedge A_2 \wedge \ldots \wedge A_m)$ must be true in U. Consequently, this wff is consistent, since it holds in some model U of **ST**. The finite set $A_1, \ldots, A_m$ thus has a model, and is therefore consistent. Since this is an arbitrary finite subset of X, the whole set X is consistent and no contradiction can ensue by a single application of our new rule.

To complete the demonstration, it suffices to observe that the proof of any wff A as a theorem of **ST** with typical ambiguity can involve only a finite number of applications of our new rule. As we observed in the footnote on page 230, we can state our new rule in the form: If $A^{++\cdots+}$, to any finite number n of type lifts, *is a theorem of* **ST**, then A is a theorem in our enlarged system. An inductive application of the foregoing proof yields that **ST** with typical ambiguity is consistent relative to **ST**.

The consistency of **ST** with typical ambiguity raises the question of whether we obtain a consistent system if we add to **ST** the stronger principle $\vdash A \equiv A^+$ for any closed wff A of **ST**. We call the system obtained from **ST** by adding this stronger principle, **ST** *with complete typical ambiguity*. **ST** with complete typical ambiguity obviously contains **ST** with typical ambiguity as a subsystem.

ST with typical ambiguity roughly corresponds to the system obtained by restricting the wffs of **NF** to stratified formulas. Clearly we have typical ambiguity in such a system. But a more thorough model-theoretic analysis is necessary to show that **ST** with typical ambiguity is essentially the same as **NF** restricted to stratified formulas. What Specker has shown is the more interesting result that **ST** with *complete* typical ambiguity is essentially equivalent to **NF** itself. More precisely, Specker proves that there is a model of **ST** with complete typical ambiguity if and only if there is a model for **NF**. Since we have already seen that **NF** has certain anomalies, this result shows that the added principle $\vdash A \equiv A^+$, A is a sentence, is very strong indeed. Intuitively, it also seems unnatural that **ST** with complete typical ambiguity is consistent, since our added principle says that everything true in one type must be true in any other type. This generally is not true of our simplified hierarchies where each set T_{n+1} is the power set of T_n. It may be that no model T of **ST** with complete typical ambiguity will be a simplified type hierarchy. T may be some nonstandard model of **ST** in which the sets T_i are related in a more "unnatural" way than the relatively simple relation of power set of a simplified type hierarchy. Yet, we have no proof that **ST** with complete typical ambiguity is contradictory. Specker's result throws light on the question of the consistency of **NF** by showing the exact relation of **NF** to **ST**, but the question of the consistency of **NF** remains open.

We close this section by briefly presenting the idea of Specker's proof that **ST** with complete typical ambiguity is consistent if and only if **NF** is. For this purpose, we suppose, for the reaminder of this section, that **NF** and **ST** are formalized without the use of a primitive term operator of abstraction. This makes it easier to visualize models of the systems, since the interpretation of the axiom of abstraction in any model is now determined by our assignment of a relation to the predicate letter "$\in$" without further specification of how to interpret the abstraction operator. We still refer to these systems as "**NF**" and "**ST**" rather than adding the asterisk to write "**NF***" and "**ST***". Moreover, since we speak not just of simplified type hierarchies, but of models of **ST** in a general sense, we should also recall the fact that we have a first-order formulation of **ST**. At this point it might be helpful for the reader to review briefly the discussion of these points in the last sections of Chapter 4.

We remarked informally that our added principle $\vdash A \equiv A^+$, A a sentence, amounts to saying that everything true in one type is true in another. Precisely, let $T = \{T_i\}$, $i \in N$, be any model of **ST** with complete typical ambiguity, and let U be the structure defined by $U_i = T_{i+1}$. (We naturally suppose that the interpretation of "$\in$" remains unchanged; that is, $a \in_U b$ holds if and only if $a \in_T b$ holds when a and b are any two elements of the domain of the interpretation U. The original model T may, of course, be any model, and not necessarily a standard one.) Then U_i is also a model of **ST** with complete typical ambiguity. In fact, U and T are elementarily equivalent as models of **ST**. To see this, let A be any sentence of our system **ST** with complete typical ambiguity. Because of our added rule, A is true in T if and only if A^+ is true in T. But A is true in U if and only if A^+ is true in T as follows from the formal relationship between A and A^+ and from the relationship between our two structures. Thus A is true in T if and only if A is true in U, and so the two structures are elementarily equivalent. Since T is a model of **ST** with complete typical ambiguity and consequently satisfies all the theorems of that system, U is also a model.

Conversely, let T be any model of **ST** which is elementarily equivalent to the structure U defined by $U_i = T_{i+1}$. Then T is a model of **ST** with complete typical ambiguity. This follows immediately from the fact that all biconditionals of the form $A \equiv A^+$, A a sentence, will hold in such a model. This, in turn, follows from the fact that A is true in U if and only if A^+ is true in T, and A is true in U if and only if A is true in T, since U and T are assumed to be elementarily equivalent.

Suppose now that we have a model T of **ST** that is not only elementarily equivalent to U as just defined, but isomorphic to it. This would mean that there is a 1–1 mapping f from the union of all the sets T_i to the union of all the sets $U_i = T_{i+1}$ that satisfies the following conditions: For all x, x is an element of T_i if and only if $f(x)$ is an element of $U_i = T_{i+1}$, and xEy holds if and only if $f(x) Ef(y)$ holds, where E is the relation that interprets the predicate letter "$\in$" of **ST** in the model T. We say informally that "there is an isomorphism between T_i and T_{i+1} for all i". Now any model T admitting an isomorphism between T_i and T_{i+1} for all i is obviously elementarily equivalent to U as previously, for the notion of isomorphism is strictly stronger than the notion of elementary equivalence. Thus, any model T admitting such an isomorphism between T_i and T_{i+1} for all i is a model for **ST** with complete typical ambiguity as follows from our previous results. What Specker succeeds in proving is that if **ST** with complete typical ambiguity is consistent, then there is some model T of **ST** admitting an isomorphism between T_i and T_{i+1} for all i (see Specker [2], p. 119). In general, elementarily equivalent models of a given theory will not be isomorphic, and so the proof of this result depends essentially on the particular structure of **ST** with complete typical ambiguity. The fact that the existence of a model for **ST** with complete typical ambiguity implies the existence of a model T with T_i isomorphic to T_{i+1} for all i shows just how strong the added rule $\vdash A \equiv A^+$, A a sentence, really is.

We can now use this basic theorem to show that there is a model of **ST** with complete typical ambiguity if and only if there is a model of **NF**. Let T be a model of **ST** with complete typical ambiguity, and let T admit an isomorphism f from T_i onto T_{i+1}. We construct a model $\langle D, g \rangle$ for **NF** by letting the domain of the model be the set T_0 of individuals. g assigns to the predicate letter "$\in$" of **NF** the set E^* of all ordered pairs $\langle a, b \rangle$, such that $\langle a, b \rangle$ is in E^* if and only if $\langle a, f(b) \rangle$ is in the relation E, which is the interpretation of "$\in$" of **ST** in the model T. We have an axiom of extensionality and an axiom of abstraction in both theories, and the nature of our isomorphism f assures us that $\langle D, g \rangle$ will be a model for **NF**.

Given now a model $\langle D, g \rangle$ for **NF**, where E^* is the relation assigned to "$\in$" of **NF** by g, we construct a model for **ST** with complete typical ambiguity. The domain of our model for **ST** with complete typical ambiguity will be the set of all ordered pairs $\langle a, n \rangle$ where a is in D and n is a natural number, i.e. the set $D \times N$. For each n, the set T_n of objects of type n will be the ordered pairs $\langle a, n \rangle$ whose second component is n. The relation E which is the interpretation of "$\in$" in **ST** with complete typical ambiguity is defined by $\langle a, n \rangle E \langle b, m \rangle$ if and only if $m = n+1$ and $\langle a, b \rangle$ is in the relation E^*. From the axioms of extensionality and abstraction of **NF** we easily obtain that this structure T is a model for **ST**. Furthermore, the mapping f defined by $f(\langle a, n \rangle) = \langle a, n+1 \rangle$, for $\langle a, n \rangle$ an element of T_n, is a 1–1 mapping from T_n onto T_{n+1}. It is easy to check that f is an isomorphism, and so it follows that our model is a model for **ST** with complete ambiguity. Thus, **NF** is consistent if and only if **ST** with complete typical ambiguity is.

Exercise. Write a short essay entitled "**NF** as a foundation for mathematics". Bring in carefully each of the positive and negative points raised in this chapter. What is the crucial open problem for **NF**?

7.5. Quine's system ML

In 1940, Quine published his work *Mathematical Logic* in which he presented a system **ML** which is related to **NF** in much the way **NBG** and **ZF** are related. That is, Quine considers the addition of proper classes to **NF**, just as we add proper classes to **ZF** to obtain **NBG**. All the sets of **NF** are classes, but there are classes of **ML** that cannot be the element of anything else. These are proper classes. Actually, the relation between **ML** and **NF** is more like the relation between **MKM** and **ZF** than that between **NBG** and **ZF**, since Quine does not restrict his axiom of abstraction to wffs predicative over the sets of **NF**.

Quine's original version of **ML** was proved inconsistent by R. Lyndon and J. B. Rosser (independently), but the inconsistency was basically attributable to a fault in exposition as Wang later showed. Wang corrected the fault and proved that the correct version of **ML** was consistent relative to **NF**. Our presentation of **ML** is based partly on the revised edition of Quine [3] and partly on Wang [1].

We formulate **ML** without use of a term operator of abstraction. **ML** is a first-order theory with one primitive predicate letter "$\in$" of degree 2. As usual, we write "$(x \in y)$" instead of "$\in (x, y)$". Our only terms are variables and consequently our wffs are all the first-order wffs definable by applying quantification and sentential connectives to our prime wffs. We begin by defining the notion "x is a set":

Definition 1. $M(x)$ for $(Ey)(x \in y)$, where y is different from x.

This is the same definition we gave for the system **NBG**. A set is something that is an element of something else.

Again, as with the system **NBG**, we define the restriction of a quantifier to sets: $(x)(M(x) \supset A(x))$ is the restriction of the quantifier (x) to sets and $(Ex)(M(x) \land A(x))$ is the restriction of (Ex) to sets. A wff in which every occurrence of every quantifier is restricted to sets is called *predicative*. We now state the axioms of **ML**.

ML.1. Same as **NF.1**. This is the axiom of extensionality.

ML.2. Where $A(x)$ is any wff of **ML** and where y is different from x and does not occur in $A(x)$, then $(Ey)(x)(x \in y \equiv M(x) \land A(x))$ is an axiom.

This is the axiom of class existence. It is clearly the same as the axiom of class existence for **MKM** of Chapter 5. The usual deduction of Russell's paradox is avoided in the same way as in that system. **ML.2** posits the existence of classes whose elements are sets, but we need an axiom of set existence. We posit:

ML.3. Let $A(x)$ be a wff of **ML** that is stratified and that is obtained from a stratified wff by restricting all bound variables to sets. $A(x)$ is a stratified, predicative wff. Let all the free variables of $A(x)$ occur in the list $x, y_1, \ldots, y_n$. Then, the following is an axiom:

$$M(y_1) \land M(y_2) \land \ldots \land M(y_n) \supset (Ew)(M(w) \land (x)(x \in w \equiv M(x) \land A(x)))$$

where w does not occur in $A(x)$.

This completes the axioms of **ML**.

ML.3 says, in effect, that the sets of **NF** are the sets of **ML**. It says intuitively that, for any condition given by a predicative, stratified wff $A(x)$ whose free variables are sets, there exists a *set* w whose elements are precisely those sets satisfying the condition. Because of **ML.2**, we also have the possibility of other classes that are not sets.

Exercise. Prove that there is a model for the axioms **ML.1** and **ML.2** in a domain with only one element.

The advantage of operating with **ML** is that certain difficulties of **NF** are avoided. For example, the principle of mathematical induction can be proved to hold for any wff, stratified or not. A theorem of infinity is also provable in **ML**. **ML** is a very liberal theory and liable to be dangerously close to the paradoxes because of its liberality. However, Wang's relative consistency proof of **ML** with respect to **NF** shows that **ML** is no more risky than **NF**. Because of this, logicians who work on Quine's systems focus their attention on **NF**, since its form is simpler and metatheorems are easier to prove for it than for **ML**. However, **ML** is more flexible as a foundation.

Exercise. Define the universal class V in **ML** and prove that $V \in V$. This differs from the universal class of **NBG**. That class is not an element of itself. In **ML**, the class of all sets is itself a set.

For a detailed development of arithmetic in **ML** (which nevertheless is quite close to our development in **NF**), consult Quine [3].

7.6. Further results on NF; variant systems

A number of interesting results have been obtained concerning various extensions and ramifications of **NF**. Perhaps the most interesting is Jensen's consistency proof for a modified version of **NF**, **NFU** (New Foundations with *Urelements*). First we describe Jensen's system.

NFU is a first-order system with equality $=$ and one binary nonlogical predicate "$\in$". Besides the logical axioms (including the axioms of equality), **NFU** has two proper axioms:

NFU.1. $(x)(y)((Ez)(z \in x) \wedge (z)(z \in x \equiv z \in y) \supset x = y)$,

where x, y, and z are all different.

NFU.2. This is identical with **NF.2***.

Thus, the only difference between **NFU** and the system **NF*** (**NF** without a *vbto* of abstraction) is the presence of the condition "$(Ez)(z \in x)$" in the extensionality axiom. This makes equality of sets extensional as before but allows for the possibility of distinct *Urelements*, that is, distinct objects having no elements. Since **NFU.1** is provable in **NF**, **NF** is at least as strong as **NFU**. In fact, as it turns out it is considerably stronger, for Jensen establishes that **NFU** is equiconsistent with the theory **ST** of simple types without **ST.3**. Thus, in particular, no theorem of infinity is provable in **NFU**.

Jensen's approach in proving the relative consistency metatheorem for **NFU** involves a modification of Specker's approach described in Section 7.4. Just as Specker proves the consistency of **NF** relative to **ST** with complete typical ambiguity (without **ST.3**) by finding appropriate models of **ST** (those with an isomorphism between T_i and T_{i+1}), so Jensen finds appropriate models of **ST** (without **ST.3**) which yield models of **NFU**.

Another consistency result involving a weaker form of **NF** is due to the Russian logician Grishin (see Grishin [1]). Let **NF**$_n$ designate the fragment of **NF** that consists of **NF.1** plus all those instances of **NF.2** for which stratification can be determined by using only the integers $0, 1, \ldots, n-1$. Grishin proves that **NF**$_3$ is consistent. His method involves examining the relationship between **NF**$_3$ and the system **ST**$_3$ (without an axiom of infinity) which is the fragment of **ST** obtained by eliminating from the language all variables x_i^n with type index $n \geqslant 3$ (thus leaving variables with type indices 0, 1, and 2). The proof again makes use of Specker's models.

In another article (see Grishin [2]), Grishin shows that **NF** is consistent relative to the fragment **NF**$_4$. Moreover, Boffa and Crabbé have established that **NF** cannot be axiomatized in **NF**$_3$ (provided **NF** is consistent), thereby proving that **NF**$_4$ is essentially stronger than **NF**$_3$ (see Boffa [1] where other references are given). Thus, there is an essential gain in complexity in the passage from **NF**$_3$ to **NF**$_4$.

These results all concern fragments of **NF**. Other results involve extensions of **NF**. One of the main extensions concerns Rosser's proposed "axiom of counting" (see Rosser [2], pp. 483 ff.). We have already noted (Theorem 17, Section 7.2) that non-Cantorian sets exist in **NF**. However, the only Cantorian sets that have ever been proved to exist are infinite (e.g. the universal class V of Theorem 17). The question naturally arises whether there are any finite non-

Cantorian sets. Rosser's axiom of counting asserts that all finite sets are Cantorian. Formally stated: $(x_1)(x_2)(x_1 \in x_2 \wedge x_2 \in N \supset \mathrm{Can}(x_1))$.

The basic result is due to Orey (see Orey [1]) who showed that the consistency of **NF** can be proved within **NF** if the axiom of counting is itself provable in **NF**. By Gödel's second theorem, we can thus conclude that if **NF** is consistent, the axiom of counting is not provable in **NF**. It is therefore either independent of **NF** or false. However, no proof of its falsity has been forthcoming and most researchers into **NF** feel that it is probably independent if **NF** is consistent.

In any case, the negation of the axiom of counting is at least consistent with **NF**, and so it is consistent to assume that there are finite non-Cantorian sets in **NF**. This is certainly very counter-intuitive since we might have thought that only very big sets could be non-Cantorian. Hensen (see Hensen [1]) has further shown that it is consistent to assume in **NF** that there are finite sets x having cardinality greater than the cardinality of their powerset $P(x)$ or else that there are sets x having cardinality greater than that of their set of unit subsets $\mathrm{Pu}(x)$. All of these possibilities exist only because the usual proof in **ZF** set theory that, for any set x, $\mathrm{Pu}(x) \, \mathrm{Sm} \, x < P(x)$ fails in **NF** because the "obvious" 1–1 correspondences involved are not stratified.

7.7. Conclusions

Quine originally propounded his systems as a basis for the logistic thesis that mathematics reduces to logic (see Quine [1] and [3]). To do this is to argue that **NF** (or **ML**) is a "naturally true" or "universally valid" system. Since some results of **NF** contradict our Cantorian intuition of sets, most mathematicians would reject such a claim. However, the paradoxes of set theory show that our original intuition is itself suspect. Consequently, something of the old intuition will probably be lost in any reformulation of set theory. In any case, the motivation of Quine's approach is clear. Recall that the notable failure of type theory in defending the logistic thesis was that the axiom of infinity, necessary for mathematics, was not a universally valid principle of type theory (it is not true in every model of the other axioms of **ST**). However, a theorem of infinity is provable in **NF**, and so any support for the view that **NF** is a "natural" system would tend to support the logistic thesis.

In recent years, Quine seems to have changed his position, at least to some degree. In any case, he clearly adopts a different philosophic attitude in Quine [4], in which he even refers to the unnatural character of **NF** (see Quine [4], p. 299). An important question for **NF** remains over the relative consistency of **NF** to **ZF**.

Chapter 8

Categorical Algebra

IN MODERN mathematics, one studies abstract structures of various kinds: groups, rings, modules, topological spaces, topological vector spaces, and the like. Categorical algebra[†] represents a basic approach to the study of such structures. In this approach it is not individual groups, rings, or spaces that are the primary interest of study. Rather, one studies various systems of functions which relate a collection of comparable structures, as well as the connections among several such systems. The study proceeds by examining how a given system of functions behaves under the operation of functional composition (applying one function to another). However, what often happens is that interesting properties of a given species of structure can be recovered solely from the known properties of the operation of functional composition within the system of functions associated with the collection of structures of the given species. In this way, one can pay less attention to the individual sets and their elements once some of the basic properties of the system of functions are established. Moreover, it can be observed that almost all interesting systems of functions satisfy certain basic and simple properties, such as the associative property of the composition of functions. This leads to the axiomatic definition of a *category* as an abstract structure of a certain kind that satisfies these basic properties. Of course, we shall have as a model for the abstract notion of a category any mathematical structure that satisfies the axioms for a category. Our systems of functions will be among these models, but they will not be the only models. In short, we can proceed, in the spirit of modern mathematics, to an axiomatic study of categories and their structure, as well as the relationships among categories.

The relevance of categorical algebra for foundations is twofold. First, categorical algebra has raised interesting questions about the commonly presumed irreducible character of the notions of set and membership. Since sets and elements are less important in category theory, and since categorical algebra has succeeded in reducing a large number of previously set-theoretic arguments to arguments about the composition of functions, the possibility of founding mathematics on notions other than set and membership is suggested. Second, there are many constructions commonly carried on in categorical algebra that cannot be done in any natural way in most set theories. In fact, some of these constructions cannot be done at all in set theories like **ZF** or **NBG**. Yet, there does not seem to be anything really contradictory or undesirable about these constructions. Thus, if one accepts these categorical constructions as natural, set theory appears to be an unnaturally restrictive and inflexible

---

[†] We will use both "categorical" and "categorial" as one-word adjectives for "category-theoretic", using "categorial" especially (though not exclusively) when there is danger of confusion with the notion "categorical" as applied to theories (cf. Chapter 1, Section 9).

language. This lack of flexibility did not appear in connection with number theory or analysis precisely because the original Dedekind–Frege constructions were designed to recover these disciplines from set theory, and the axioms of our set theories have, in turn, been designed to accommodate these constructions or something like them. Thus, categorical algebra suggests that there might be other constructions germane to the scope of modern mathematics, especially in view of the present emphasis on the study of a wide variety of different abstract structures, which cannot be naturally represented in our presently conceived foundational systems.

Of course one has the obvious alternative of refusing to acknowledge the validity of those constructions of category theory which do go beyond present set theory. Choosing this alternative amounts to taking the "believer" position that some particular system of set theory such as **ZF** is *the* right one, and that intuitively conceived constructions which cannot be expressed in such a system are therefore illegitimate. However, if constructions in category theory, or in any yet-to-be-invented mathematical discipline, are not in fact contradictory, then it seems rather arbitrary to refuse to contemplate the use of these constructions or to ignore the problem of building a more flexible foundation in order to accommodate them. After all, the primary purpose of a foundational system is to furnish a foundation for mathematics, not to restrict mathematical inquiry by setting up arbitrary limits on what should be considered legitimate mathematics. Such paradoxes as Russell's show that some sort of restriction on intuition is necessary, but the precise problem that studies of foundations attempt to solve is that of discovering what the necessary restrictions are. It therefore seems more reasonable to take the position that any conceivable type of construction or method of reasoning that does not in fact lead to contradiction is potentially worthy of incorporation into mathematics. This is particularly true if the construction yields interesting results.

More will be said about the foundational problems of category theory in later sections. We turn now to the problem of clarifying the basic notions, which we have only discussed in a general way here. Our method will be to take a simple example leading up to the abstract definition of a category by means of a first-order language of abstract categories.

8.1. The notion of a category

In 1945, Eilenberg and MacLane defined the notion of a category. Their immediate objective was a simplification of certain aspects of algebraic topology, but the concepts of categories, functors between categories, and natural transformations between functors have since proved to be of significance as unifying concepts in other branches of mathematics. We discuss these notions by using, for purposes of illustration, the example of the theory of groups. We have chosen group theory because of its simplicity and accessibility as an example of categorical ideas. However, the reader should not suppose that the most significant contribution of category theory has been to the theory of groups.

In abstract algebra, a structure is defined as a set S together with certain operations on the set. These operations are mappings from S^n to S, where n is a natural number. We allow $n = 0$, and we intend that an operation of degree 0 on S is simply a constant element of S.

For example, consider the well-known structure of a group. This consists of a set S together with three operations, $+$, $'$, and e, of degree 2, degree 1, and degree 0 respectively. The operation of degree 2 will be called addition, the operation of degree 1 will be called inversion, and the constant will be called the neutral element. What distinguishes a group from other similar structures (i.e. structures with the same number of operations of the same degree) is that we require the group operations to satisfy certain basic properties or axioms. These are:

$$(1) \quad +(+(x, y), z) = +(x, +(y, z)); \qquad (2) \quad +(e, x) = +(x, e) = x;$$
$$(3) \quad +(x, '(x)) = +('(x), x) = e.$$

From now on, we follow the more traditional way of writing the function symbol "$+$" between the arguments, of writing x' instead of $'(x)$, and of suppressing unnecessary parentheses.

By a group we mean any set S together with the above-defined operations that satisfy the three properties we have specified. We can easily express the three axioms of a group in a first-order language with equality, one constant letter and two function letters. An equivalent definition of a group would then be the following: A group is any model for the first-order language of group theory.

However, most algebraists do not study first-order group theory. What is more frequently done is to define within set theory (the particular set theory in which one is working) an appropriate predicate $G(x, y, z, w)$ where x is a set, y is a function from x^2 to x, z is a function from x to x, and w is a constant. Any ordered quadruple satisfying the predicate G is a group.

The study of groups continues with the study of the "inside" structure of a group. That is to say, one examines the consequences for the structure of a group of various conditions. How many elements does it have? What elements are generators? What subgroups does it have, what quotient groups does it have, and so on?

The relationship between groups is also important. The basic tool for this study is the notion of a *homomorphism* between two groups S and S^*. This is a function h from S to S^* that "preserves" the group structure, i.e. such that $h(x+y) = h(x)+h(y)$, $h(x') = (h(x))'$, and $h(e) = e^*$. (As it turns out, the latter two conditions can be proved to hold if the first condition does.) The image $h(S)$ under a homomorphism may not be the whole set S^*, but it is easily shown that $h(S)$ is a subgroup of S^*. Thus, there may be many different homomorphisms from S to S^*.

Given a group S, a *homomorphic image* of S is any group S^* for which there exists a homomorphism h from S to S^* such that $h(S) = S^*$ (i.e. h is onto). If S^* is a homomorphic image of S by h and if h is an injection, that is, a 1–1 function, then we say that S and S^* are *isomorphic* and that h is an *isomorphism*. In group theory, two isomorphic groups have the same algebraic structure even though the actual sets S and S^* may have different elements. Obviously, isomorphic groups must have the same number of elements, since an isomorphism is a bijection.

Now, a given group S will only admit certain groups as homomorphic images, and it will be the homomorphic image only of certain groups. Exactly which groups are so related to S depends on the internal structure of S. More generally, we might think of the internal structure of S as being determined by all the homomorphisms h with domain S and all the homomorphisms h^* going to S from some group S^*. For example, any one-element group $\{e\}$ has exactly one homomorphism to it from any other group, and it has exactly one homomorphism from it to any other group. This fact uniquely characterizes one-element groups $\{e\}$ up to isomorphism, which means that any two groups having this property admit of an isomorphism between them.

Given any two groups S and S^*, let $\mathrm{Hom}(S, S^*)$ be the set of all homomorphisms from S to S^*. Hom is a function which associates a set with every ordered pair $\langle S, S^* \rangle$ of groups.

Let G be the class of all groups and V the class of all sets. Then Hom is a *heterogeneous* operation, namely a mapping from $G \times G$ to V.[†] Thus, the class of all groups has, itself, an algebraic structure.

Associated with this algebraic structure on the class of groups is the *partial* (i.e. not everywhere defined) operation of the composition of homomorphisms. For a function f from a set S to a set S^*, and a function g from a set S^* to a set S^{**}, the composition gf of f with g is defined by the relation $g(f(x)) = gf(x)$ for all $x \in S$. It is easy to show that composition is associative (when it is defined). If S, S^*, and S^{**} are groups, and if f and g are homomorphisms, then it is easy to show that gf is also a homomorphism from S to S^{**}. Thus, the composition of two homomorphisms is a homomorphism. It follows that homomorphism composition is defined under the same conditions as functional composition and that homomorphism composition is associative when defined.

We make this a bit more precise. If $f \in \text{Hom}(S, S^*)$, we speak of S as the *domain* of the homomorphism and S^* as the *codomain* (the range being $f(S) \subset S^*$). We define the composition only when the domain of g is the codomain of f. Thus, composition is properly viewed not as a mapping from $H \times H$ into H (where H represents the class of group homomorphisms) but rather from some subclass K of $H \times H$ to H.

Let us now call the structure presently elaborated, that is to say the class of G of groups together with the operations Hom and composition of homomorphisms, the *category* of groups. There are two basic things that are unwieldy about this structure. One is that our operation of composition is defined on a class, the class of homomorphisms, which is different from our basic class, the class G of groups. The other difficulty is that the operation of composition is not everywhere defined.

The first of our two difficulties can be overcome in the following way. We observe that each group S has a uniquely associated identity homomorphism from S to itself. (Notice that there are, in general, many other homomorphisms from S to S that are not the identity.) This identity mapping is a member of our class H of homomorphisms. Moreover, an identity homomorphism acts as a neutral element under composition: $ih = h$ and $gi = g$, where i is an identity homomorphism and g and h are any homomorphisms for which the indicated composites are defined. We can thus "throw away" the class G of objects by identifying a group S with its identity homomorphism. The domain and codomain of homomorphisms now become other homomorphisms of a special kind, the identity homomorphisms. We can consider the category $\mathcal{G}$ of groups to consist of the class H of homomorphisms, together with the partially defined operation of composition. By doing this, we obtain an algebraic structure defined on one class, the class of homomorphisms, having a partial binary associative operation of composition and certain identity elements. The only thing that prevents this structure from being a semigroup is the fact that our operation of composition is not everywhere defined.

This leads us to our second difficulty which was the fact that the operation of composition is not everywhere defined. This difficulty will not be overcome. Of course, it is possible to extend trivially any operation defined on a given class. We could, if we chose, add some "dummy" element to the category and consider the composition of any two morphisms which was previously undefined as now defined to be the dummy element. It is easy to see that this

---

[†] The general theory of heterogeneous algebras, i.e. those for which the operations are heterogeneous, has been developed in Birkhoff and Lipson [1] and [2].

would preserve associativity. We would thus obtain a semigroup with certain identities which we have previously described. However, we shall not do this.

Before continuing our discussion, we wish to introduce a notational convention. In analysis and in set theory, it is the usual practice to write the composition of the function f with the function g as gf in order that $gf(x) = g(f(x))$. From now until the end of this book, we write the composition of f with g as fg rather than gf, and we will write $(x)f$ instead of $f(x)$. Thus, $(x)fg = ((x)f)g$. The convention is to apply in both informal discussion of functions and in formal systems (to be presented). The reason for this change is that we are now considering the operation of composition as an algebraic operation, and it is customary to write algebraic operations from left to right rather than right to left. We now continue with our discussion.

We have considered the category $\mathcal{G}$ of groups as an example of a category. Yet we have not said what we mean by a category in general. Intuitively, what we mean by a category is a class of *sets with similar* (i.e. comparable) *structure*, which we will call *objects*, together with all the *structure-preserving functions*, which we will call *maps, mappings* or *morphisms*, between them. Thus, other examples of systems we will wish to call categories are the following: The class of topological spaces as objects with continuous functions as mappings; the class of vector spaces as objects with linear transformations as mappings; the class of sets (just plain sets) as objects with functions (just plain functions) as mappings; the class of rings as objects with ring homomorphisms as mapping; the class of abelian groups as objects with group homomorphisms as mappings; the class of modules as objects with module homomorphisms as mappings.

In each of these examples, we can see that the identity function on a given object of the category is trivially structure preserving and thus a map in the category. The identity function on a topological space is continuous and hence a map in the category of topological spaces. Similarly, the identity function on a set is a function, the identity function on a vector space is a linear transformation, and so on. Therefore, we can, in every case, assimilate the class of objects as a subclass of the class of maps by identifying the objects with their identity maps, just as we did with groups and group homomorphisms. In every case we can see that the composition of two maps of the category is a map of the category, and that composition is associative when it is defined. In short, for each of these examples we obtain the same kind of algebraic structure. This structure consists of a basic class of maps with a partially defined operation of composition that is associative when defined and with certain identity maps which intuitively replace objects. It is this common algebraic structure that constitutes the abstract notion of a category.

By an *abstract category*, we thus understand some class M whose elements are called *morphisms (maps, mappings)*. This class has a partially defined binary operation called *composition*, the operation being associative when it is defined. There are certain identity morphisms which are called the *domain* of morphisms for which they are the left identity, and the *codomain* of morphisms for which they are the right identity. Every morphism has a unique domain and a unique codomain; every identity morphism is its own domain and codomain. The composite ab of two morphisms a and b is defined exactly when the codomain of a is the domain of b. The domain of the composite ab is the domain of a and the codomain is the codomain of b.

Notice that we use the three terms "morphism", "map", and "mapping" with complete synonymity to designate the elements of the basic class of a category. This is current in the literature. Another current notational convention is the following: For two identity morphisms A and B, $\operatorname{Hom}(A, B)$ represents the class of all morphisms with domain A and codomain B. When $x \in \operatorname{Hom}(A, B)$ we say that x is a map *from A to B*. The identity morphisms of a category are called *objects*.

We will wish to make the concept of a category even more precise by exhibiting a formal, first-order language of abstract categories. Before doing this we wish to motivate further the abstract study of categories.

An essential idea of category theory is that every particular category has a characteristic structure which distinguishes it from other categories. The category of groups has a different structure from that of rings or of abelian groups. The very notions "group", "ring", and "abelian group" can thus be thought of as characterized by their respectively associated categories. Let us take a simple example.

By a *terminal object* in a category, we shall mean an identity morphism A such that for every other identity morphism B there is a unique morphism f from B to A. In the category of groups, any one-element group will be a terminal object, since there will be one and only one homomorphism from any group into the one-element group. In the category of sets and functions, any one-element set will be a terminal object.

Dually, we define an *initial object* as an identity morphism that is the domain of exactly one morphism f to each identity morphism in the category. *A zero object* is defined as an identity morphism that is both initial and terminal. In the category of groups, any one-element group $\{e\}$ is a zero object, since $\{e\}$ will have exactly one morphism to and from every other group. However, in the category of sets and functions, one-element sets are *not* initial, as there are in general many different mappings from a one-element set into another set. The category of sets has no zero object whereas the category of groups does have zero objects, namely, one-element groups. This fact structurally distinguishes the two categories to some degree.

Of course the foregoing example is trivial, and more interesting structural properties of categories will be forthcoming.

Exercise. Does the category $\mathcal{T}$ of topological spaces and continuous functions have initial objects? Terminal objects? Zero objects?

Three paragraphs ago, we spoke of the definition of "initial object" as one that is defined "dually" to that of terminal object. Informally, the reason is clear: Just take the definition for "terminal object" and substitute "domain" for "codomain" and "codomain" for "domain" and we obtain the definition for "initial object". The following question is obvious: Is there some general notion of duality applicable to categories? The answer is "yes".

Given any category $\mathcal{C}$, let us denote by $\mathcal{C}^{op}$ the category having the same class of morphisms, but such that composition of morphisms is defined as follows: Where a and b are any two morphisms of the class, $c = ab$ in $\mathcal{C}^{op}$ if and only if $c = ba$ in $\mathcal{C}$. Thus, in the dual category, everything is "turned around", domains and codomains switch places and the maps "go the other way". The *dual* of a statement about category $\mathcal{C}$ is the same statement made about category $\mathcal{C}^{op}$. This also will be made more precise when we develop the first-order language of categories. Before proceeding to consider this language, we need to discuss one further question.

In Chapter 5, we defined a function from a set A to a set B as a certain kind of subset of $A \times B$. We now alter this definition slightly. By a function, we now mean an ordered triple

$f = \langle A, B, X \rangle$ of sets where $X \subset A \times B$ is a functional relation in which every element of A occurs as a first element in a pair of X. We call A the *domain* of the function, B the *codomain*, and X the *graph* of the function. It is the graph X that we have previously identified with the function itself. The domain A is still defined in the same way as the set of first elements in the graph. It is the codomain that is really new here, for the set $(A)f$ which is the range of the function may not be equal to the codomain. Since the domain can be recovered from the graph X, it would be possible to make do with ordered pairs $\langle B, X \rangle$ in our extended definition of function, but we shall maintain the ordered triple notation.

Our reason for the extended definition is the following: In the category of sets we want the composition of f and g to be defined only when the codomain of f is the domain of g. However, the notion of a function as a functional relation does not involve the notion of codomain. Moreover, relational composition is defined for any two sets of ordered pairs (although the composite may be the null relation), whereas we wish to restrict the composition of functions to certain cases involving domain and codomain.

In terms of our new definition of function, there is a function whose codomain is any given set B and whose domain is the null set Λ. This is the function $\langle \Lambda, B, \Lambda \rangle$. Notice that the triple $\langle A, \Lambda, \Lambda \rangle$ is not a function unless $A = \Lambda$, since A must be the set of first elements of the graph. By the *null function*, we shall always mean the unique function $\langle \Lambda, \Lambda, \Lambda \rangle$.

Exercise. Show that in the category of sets, the null function is the identity function associated with the null set. Show that the null set is an initial object but not a terminal object of the category of sets.

8.2. The first-order language of categories

We now present a first-order language of categories that bears a relation to the abstract notion of a category similar to that born by the first-order language of group theory to the abstract notion of a group. Our presentation is based essentially on Lawvere [3].

Our system **C** is a first-order theory with equality and one other primitive predicate letter "K" of degree three. $K(x, y, z)$ means intuitively "z is the composition of x with y". There are two primitive singulary function symbols "D" and "C". $D(x)$ and $C(x)$ mean intuitively the "domain of x" and "codomain of x" respectively. The proper axioms of our language, besides the axioms of equality (see Chapter 1), are as follows:

C.1. $(x_1)(D(C(x_1)) = C(x_1) \wedge C(D(x_1)) = D(x_1))$.

The domain of the codomain of x_1 is the codomain of x_1, and the codomain of the domain of x_1 is the domain of x_1.

As usual, our intuitive explanation of the axioms is not part of the formal language.

C.2. $(x_1)(x_2)(x_3)(x_4)(K(x_1, x_2, x_3) \wedge K(x_1, x_2, x_4) \supset x_3 = x_4)$.

The composition of x_1 with x_2 is unique when it is defined.

C.3. $(x_1)(x_2)((Ex_3) K(x_1, x_2, x_3) \equiv (C(x_1) = D(x_2)))$.

The composition of x_1 with x_2 is defined if and only if the codomain of x_1 is the domain of x_2.

C.4. $(x_1)(x_2)(x_3)(K(x_1, x_2, x_3) \supset (D(x_3) = D(x_1) \wedge C(x_3) = C(x_2)))$.

If x_3 is the composition of x_1 with x_2, then the domain of x_3 is the domain of x_1, and the codomain of x_3 is the codomain of x_2.

C.5. $(x_1)(K(D(x_1), x_1, x_1) \wedge K(x_1, C(x_1), x_1))$.

For any x_1, the domain of x_1 is a left identity for x_1 under composition, and the codomain is a right identity.

C.6. $(x_1)(x_2)(x_3)(x_4)(x_5)(x_6)(x_7)((K(x_1, x_2, x_3) \wedge K(x_2, x_4, x_5)$

$$\wedge K(x_1, x_5, x_6) \wedge K(x_3, x_4, x_7)) \supset x_6 = x_7).$$

Composition is associative when it is defined.

The foregoing are the axioms for the theory **C**.

We can now define a category as any structure that is a model for **C**. The elements of the class which is the domain of such a model are the morphisms of the category. We continue the practice of using the terms "morphism", "map", and "mapping" synonymously to refer to elements of our universe of discourse in the same manner that we used the term "set" when we worked with systems such as **ZF**. The reader will recognize this as simply a metalinguistic convention involving no extension of our language.

We could develop the general theory of abstract categories by proceeding to deduce theorems in this system. However, our basic interest is in applying the notions of category theory, and **C** represents a basic language to which various other axioms will be added to obtain a stronger theory. Some general development is desirable for future use.

Definition 1. $\mathrm{Ob}(x)$ for $x = D(x) \wedge x = C(x)$.

An object is a morphism which is equal to its own domain and codomain. Thus, objects appear as special kinds of morphisms, which is in line with our introductory discussion in Section 1. Axiom **C.5** tells us that these objects are identity morphisms of a certain kind. In the following exercise, we ask the reader to recover some desirable properties of objects.

Exercise.   Prove

(1) $\vdash (x_1)(\mathrm{Ob}(x_1) \equiv (x_1 = D(x_1)))$;

(2) $\vdash (x_1)(\mathrm{Ob}(x_1) \equiv (x_1 = C(x_1)))$;

(3) $\vdash (x_1)(\mathrm{Ob}(x_1) \equiv ((Ex_2)(x_1 = D(x_2)) \vee (Ex_2)(x_1 = C(x_2))))$;

(4) $\vdash (x_1)(\mathrm{Ob}(x_1) \equiv ((x_2)(x_3)(K(x_2, x_1, x_3) \supset x_2 = x_3)$

$$\wedge (x_2)(x_3)(K(x_1, x_2, x_3) \supset x_2 = x_3)))$$;

(5) $\vdash (x_1)(D(D(x_1)) = D(x_1) \wedge C(C(x_1)) = C(x_1))$.

Our exercise gives different sets of equivalent conditions for the property of being an object and also gives us another property of the domain and codomain operations, namely that they are idempotent.

From now on, we shall use capital Latin letters as abbreviations for variables restricted to objects. Thus, where $W(x)$ is some wff, we shall abbreviate $(x)(\mathrm{Ob}(x) \supset W(x))$ by $(A)W(A)$ where A is some capital Latin letter not appearing in $W(x)$, and where $W(A)$ is the result of substituting A for x in $W(x)$ for all free occurrences of x. Similarly, $(EA)W(A)$ will be an abbreviation for $(Ex)(\mathrm{Ob}(x) \wedge W(x))$ where A and x are related in the same way.

Departing from the usual way of abbreviating formulas of first-order systems, we introduce a two-dimensional notational abbreviation.

Definition 2. $A \xrightarrow{x} B$ for $D(x) = A \wedge C(x) = B$.

The arrow notation is read "x is a morphism with domain A and codomain B" or "x is a map (mapping) from A to B". We will sometimes render this same notion by the strictly linear notation $x : A \longrightarrow B$.

Definition 3. $(xy) = z$ for $K(x, y, z)$.

In this chapter, we do not bother to state explicitly all the various restrictions on the meta-variables "x", "y", and the like when we state definitions. The reader should now be so familiar with the usual type of restrictions necessary that he can supply them from context. Likewise, as the vernacular will be increasingly used even in the statement of theorems, we shall sometimes omit the asserter sign "$\vdash$", which has a rather formal character and which is not usually employed with vernacular statements. (There is no logical point involved here, only a linguistic one.)

Definition 4.

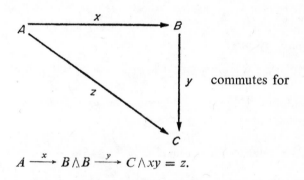

commutes for

$$A \xrightarrow{x} B \wedge B \xrightarrow{y} C \wedge xy = z.$$

The display of Definition 4 is called a *triangle*. In general we can make much more complicated displays by iterating triangles. These displays are called *diagrams*. Any statement (closed wff) of **C** will usually assert that some morphisms are the composition of others (unless the statement contains no occurrences whatever of the primitive predicate K). Every such statement will have an associated diagram. For example, the following diagram is associated with axiom **C**.6:

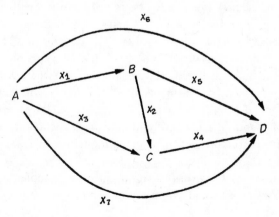

To say that such a diagram *commutes* means that all possible triangles that are parts of the diagram commute. **C.6** says that if certain particular triangles commute in the foregoing diagram, then the whole diagram must commute. Referring to **C.6**, and supposing the hypotheses to be fulfilled, we have the following equalities: $x_3 = (x_1x_2)$, $x_5 = (x_2x_4)$, $(x_1x_5) = x_6 = x_7 = (x_3x_4)$. From this, and using Definition 3, we obtain the usual form of associativity whenever the necessary compositions are defined: $(x_1x_2)x_4 = x_3x_4 = x_7 = x_6 = x_1x_5 = x_1(x_2x_4)$.

As we discuss categories and make statements, we shall often supply an associated diagram. When we do not, the reader is invited to supply it himself. It is the "visual" aspect of category theory that accounts for some of its appeal (although this same aspect has caused some to label category theory "abstract nonsense"). When we are writing systems of equations involving the composition of many morphisms, we shall often insert or omit parentheses at will, as is usual when dealing with an associative operation.

Definition 5. $\forall x$ for (x) where x is any variable.

We introduce the usual sign for the universal quantifier in informal mathematical literature. This will be helpful, since our parentheses are becoming dangerously overworked.

Definition 6. Mono(f) for $\forall x \forall y(xf = yf \supset x = y)$.

The morphism f is a monomorphism if it can be right cancelled under composition.

Exercise. Prove that in the category of sets and functions (recall our extended definition of this latter term) monomorphisms are 1–1 functions.

Definition 7. Epi(f) for $\forall x \forall y(fx = fy \supset x = y)$.

An epimorphism can be left cancelled.

Exercise. Prove that in the category of sets, epimorphisms are onto functions, i.e. functions whose range is the whole codomain.

Definitions 6 and 7 present us with another occasion of duality, for the notions of epimorphism and monomorphism are dual. Now that we have made our first-order language **C** quite precise, let us pause here to make the notion of duality equally precise.

Suppose that we have some model $\mathcal{M}$ of our first-order language **C**. Suppose we construct from $\mathcal{M}$ another model $\mathcal{M}^{op}$, in the following way: The domain of $\mathcal{M}^{op}$ is the domain of $\mathcal{M}$. For every triple $\langle x, y, z \rangle$ of elements of $\mathcal{M}^{op}$, the relation $K'(x, y, z)$ holds (where K' is the interpretation of K in $\mathcal{M}^{op}$) if and only if $K\#(y, x, z)$ holds in $\mathcal{M}$ (where $K\#$ is the interpretation of K in $\mathcal{M}$). For every pair $\langle x, y \rangle$ of elements of $\mathcal{M}^{op}$, $D'(x) = y$ if and only if $C\#(x) = y$, and $C'(x) = y$ if and only if $D\#(x) = y$, where D', C', $D\#$, and $C\#$ represent the interpretations of D and C in $\mathcal{M}^{op}$ and $\mathcal{M}$ respectively. By checking the axioms of **C** one by one, it is easy to verify that such a structure $\mathcal{M}^{op}$ is indeed a model for **C** if $\mathcal{M}$ is itself a model. $\mathcal{M}^{op}$ is called the *dual category* of $\mathcal{M}$.

As we check the axioms of **C** to verify that $\mathcal{M}^{op}$ really is a model if $\mathcal{M}$ is, we can observe that there is a certain kind of formal symmetry in each axiom. For example, if we take **C.1** and permute the role of C and D, we obtain **C.1** again. If in **C.2** we permute the occurrences of the variables x_1 and x_2 (intuitively changing the order of composition), we obtain, in a purely formal way, the axiom **C.2** again. If, in **C.3**, we permute the occurrences of x_1 and x_2 and the occurrences of C and D, we obtain **C.3** again. Similar statements hold for **C.4**, **C.5** and **C.6**. Thus, if we take an axiom of **C**, interchange everywhere the role of C and D and permute everywhere the order of composition, we obtain an axiom of **C** again.

We can sum these observations up as follows: For a wff A of **C**, the dual $\hat{A}$ of A is the wff of **C** obtained by interchanging everywhere in A the occurrences of the function letters D and C, and by permuting everywhere the order of compositions, i.e. by replacing every atomic wff of the form $K(x, y, z)$ by the atomic wff $K(y, x, z)$, where x, y, and z are any terms.

On the basis of this precise notion of duality, we can prove the following metatheorem:

THEOREM 1. *Let A be any wff provable in* **C**, *then the dual $\hat{A}$ of A is also provable in* **C**.[†]

Proof. Clearly the formal operation of duality preserves the quantifiers and sentence connectives. Taking our underlying logic to be our original logical axioms with just the rules MP and UG, it follows immediately that the dual of a logical axiom is a logical axiom. Moreover, duals preserve the rules MP and UG. Since duality preserves the proper axioms of **C**, the dual of a theorem of **C** is a theorem of **C**. Indeed, the dual of a proof is the proof of the dual.

Notice that a wff is provable if and only if its dual is provable, since the dual of the dual of A is A itself. We immediately obtain the following corollaries of Theorem 1 by using our general theorems of first-order logic.

COROLLARY 1. *If the closed wff B of* **C** *holds in all categories (i.e. all models of* **C***) for which the closed wff A is true, then $\hat{B}$ holds in all categories in which $\hat{A}$ is true.*

Proof. If B holds in all categories in which A is true, then the wff $A \supset B$ is true in every model of **C**. Thus, by the completeness theorem of first-order logic, $\vdash_C A \supset B$, and so $\vdash_C \hat{A} \supset \hat{B}$ by Theorem 1. But this means that the conditional $\hat{A} \supset \hat{B}$ holds in every category. Consequently, if $\hat{A}$ is true in such a model, so is $\hat{B}$.

COROLLARY 2. *For closed wffs A and B, B holds in all categories in which A is true if and only if $\hat{B}$ holds in all categories in which $\hat{A}$ is true.*

Proof. The reader will prove Corollary 2 as an exercise.

These metatheorems are obviously useful and underlie all our discussion of category theory.

We can see immediately from Definitions 6 and 7 that epimorphism and monomorphism are dual. Thus, the dual of any statement provable about monomorphisms is provable about epimorphisms and vice versa. The reader can also check that the notions of terminal object and initial object informally discussed in Section 8.1 are dual in the precise sense of our meta-definition when the respective definitions of these notions are formulated in our language **C**.

[†] It should be clear that the duality involved here has nothing to do with the Boolean duality of our logical connectives, which is not relevant to the present discussion.

From the way we have defined the dual category $\mathcal{M}^{op}$ of a category $\mathcal{M}$, we can also see that the dual of any statement that is true in $\mathcal{M}$ is true in $\mathcal{M}^{op}$ and vice versa. However, it is certainly *not* the case that the dual of any statement true in $\mathcal{M}$ is also true in $\mathcal{M}$. Suppose $\mathcal{M}$ has a terminal object but no initial object. Then the statement

$$(Ex_1)(Ob(x_1) \wedge (x_2)(Ob(x_2) \supset (E! x_3)(D(x_3) = x_2 \wedge C(x_3) = x_1)))$$

is true in $\mathcal{M}$, but its dual is not. However, it is trivial to see that its dual is true in $\mathcal{M}^{op}$. Notice that the foregoing statement is not provable in our language **C**.

So much for our duality principle.

Definition 8. Iso(f) for $(Eg)(fg = D(f) \wedge gf = C(f))$.

An isomorphism is a map f for which there is an "inverse" mapping g from $C(f)$ to $D(f)$ and such that fg and gf are equal to the appropriate identities.

Exercise. Prove that $\vdash$ Iso(f) $\supset$ (Mono(f)$\wedge$Epi(f)), but not conversely. For a counter-example to the converse, try the category of semigroups and semigroup homomorphisms, and let f be the inclusion map from N (as the semigroup of natural numbers under addition) to Z (as the semigroup of integers under addition).

The next definition is natural.

Definition 9. $A \cong B$ for $(Ef)(A \xrightarrow{f} B \wedge \text{Iso}(f))$.

Two objects A and B are isomorphic, written $A \cong B$, if there is an isomorphism from A to B.

Exercise 1. Prove that two groups S and S' are isomorphic in the sense of Definition 9 if and only if there is a bijective homomorphism h from S to S' (which is the way we defined the isomorphism of groups in our discussion of Section 8.1).

Exercise 2. Prove that "$\cong$" is an equivalence relation (reflexive, symmetric, and transitive).

Exercise 3. Prove that any two terminal objects in a given category are isomorphic. Prove the same for any two initial objects.

It is the essence of category theory that isomorphic objects can be regarded as practically identical. In the category of groups, algebraists generally consider two isomorphic groups as identical. It is true that occasionally some particular representation of a group has more practical value than another (when a group is presented recursively by generators and relations, for example). But for the purposes of abstract algebra, these differences may be ignored, since they are not part of the theory. In short, abstract algebra studies the isomorphic-invariant properties of groups, other properties being regarded as nonessential to purely algebraic matters.

Similarly, two isomorphic objects in the category $\mathcal{T}$ of topological spaces and continuous functions are homeomorphic spaces. Topology is often defined as the study of properties invariant under homeomorphisms. Analogous remarks hold for the category $\mathcal{V}$ of vector spaces and linear transformations, and indeed, for every other example we have given.

Despite this essential sameness of two isomorphic objects, it is possible in most categories for two different objects to be isomorphic. We say that a category is *skeletal* when this never happens. Precisely, a category (a model of **C**) is skeletal if and only if the wff $\forall A \forall B (A \cong B \supset A = B)$ is true in the category. All isomorphic objects are equal in any skeletal category.

In every category, skeletal or not, the relation "$\cong$" partitions the objects of the category into equivalence classes. Intuitively, we can pick a subcategory from the original category by choosing exactly one object from each equivalence class while keeping all morphisms except those we have to throw out because of the eliminated objects. Such a subcategory is said to be a *skeleton* of the original one, and it is equivalent to the original one in a sense of equivalence to be defined later. Isbell and Wright have shown that, in the proper set-theoretic framework, the existence of a skeleton for an arbitrary category is equivalent to the axiom of choice.

In the category $\mathcal{S}$ of sets and functions, isomorphisms are simply bijective functions, and so sets are isomorphic precisely when they have the same cardinality. Hence, from the point of view of the category of sets, the only relevant thing about a set is its cardinality, since the theory of cardinals is the study of properties of sets that are invariant under bijections. Though it may seem paradoxical, even the category of sets can be made skeletal, and we can still speak of elements of sets. This will be done in one system of this chapter.

Definition 10. Endo(f) for $D(f) = C(f)$.

The morphism f is an endomorphism if the domain and codomain of f are equal.

THEOREM 2. $\vdash (A \xrightarrow{A} A)$.

Proof. Since A is an object, we have by Definition 1 that
$$D(A) = A = C(A).$$

Theorem 2 tells us that every object A has an identity endomorphism, which is A itself. It may be that an object A in a category has only the identity endomorphism, but in general one must guard against the assumption that the class of endomorphisms from A to A is necessarily simple. The class may consist of only one member or it may consist of many different morphisms. An infinite group, for example, will generally have many different homomorphisms into itself.

Definition 11. Auto(f) for Endo(f)$\wedge$Iso(f).

An automorphism is an endomorphism that is an isomorphism.

Exercise. Prove that, in any category whatever, the class of automorphisms of a given object A is a group whose identity is A.

Definition 12. Proj(x, y) for
$$D(x) = D(y) \wedge \forall f \forall g((D(f) = D(g) \wedge C(f) = C(x) \wedge C(g) = C(y)) \supset (E!z)(zx = f \wedge zy = g)).$$

The maps x and y are *projections* if they have a common domain, and if, for every ordered pair of maps f and g with a common domain, and such that $D(f) \xrightarrow{f} C(x)$ and

$D(g) \xrightarrow{g} C(y)$, there exists a *unique* map z such that $zx = f$ and $zy = g$. We sometimes say that f and g "factor through" the projections x and y. Given a pair of projection maps x and y, the unique map z associated with a given pair f and g which factor through the projections is called the *product map* of f and g relative to the given pair of projections x and y. We denote the product map of f with g by $\langle f, g \rangle$.

Dual to the notion of a pair of projection maps is the notion of a pair of injections.

Definition 13. Inj(x, y) for
$$C(x) = C(y) \wedge \forall f \forall g (C(f) = C(g) \wedge D(f) = D(x) \wedge D(g) = D(y) \supset (E! z)(xz = f \wedge yz = g)).$$

The maps x and y are *injections* if they have a common codomain, and if, for every ordered pair of maps f and g with a common codomain, and such that $D(x) \xrightarrow{f} C(f)$ and $D(y) \xrightarrow{g} C(g)$ there is a unique map z such that $xz = f$ and $yz = g$. We call z the *sum map* of f with g relative to the given pair of injections, and we write $[f, g]$ to denote the unique sum map of f and g relative to a given pair of injections.

Since projections and injections are dual, any theorem proved for projections gives rise immediately to a dual theorem for injections and vice versa.

We now use projections to define *products* and *sums* of objects.

Definition 14. Pr(P, A, B, x, y) for
$$\text{Proj}(x, y) \wedge P = D(x) = D(y) \wedge A = C(x) \wedge B = C(y).$$

P is a *product object of A and B relative to the projections x and y* if P is the common domain of the projections x and y, and if A is the codomain of x and B the codomain of y. Notice, we say that P is "a product" of A and B relative to the projections, since nothing excludes P from being a product of A and B relative to different pairs of projections, and nothing excludes the same pair of objects A and B from having two different product objects P and P' relative to different pairs of projections.

When there is no danger of ambiguity, we will omit the modification "object" and speak simply of a product P of A and B, and the projections in question will be noted $p_A: P \longrightarrow A$ and $p_B: P \longrightarrow B$, called "projection on A" and "projection on B" respectively. Using this symbolism, we have the following diagram associated with the definition of product objects and product maps:

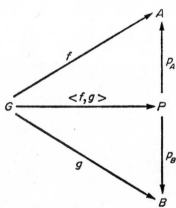

where G is the common domain of f and g.

P is a product of A and B relative to the projections p_A and p_B if and only if the above diagram commutes for all possible maps f and g satisfying the conditions of Definition 12. The map $\langle f, g \rangle$ is our unique product map associated with f and g. The uniqueness of the map $\langle f, g \rangle$ in the definition of projections is important, for it allows us to prove the following simple but important theorem that states that any two product objects P and P' of the same objects A and B are isomorphic:

THEOREM 3. $\vdash \mathrm{Pr}(P, A, B, p_A, p_B) \wedge \mathrm{Pr}(P', A, B, p_A', p_B') \supset P \cong P'$.

Proof. Assuming the hypotheses, we observe that p_A and p_B are maps with a common domain and with codomain A and B respectively. There is thus a unique map $P \xrightarrow{\langle p_A, p_B \rangle} P'$, since P' is a product of A and B. Similarly, there is a unique map $P' \xrightarrow{\langle p_A', p_B' \rangle} P$, and the following diagram commutes:

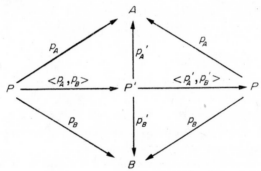

Now, Definition 12 tells us that there is one and only one map z such that

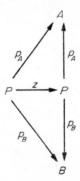

But $Pp_A = p_A$ and $Pp_B = p_B$ since P is the identity. The uniqueness of z implies that $z = P$ and $z = \langle p_A, p_B \rangle \langle p_A', p_B' \rangle$ and so $P = \langle p_A, p_B \rangle \langle p_A', p_B' \rangle$. By a similar argument, $\langle p_A', p_B' \rangle \langle p_A, p_B \rangle = P'$, and so $\mathrm{Iso}(\langle p_A, p_B \rangle)$ and $P \cong P'$.

Theorem 3 shows us that any two product objects of the same pair of objects in a category are unique up to isomorphism, regardless of the projections relative to which the product objects are defined. If the category in question is skeletal, we obtain immediately that any pair of objects has at most one product. Of course it is not necessary that there are any product objects at all in a category. We say that *products exist* in a category X if and only if there is a product object P in X for every pair of objects A and B in X.

Exercise. Show that a lower semilattice is a category in which the objects of the category are the elements of the lattice, and a morphism from A to B is the ordered pair $\langle A, B \rangle$ if $A \leqslant B$ holds in the lattice (thus, $\text{Hom}(A, B)$ has at most one element for all pairs A and B of objects in the category). Show that products exist and are identical with the meet operation in the semilattice. Show that the category is skeletal and that products are therefore unique.

The method of proof of Theorem 3 is typical of what often occurs in category theory in which a uniqueness condition allows us to prove certain equalities. Theorem 3 exemplifies the informal way in which we shall present proofs in this chapter. This relaxing of purely formal arguments was already begun in Chapter 5, and it is particularly suited to category theory in which the use of diagrams is frequent.

In the category of sets and functions, products exist. They are cartesian products, and the projection maps are the usual projection functions. Products also exist in the category of groups. They are the usual notion of the direct product of two groups. In the category of topological spaces, products exist and are represented by the product space of two spaces. We now define the notion of a *sum* dual to that of products.

Definition 15. $\text{Sm}(S, A, B, x, y)$ for

$$\text{Inj}(x, y) \wedge S = C(x) = C(y) \wedge A = D(x) \wedge B = D(y).$$

S is a *sum object of A and B relative to the injections x and y* if S is the common codomain of the injections x and y, and if A is the domain of x and B the domain of y. As with products, we omit the modification "object" and speak of the sum of two given objects. If S is the sum of A and B, we note the injections relative to which S is a sum of A and B by "i_A" and "i_B". We think of these as standing for "injection from A" and "injection from B" respectively.

Exercise. Draw the diagram associated with sums by dualizing the diagram for products.

THEOREM 4. *Any two sums of the same objects are isomorphic.*

Proof. Immediate from our duality principle, Theorem 1, and Theorem 3.

Sums are often called *coproducts* because they are defined dually with respect to products. It is a frequent practice in category theory to introduce a term for a dual notion by putting the prefix "co" in front of a term for the notion of which it is the dual. However, the practice is irregular and unsystematic on many counts. Why, for example, not "sums and cosums" instead of "products and coproducts"? And one never hears "projections and coprojections". We shall simply follow the practice of introducing the "co" terminology where it has become traditional and skip it elsewhere. For coproducts, the terminology is fairly well established.

In the category of sets, coproducts are disjoint sums. In the category of abelian groups, coproducts are direct sums. Of course, for abelian groups, direct sums are direct products, and so products and coproducts are equal in the category of abelian groups. Coproducts also exist in the category of groups. The coproduct of two groups is the so-called *free product* of the groups. However, products and coproducts are not the same thing in the category of groups. Again, we note that there are many categories which do not have a product object for every pair of objects or a coproduct object for every pair of objects.

Theorems 3 and 4, as well as the definitions on which they are based, illustrate one of the key notions of category theory, namely that of *universality*. The proper generality for discussing this notion involves more of the theory of functors (to be defined) than is really necessary for the purposes of this work. However, the basic idea is easily accessible at the present level of discussion.

Consider our definition of the product of two objects in a category. Given any category $\mathcal{C}$ and any two objects A and B, we can form a new category $\mathcal{D}$ of diagrams whose objects are all pairs of morphisms $f : X \longrightarrow A$ and $g : X \longrightarrow B$ in $\mathcal{C}$. The maps in the new category will be morphisms $h : X \longrightarrow X'$ such that $hf' = f$ and $hg' = g$, where $f' : X' \longrightarrow A$ and $g' : X' \longrightarrow B$ are another object in $\mathcal{D}$ (= appropriate diagram of morphisms of $\mathcal{C}$). It is easy to verify that $\mathcal{D}$ does satisfy the axioms of a category. Now, the product P of A and B with projections p_A and p_B is an object of $\mathcal{D}$ by definition. In fact, it is the (necessarily unique up to isomorphism) terminal object of $\mathcal{D}$. It is *universal* among those diagrams of $\mathcal{C}$ which are objects of $\mathcal{D}$. Thus, a universal diagram is one which is a terminal object in an appropriate category of diagrams. Dually, a diagram is *couniversal* if it is an initial object in an appropriate category of diagrams.[†]

Exercise. Given a category $\mathcal{C}$, formulate the appropriate category of diagrams and prove that sums are couniversal relative to it.

It is important to realize that universality and couniversality are always defined relative to an appropriate category of diagrams. Any diagram in any category can be considered universal: just take the category having that diagram as unique object (identity morphism) and with no other morphisms. In the case of products, however, the category $\mathcal{D}$ was the category of *all* diagrams having a certain form. A similar remark holds for sums as well. The only way to make all of this precise is by moving the whole discussion to the realm of functors and natural transformations. Since this is not a book on general category theory, we will not go further in our explanation of universality. However, the elements we have given should be sufficient for the reader to see how most of the constructions of category theory arise in a natural way as solutions to a universal problem. In other words, such constructions amount to seeking a universal or couniversal object in an appropriate category of diagrams.

We now pass on to a consideration of the foundational problems related to category theory.

8.3. Category theory and set theory

So far, we have avoided a detailed discussion of the relationship between category theory and set theory in order to allow for a natural and uncomplicated introduction to the basic notions of category theory.

We must now be a bit more sophisticated. We have defined a category as any model of our theory **C**. Yet, the notion of a model is relative to the set theory in which we choose to work. Except on an intuitive basis, there is no such thing as *the* category of groups or *the* category of sets. There is the category of groups in **NF** or **ML**, or in **ZF** or **NBG**. In fact, since a group can be defined on any set, the category of groups contains as many objects as the

---

[†] Some authors use the reverse terminology, calling "universal" what I have called "couniversal" and vice versa. Still others use "universal" to cover both cases, eliminating the term "couniversal" altogether.

category of sets. Consequently, the troublesome notion of "all sets" is involved in the very notion of the category of groups, as well as the category of sets.

In some set theories we have studied, the notion of "all sets" is not even definable. This is true for **ZF** in particular. In Quine's theories, and in **MKM**, a universal class V is definable. In **NBG**, we have a universal class V that is not a set. A category of sets is definable in all of these theories, but not in **ZF**.

Since **ZF** is a size-limitation theory, what can be done is to pick some fixed infinite cardinal number α and consider some category of sets and functions in which the sets chosen have cardinality less than α. But there is nothing to motivate the particular choice of one cardinal over another. How high a cardinal do we pick? There does not seem to be any natural stopping place. Moreover, certain natural operations, such as power set, will not be generally defined in such a category. In this way, such a category of sets in **ZF** will differ in structure from the category of sets defined in some of our other theories such as **NF** or **NBG**.

Another interesting approach is due to Isbell [1]. There all categorical constructions are performed relative to a fixed inaccessible cardinal α. As we saw in Chapter 5, there is a model M for **ZF** among the sets smaller than α; M can be taken as the class of all sets of rank less than α. M is thus a category of sets closed under the various operations of **ZF**. Of course, working with inaccessible cardinals amounts to working outside of **ZF**, and so we really have no satisfactory definition of the category of sets *within* **ZF**.

In the future we will speak of "the category of sets and functions *of* **ZF**". We will mean by this the intuitively conceived category $\mathcal{S}$ whose objects are the sets of **ZF** and whose functions are the functions in **ZF**. This category is not definable *within* **ZF**, but it is the category of sets definable within **NBG**.

Defining a category of sets presents obvious difficulties for type theory as well, for in **ST** we have no single universal class, but only a universal class for each type. Furthermore, many set-theoretic operations cause us to change types when performed within **ST**. If types are cumulative, then the universal class of a given type is a subset of the universal class of any higher type, which is of some help. But, as we have already seen, **ST** has a model in **ZF**, and so any attempt to define satisfactorily a category of sets in **ST** will encounter difficulties at least as great as those involved with **ZF**.

We might formulate the general problem at hand in the following way: We intuitively feel that the category of groups or the category of sets ought to be a fixed structure with a definite theory. Yet these structures will have different theories according to the various foundational systems in which we may choose to work. This same relativity will obtain for other examples of categories which we have discussed in Section 8.1.

It might appear at first that this problem is really not different from the situation that obtains for any other notion, such as "real number" or "natural number". The set of natural numbers was, after all, differently defined in **ZF** than in **NF**, for example. Yet we certainly feel, in an intuitive manner, that there is only one set of natural numbers in the "real world". The difference here is that each of these various constructions amounts only to a different way of obtaining the same theory. We proved the Peano postulates in each of these theories to illustrate this fact. As we pointed out in our detailed discussion of the matter in Chapter 3, once the Peano postulates and a few other basic tools (such as a theorem of recursion) are at hand, the constructions are actually dispensable in favor of the proved properties. And these proved properties are the same in each theory.

In other words, our various foundations differ primarily at their "outer reaches". They all share a common core theory, constructible in one way or another. This common core theory is essentially classical analysis. But category theory is a new theory which goes beyond the classical approach and actually involves these outer fringes of set theory to a considerable extent. In fact, the question of the structure of the category of sets really is just another way of posing the question of the relative consistency of the various set theories. Whether or not the category of sets in **NF** is definable in **NBG** is practically the same question as whether or not there is a model of **NF** in **NBG**, a question which is quite unresolved as of the present writing. But there certainly is a model in **NBG** for, let us say, elementary number theory of **NF**.

Traditional foundational systems were constructed with the express aim of furnishing a basis for classical notions. It is thus hardly surprising that basic theorems relating to classical notions are the same in all such foundations. If we obtained a nonclassical theory of real numbers, we would probably reject the foundation.[†] On the other hand, it is not yet exactly clear what the "classical" theory of the category of all groups really is or should be.

8.4. Functors and large categories

One way out of the dilemma of the relativity of the notion of such structures as "category of groups" and "category of sets" is to choose a fixed foundational system as *the* foundational system, at least for the purpose of category theory. By "category of sets" we would then mean the category of sets of that system and no other. The same would apply for groups and the like. Nothing would prevent the comparative study of the notions in different theories if this proved desirable, but the given system would be the main one, serving as a standard of comparison to other possible theories.

In a way, this is what has come about in the approach to category theory by many authors. The system **NBG** is very often considered to be the basic theory underlying most work in category theory. Of course, authors, for the most part, do not bother to make everything formal or explicit, but they assume in a general sort of way that they have available the apparatus of **NBG**. In **NBG**, such large collections as the class of all groups or the class V of all sets do exist, but these are, as we saw in Chapter 5, *proper* classes and cannot themselves be elements of another class.

We suppose, for the moment, that we are working in **NBG**.

Let us call a category a *large* category if its class of morphisms is a proper class of **NBG**. Otherwise the category will be called *small*. For both large and small categories, it is the usual practice to assume that Hom(A, B), the class of all morphisms with domain A and codomain B, is a set. Thus, even though the class of all morphisms of a large category is a proper class, the class Hom(A, B) is usually assumed to be a set. Notice that if a category is small, its class of objects must also be a set, since this is a subclass of the morphism class.

Most of the examples we have given of categories are obviously large categories. There are, however, many examples of small categories. Any one-element group $\{e\}$ is a one-object category. Any quasi-ordered set (the relation is reflexive and transitive) is a category, the

---

[†] This remark should not be confused with the question of nonstandard analysis, which studies nonstandard models for the real numbers. It is not the possibility of different, nonclassical models that would cause us to reject a theory, but rather a theory in which the classical theorems could not be proved.

morphisms being the ordered pairs of the relation. The reflexivity of the relation assures us of identity morphisms, and transitivity yields the composition of morphisms. Thus, partially, ordered sets, Boolean algebra, and lattices are all examples of small categories.

We shall see that this distinction between large and small categories, which has become current in the literature of the subject, leads to new foundational problems. To understand the exact nature of the difficulty, we introduce the notion of a *functor*.

One aim of category theory is the study of the relationship between categories. We need the notion of a "homomorphism" between two categories, i.e. a structure-preserving function from one category to another. This is provided by the notion of *functor*. By a functor F from a category $\mathcal{C}$ to a category $\mathcal{C}^*$, we mean a function which associates with every morphism x of $\mathcal{C}$ a morphism $(x)F$ of $\mathcal{C}^*$ such that (i) if $\mathrm{Ob}(x)$, then $\mathrm{Ob}((x)F)$, and (ii) $(xy)F = (x)\,F(y)F$, i.e. the composition of morphisms is preserved.

An example of a functor is the first homology functor, which associates a certain group with every topological space and a group homomorphism with every continuous map. Another example is the free group functor, which associates with every set x the free group generated by x, and which associates with every function between two sets the unique group homomorphism which is the extension of the function from generators to generators of the respectively associated free groups.

Although we shall not have much need for the concept in our discussion, the notion of a *contravariant* functor is also used in category theory. This is a function F from a category $\mathcal{C}$ to another category $\mathcal{C}^*$ such that (i) if $\mathrm{Ob}(x)$, then $\mathrm{Ob}((x)F)$, and (ii) $(xy)F = (y)F(x)F$. To distinguish them from contravariant functors, the functors we first introduced are often termed *covariant*. Whenever we use the term "functor" without modification, we mean covariant functor. In fact, a contravariant functor from $\mathcal{C}$ to $\mathcal{C}^*$ is the same thing as a covariant functor from $\mathcal{C}^{op}$ to $\mathcal{C}^*$.

Now, when given several categories, we can associate with them an intuitively conceived category whose objects are the given categories and whose morphisms are the functors between the given categories. Every category has an identity functor and the composition of two functors is easily seen to be a functor. We thus have an "operation" that associates with given categories a new category of categories. There are, in fact, a large number of standard ways of constructing categories from given categories. The theory of categories makes it desirable that these constructions be indiscriminately applicable to categories large and small. This is not possible on the basis of **NBG**, however, and herein lies the basis for a second foundational problem concerning categories. Let us see just how much flexibility we really have in **NBG**.

For any large category, the class of morphisms is not a set. Consider how such a category, say the category $\mathcal{S}$ of sets, would be defined in **NBG**. The objects of $\mathcal{S}$ are the elements of the class V of all sets. Since we have chosen to identify objects with identity morphisms, the class of morphisms is really the class F of all functions (defined as ordered triples of *sets*). A set x is then identified with the function $\langle x, x, I \rangle$ where I is the identity relation on x. Intuitively the operation K of composition of functions is a partially defined function from $F \times F$ to F. Since the codomain F of K is given, we can identify K with its graph $R \subset F^3$. In short, the category of sets, defined in **NBG**, should be an ordered pair $\langle F, R \rangle$ where F is the class of all functions and $R \subset F^3$. But is this really legitimate in **NBG**? Not quite, for an ordered pair $\langle F, R \rangle$ is the class $\{\{F\}, \{F, R\}\}$ by definition. But F is a proper class and thus is an element of no other class. Thus, the class $\{F\}$ does not exist in **NBG** and the ordered pair $\langle F, R \rangle$ is not definable in the usual manner.

What can be done is to introduce into **NBG** a predicate $\mathrm{Cat}(X, Y)$ for "X is the morphism class of a category whose composition is $Y \subset X^3$". (Let us recall that a proper class *can* be a subclass of another class. It is only class membership that is excluded.) Thus, through notational identification of a category with its morphism class, we can speak of the category F of sets by proving in **NBG** that $\mathrm{Cat}(F, R)$ holds.

We can speak of a functor from a large category $\mathrm{Cat}(X, Y)$ to another large category $\mathrm{Cat}(X^*, Y^*)$ by defining a predicate $\mathrm{Funct}(t_1, t_2, t_3, t_4, t_5)$, meaning "$t_1$ is a functor from the category with morphism class t_2 and composition class t_3 to the category with morphism class t_4 and composition class t_5", and then proving $\vdash \mathrm{Funct}(t, X, Y, X^*, Y^*)$ for some appropriate term t. In our definition of the relation Funct, t_1 would be the graph of the functor, a certain subclass of $t_2 \times t_4$.

An alternative device for defining large categories in **NBG** is the following: If X is the morphism class and Y the composition class of the category, define the category Z to be the *disjoint union* of X and Y. This is defined as $Z = X \times \{0\} \cup Y \times \{1\}$. Z is a well-defined class even if X and Y are not sets. Z is a proper class if either X or Y is proper. Rubin [1] uses the device of disjoint union to define the notion of an ordered pair of proper classes in a set theory like **NBG**. This is certainly legitimate, since we can prove that

$$\vdash_{NBG} X \times \{0\} \cup Y \times \{1\} = X^* \times \{0\} \cup Y^* \times \{1\} \equiv (X = X^* \wedge Y = Y^*),$$

where X, Y, X^*, and Y^* are classes (possibly proper) of **NBG**.

Using disjoint union, we can thus define within **NBG** a proper class Z. We call Z the large category with morphism class X and with composition class Y, where one or both of these classes is proper, and $\mathrm{Cat}(X, Y)$ holds. Similarly we can define a functor from a large category Z to a large category Z^* to be the disjoint union of the various classes t_1, t_2, t_3, t_4, and t_5, where $\mathrm{Funct}(t_1, t_2, t_3, t_4, t_5)$ holds, and where $Z = t_2 \times \{0\} \cup t_3 \times \{1\}$, $Z^* = t_4 \times \{0\} \cup t_5 \times \{1\}$ and t_1 is the graph of the functor.

However we choose to deal with the definition of large categories in **NBG**, we shall have the following problem: Given some large categories, how can we form in **NBG** the intuitively conceived category consisting of the given categories as objects and the functors between them as morphisms? We cannot collect the given categories into a class, since proper classes are not elements of any other class. We cannot form the morphism class of such a category in **NBG**. Unless we can find some novel device for forming such a category, without having to consider the given categories as elements of a class, we can only study the functors between the various given categories "one by one", i.e. study the properties of one particular functor and then another, and so on.

On the other hand, observe that no such difficulties obtain for small categories. They are defined on sets, just like groups or rings or any other algebraic structure. Thus, we can form categories whose objects are small categories. In fact, one well-defined, large category is the category of all small categories and functors between them.

Our second foundational problem of categories can be stated as follows: Certain intuitively natural constructions on categories cannot be performed indiscriminately for large and small categories. We have given one such example, that of forming a category of categories. Let us consider one more, that of forming *functor categories*.

Consider two categories $\mathcal{C}$ and $\mathcal{C}^*$ and let F and G be functors from $\mathcal{C}$ to $\mathcal{C}^*$. By a *natural transformation* from F to G, we mean a function π which assigns a morphism $(A)\pi$ of $\mathcal{C}^*$ to every *object* A of $\mathcal{C}$ such that (1) $(A)\pi \in \mathrm{Hom}((A)F, (A)G)$, and (2) for every $A \xrightarrow{\ x\ } B$ in $\mathcal{C}$, the diagram

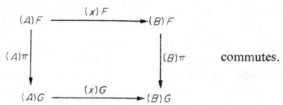

commutes.

If, further, for every A in $\mathcal{C}$, the morphism $(A)\pi$ is Iso, then we say that F and G are *naturally equivalent*.

Exercise. Prove that the relation of natural equivalence is an equivalence relation on the class of all functors from $\mathcal{C}$ to $\mathcal{C}^*$.

Two categories $\mathcal{C}$ and $\mathcal{C}^*$ are said to be *isomorphic* if there is a functor F from $\mathcal{C}$ to $\mathcal{C}^*$ and a functor G from $\mathcal{C}^*$ to $\mathcal{C}$ such that the composition of functors FG and GF are the identity functors on $\mathcal{C}$ and $\mathcal{C}^*$ respectively. We say that $\mathcal{C}$ and $\mathcal{C}^*$ are *equivalent* if there exist F from $\mathcal{C}$ to $\mathcal{C}^*$ and G from $\mathcal{C}^*$ to $\mathcal{C}$ such that FG is naturally equivalent to the identity functor on $\mathcal{C}$ and GF is naturally equivalent to the identity functor on $\mathcal{C}^*$. The notion of equivalence is clearly more general than isomorphism, since equality with the identity functor certainly is a special case of natural equivalence. Again, the philosophy of category theory wills that the relevant structure of categories be invariant under isomorphism (and, generally speaking, under equivalence as well).

Now, given two categories $\mathcal{C}$ and $\mathcal{C}^*$, we can form from them the intuitively conceived *functor category*, consisting of all functors from $\mathcal{C}$ to $\mathcal{C}^*$ as *objects* and all natural transformations between the functors as morphisms (the composition of two natural transformations is a natural transformation). However, if $\mathcal{C}$ and $\mathcal{C}^*$ are large, then a functor from $\mathcal{C}$ to $\mathcal{C}^*$ will be a proper class, and so we cannot collect the functors to form the object class of the (intuitively conceived) functor category in **NBG**.

On the other hand, if $\mathcal{C}$ and $\mathcal{C}^*$ are small categories, then the respective morphism classes M and M^* are sets and the set of all functors from $\mathcal{C}$ to $\mathcal{C}^*$ is a subset of M^{*M}. Moreover, the class of all natural transformations is a set, as is clear. We can form the functor category.

Again, as with categories of categories, we see that we cannot form functor categories indiscriminately, but only under certain conditions.

Because of these and other cases when classes of classes are needed, category theorists have been led to seek some kind of enlargement of set theory in order to be free of these nagging foundational questions. One idea, due to Grothendieck and others, is that of considering an extension of set theory by adding "universes" to **NBG**. A universe is a set which is itself closed under all of the set-theoretic operations and relations of **ZF**. A natural way of obtaining such a set is to add an inaccessible cardinal α to the system. A universe will then be the set of all sets of rank less than α.

The approach of Grothendieck is to add not one but an arbitrary number of new universes to our system. This can be accomplished by adding the simply stated axiom that every set is an element of some universe. Thus, any universe U_1 must itself be an element of some universe $U_2 \neq U_1$ which must, in turn, be an element of a universe U_3 different from both U_2 and U_1, etc.

Because a universe is closed under the usual set-theoretical operations of **ZF**, all constructions performable in **ZF** can be done within any universe. And those constructions of category theory which take one outside of a given universe can be done in another universe. There is, in fact, no real need for classes at all since what was done before with classes can now be done with sets from a larger universe.

As smooth as this approach is in many respects, it carries with it the intuitively repugnant feature of a continuous switching back and forth from one universe to another. There is no universe of all sets, but rather a plethora of universes for each of which the basic categories, such as groups, topological spaces, etc., can be defined and on each of which the basic constructions of category theory can be effected.

A more recent "one-universe" approach is the closest to anything like a "standard" foundation for category theory (cf. MacLane [3], and Herrlich and Strecker [1]). Here we assume the existence of exactly one inaccessible cardinal α. As before, all the sets of rank less than α form a model for **ZF**. These are called *small* sets. Collections of small sets are just *sets* (these correspond to the classes of **MKM**). Finally, any collection of objects is called a *class*. The advantage is that all the set-theoretical operations of **ZF** can be performed on the classes.

Thus, in this system categories formed from small sets are sets (and thus classes). We can further form any collection of such sets to obtain a class and we can operate on such classes as freely as we can with the sets of **ZF**. Thus, for such large categories as the category of small sets, small groups, etc., we can form functor categories indiscriminately and we can form categories of categories as much as we need.

We describe briefly a formal language for the one-universe foundation (cf. Fraenkel, Bar-Hillel, and Levy [1], pp. 142–143). We start with the language and theory **ZF**. These are the classes. We add a unary predicate "M" and understand that $M(x)$ means intuitively "x is a small set". We introduce restricted quantification over small sets and we add the full axioms of **MKM** with the small sets here playing the role of the sets of **MKM**. Formally, these additional axioms here are identical with the axioms of **MKM** given the difference in the status of the predicate M: in **MKM** the predicate is defined whereas here it is primitive.

Thus, the objects of our present system are all classes. Those classes which satisfy the predicate M are small sets and classes which have only small sets as members are sets.

There have also been approaches which try to combine **ZF** set theory with some version of type theory or of stratification à la Quine's NF. The systems of da Costa [1] and da Costa and Dias [1] are perhaps the most successful in this direction.

The way in which category theory becomes so easily involved in foundational questions and the very generality and perspective derived from the "categorial viewpoint" on mathematics naturally leads one to wonder whether it might not be possible to found mathematics on category theory itself. At the very least, one would expect that an analysis of set theory and of other traditional foundational systems from the standpoint of category theory would lead to interesting new insights into their structure. This expectation has been largely justified by the development of several different foundational systems based on category theory.

One of the prime movers of these developments has been F. W. Lawvere and his collaborators, notably M. Tierney. However, there have been significant contributions from other sources, especially from the French school of algebraic geometers comprising A. Grothendieck, J. Giraud, and J.-L. Verdier.

The first step was the development of a categorial set theory by Lawvere in 1964. This system was subsequently generalized and cast in a better form, called a *topos*. The initial work by Lawvere and Tierney on toposes was done during the year 1969–1970. Since that time, the theory of toposes has been developed to a considerable extent.

Our approach in the present work will be to present first the original system of the category of sets of Lawvere. We will then turn to a consideration of topos theory, showing clearly and explicitly the relationship between the two systems.[†] The pedagogical advantage in this approach is that the axioms of the system **CS** of the *category of sets* are easily seen to be direct categorial (usually universal or couniversal) formulations of simple structural properties of the category of sets of **ZF**. This facilitates a transition from set-theoretical to categorial thinking. Once this transition is accomplished, the consideration of topos theory will be quite direct and uncomplicated. Moreover, the notions and techniques we will have acquired in the course of formulating and studying the language and theory **CS** will be useful in the study of topos theory.

8.5. Formal development of the language and theory CS

As we develop the theory **CS**, we will have continually in mind the category of sets of **ZF** (thus the class V of sets, together with all functions between them, definable in **NBG**). This will serve us as a guiding, intuitive model for our axioms.

The language **CS** has the same symbols and wffs as the language **C** of Section 8.2. Moreover, all the axioms of **C** are axioms of **CS**. We now assimilate these, as well as Definitions 1 through 15 and Theorems 1 through 4. We number our new axioms as **CS**.1, **CS**.2, and so on, but we maintain numberings consecutive with those of Section 8.2 for definitions and theorems. All our other conventions, such as the use of capital letters for object variables, will also be retained.

As we state each axiom of **CS**, we shall give an intuitive interpretation of the axiom to show that it is "true" in the category of sets. It must be clear that this is an aid to the understanding of the model we have in mind, but it is not relevant to proofs of theorems. These are, as always, forthcoming by deduction. As with **ZF** and Quine's systems of previous chapters, we shall not exhibit detailed formal proofs of theorems, but shall give only an informal discussion, which nevertheless will be sufficient to allow the serious reader to recover the full proof.

CS.1. $(EA)\forall B(E!\,x)(B \xrightarrow{\ x\ } A)$.

A is a terminal object. Intuitively, this is a one-element set.

CS.2. $(EA)\forall B(E!\,x)(A \xrightarrow{\ x\ } B)$.

A is an initial object. Intuitively A is the null set.

[†] Another foundational system, the category of categories, presented in Lawvere [3], has been less successful. It turned out to have some defects of detail, though not of conception (cf. Isbell [2] and Blanc and Preller [1]). A more recent version (cf. Blanc and Donnadieu [1]) removes some of these defects, but we will not consider this system in the present work.

THEOREM 5. *Any two initial objects are isomorphic, and any two terminal objects are isomorphic.*

Proof. The reader has already proved this in Exercise 3 following Definition 9, Section 8.2.

In this chapter we go a step further toward an informal presentation by stating theorems in the vernacular. The reader may take it as an exercise to translate the vernacular into the formal language.

We have already discussed products and sums in Section 8.2. We posit:

CS.3. $\forall A \forall B(EP)(Ex)(Ey)\Pr(P, A, B, x, y)$.

CS.4. $\forall A \forall B(ES)(Ex)(Ey)\mathrm{Sm}(S, A, B, x, y)$.

Theorem 3 and Theorem 4 tell us that products and sums of two given objects are unique up to isomorphism. The same is true for initial objects and terminal objects. **CS.3** and **CS.4** tell us that, for any two given objects A and B, there is a product of A with B and a sum of A with B.

We have already explained that products represent cartesian products in the category of sets and sums represent disjoint sums.

Before continuing with our axioms, we might pause to get a better idea of the model we have in mind. In **ZF**, the ordinal structure was the most important. We have a model of **ZF** in which every set in the model has an ordinal number as its rank. Since the ordinals are defined in **ZF** by means of the "$\in$" relation, one might say that the important thing about a set of **ZF** is its structure in terms of the $\in$ relation. From the point of view of category theory, on the other hand, two isomorphic sets (thus with the same cardinality) are practically the same. Though an $\in$ relation will be forthcoming in **CS**, its importance will be much less than it is in **ZF**.

In **ZF**, we have only one no-element class, which was 0. We have many different one-element classes. We need the existence of these different one-element classes in order to construct the ordinals in terms of the $\in$ relation, for our successor function in **ZF** is $x \cup \{x\}$, and so there is an infinity of different one-element sets: $\{x\}$ is different from $\{y\}$ whenever x is different from y.

In **CS**, cardinality is the important concept. We seek axioms that will intuitively guarantee us a representative for every cardinal class. But it does not matter *which* representative or *how many* representatives are chosen from each such class. Those operations we constructed in **ZF** by complicated membership relations are the object of special postulation in **CS**. For example, **CS.3** posits the existence of products. These are constructed in **ZF** as sets of ordered pairs, an ordered pair $\langle x, y \rangle$ being a set of the form $\{\{x,\} \{x, y\}\}$ that involves a complicated membership structure (and in particular the existence of the one-element set $\{x\}$ as well).

Since the number of representatives of each cardinal class is not relevant in **CS**, we can assume if we wish that we have exactly one. This amounts to assuming that the category of sets we are describing is skeletal, since all members of any cardinal class are isomorphic. We make this specific assumption in **CS** as our next axiom.

CS.5. $\forall x_1 \forall x_2((\mathrm{Ob}(x_1) \wedge \mathrm{Ob}(x_2) \wedge x_1 \cong x_2) \supset x_1 = x_2)$.

CS.5 plays very much the same role in **CS** as the axiom of restriction does in **ZF**. The axiom of restriction serves to eliminate models with certain undesirable (or undesired features); for example, models with sets containing themselves as elements. **CS.5** does the same here, for it eliminates all those models of **CS** in which we choose more than one representative for each cardinal class.

We now know by Theorems 3 through 6 and **CS**.5 that there is exactly one terminal object, exactly one initial object, and exactly one sum object and exactly one product object for each pair of objects. This motivates the following definitions:

Definition 16. 0 for the unique initial object.

Definition 17. 1 for the unique terminal object.

Definition 18. $(A \times B)$ for the unique product object of A and B.

Definition 19. $(A + B)$ for the unique sum object of A and B.

Exercise 1. Prove that products and sums are associative and commutative. (Commutativity follows almost immediately from the definition. Associativity requires an argument. In each case prove isomorphism and appeal to **CS**.5.)

Exercise 2. Prove that products and projections are totally determined by two binary operations $\times$ and $\langle , \rangle$, the first defined for all pairs of objects and the second for all pairs of morphisms with common domain, such that the following identities hold: (1) $\langle a, b \rangle p_A = a$, (2) $\langle a, b \rangle p_B = b$, (3) $\langle dp_A, dp_B \rangle = d$, where these mappings have the obvious domains and codomains (and thus involve the object operation $\times$). Note in particular that $\langle p_A, p_B \rangle = A \times B$. Formulate and state an appropriate dual set of conditions for sums and injections.

The above identities (and the dual ones for sums and injections) will be frequently used in the remainder of this chapter. We now prove some other extremely useful properties of products, giving both diagrammatical and equational proofs of them. In each case, we leave as an exercise to the reader the task of formulating the dual result for sums and injections.

THEOREM 6. *Let* $X \xrightarrow{\langle a, b \rangle} A \times B$ *be the product mapping of* $X \xrightarrow{a} A$ *and* $X \xrightarrow{b} B$, *and let* $X' \xrightarrow{z} X$, *then* $z \langle a, b \rangle = \langle za, zb \rangle$.

Proof. Consider the following diagram:

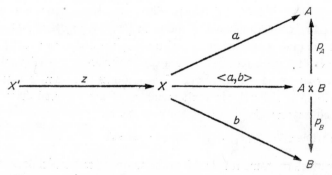

By the definition of products, this diagram commutes. Hence, by the uniqueness of the product map $\langle za, zb \rangle$ we have $z \langle a, b \rangle = \langle za, zb \rangle$.

Equationally, we have: $z\langle a, b\rangle = \langle z\langle a, b\rangle p_A, z\langle a, b\rangle p_B\rangle = \langle za, zb\rangle$.

In the next definition, we extend the notion of product mapping in an obvious way.

Definition 20. Given $A \xrightarrow{y} B$ and $C \xrightarrow{z} D$, then we denote by $(y \times z)$ the uniquely defined map $\langle p_A y, p_C z\rangle$ from $(A \times C)$ to $(B \times D)$. We shall also refer to $(y \times z)$ as the *product map* of y and z. Note that the product map of

$$A \xrightarrow{A} A \quad \text{and} \quad B \xrightarrow{B} B \quad \text{is} \quad \langle p_A A, p_B B\rangle = \langle p_A, p_B\rangle = A \times B.$$

Thus, the extension of the "$\times$" operation to arbitrary maps is consistent with the notation already introduced.

The essential difference between the product map $\langle y, z\rangle$ and the product map $(y \times z)$ is that, in the first instance, the maps y, z and $\langle y, z\rangle$ all have a common domain whereas in the second case this is not necessarily so. Even if y and z have a common domain X, the mapping $(y \times z)$ is, in general, different from the mapping $\langle y, z\rangle$. The latter has X as a domain whereas the former has $(X \times X)$ as a domain, and $(X \times X)$ is not in general equal to X. On the other hand, observe that if $A = A \times C = C$, $p_A y = y$, and $p_C z = z$, then $(y \times z) = \langle y, z\rangle$.

Although, by **CS.5** and Theorem 3 the product $(A \times B)$ of two objects A and B is uniquely determined, the projections p_A and p_B are not. That is, nothing prevents the same object from being a product of the same two given objects in different ways, with different projection maps in each case. Since this is so, the notations "$(y \times z)$" and "$\langle y, z\rangle$" are always understood to be relative to any particular pair of projection maps under consideration. A similar remark holds for the sum map of two maps.

Exercise. Define a generalized sum map $(x + y)$ of two arbitrary maps in analogy with our foregoing definition of $(x \times y)$. (*Hint*: Dualize Definition 20.)

As a simple illustration of the way in which the same object can be the product of the same two objects with different projections, consider the arithmetic product $2 \times 3 = 6$ illustrated by the rectangular display

$$
\begin{array}{c}
\xleftarrow{\quad p_2 \quad}\ (6) \\[4pt]
\begin{array}{cccc}
 & 0 & \;0\;\;1\;\;2 \\
(2) & 1 & \;3\;\;4\;\;5
\end{array} \Bigg| \; p_3 \\[10pt]
\begin{array}{c}
0\;\;1\;\;2 \\
(3)
\end{array}
\end{array}
$$

and also the display

$$
\begin{array}{c}
\xleftarrow{\quad p_2' \quad}\ (6) \\[4pt]
\begin{array}{cccc}
 & 0 & \;0\;\;3\;\;5 \\
(2) & 1 & \;1\;\;2\;\;4
\end{array} \Bigg| \; p_3' \\[10pt]
\begin{array}{c}
0\;\;1\;\;2 \\
(3)
\end{array}
\end{array}
$$

18

In the first instance $(5)p_2 = 1$, and in the second $(5)p_2' = 0$. We can obtain a different projection onto one of 2 or 3 for every permutation of the elements of 6, and so there are 720 different projection pairs associated with this product. The product 3×2 will be represented in a different way again, even though commutativity holds, that is, $2\times3 = 3\times2$, no matter what the projections.

Exercise. Make a display for 3×2 like the one for 2×3.

THEOREM 7. *For any mappings a, b, c, and d, we have*

$$(ab\times cd) = (a\times c)(b\times d).$$

Proof. This follows immediately from the fact that the following diagram commutes by Definition 20, and from the uniqueness of the product map $ab\times cd$:

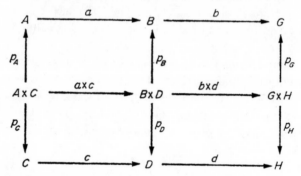

Equationally, we have: $(a\times b)(b\times d) = \langle p_A a, p_C c\rangle\langle p_B b, p_D d\rangle$

$$= \langle\langle p_A a, p_C c\rangle p_B b, \langle p_A a, p_C c\rangle p_D d\rangle = \langle p_A ab, p_C cd\rangle = (ab\times cd).$$

THEOREM 8. *If* $X \xrightarrow{a} A$, $X \xrightarrow{b} B$, $A \xrightarrow{c} C$, $B \xrightarrow{d} D$, *then*

$$\langle a, b\rangle(c\times d) = \langle ac, bd\rangle.$$

Proof. This results from the relevant definitions, the commutativity of the following diagram, and the uniqueness of the product map $\langle ac, bd\rangle$:

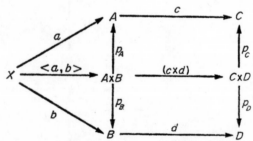

Equationally: $\langle a, b\rangle(c\times d) = \langle a, b\rangle\langle p_A c, p_B d\rangle = \langle\langle a, b\rangle p_A c, \langle a, b\rangle p_B d\rangle = \langle ac, bd\rangle$.

THEOREM 9. *If* $A \xrightarrow{a} B\times C$, $B \xrightarrow{b} D$, $C \xrightarrow{c} G$, *then*

$$a(b\times c) = \langle ap_B b, ap_C c\rangle.$$

Proof. We give only an equational proof: $a(b \times c) = a \langle p_B b, a p_C c \rangle = \langle a p_B b, a p_C c \rangle$.

In the future, rather extensive use will be made of the results of Theorems $6-9$ and of Exercise 2 immediately preceding Theorem 6. When we do so, we shall offer the justification "by the properties of products" instead of citing each separate use of each theorem. In fact, in most instances, we will be using somewhat complicated chains of equations mainly involving the identities of Theorem 6 and of Exercise 2 preceding it.

Since our operations of product and sum correspond to cartesian product and disjoint sum, these operations amount to cardinal multiplication and cardinal addition in **CS**. Thus, we should expect that the usual properties of these operations will be forthcoming. We have already proved in an exercise that these operations are associative and commutative. Another property is that 1 is a neutral element under multiplication.

THEOREM 10. *For any object A, $(A \times 1) = A$.*

Proof. Let $A \xrightarrow{\ t\ } 1$ be the unique terminal map with domain A. We prove $(A \times 1) \cong A$ by proving that $\mathrm{Iso}((A \times 1) \xrightarrow{\ p_A\ } A)$. $\mathrm{Iso}(p_A)$ means that p_A has an inverse. We show that $A \xrightarrow{\langle A, t \rangle} A \times 1$ is an inverse of p_A, i.e. $p_A \langle A, t \rangle = (A \times 1)$ and $\langle A, t \rangle p_A = A$.

In fact, this last equation is immediate by the properties of products. As for the other one, observe that $(p_A \langle A, t \rangle) p_1 = p_1$ since there is only one map $A \times 1 \longrightarrow 1$. Also,

$$(p_A \langle A, t \rangle) p_A = p_A (\langle A, t \rangle p_A) = p_A A = p_A.$$

Thus,

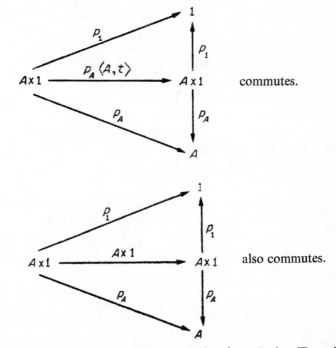

commutes.

But

also commutes.

Consequently, the uniqueness of products gives us $p_A \langle A, t \rangle = A \times 1$. (Equationally, $p_A \langle A, t \rangle = \langle (p_A \langle (A, t) \rangle) p_A, (p_A \langle A, t \rangle) p_1 \rangle = \langle p_A, p_1 \rangle = A \times 1$.) Thus, $A \times 1 \cong A$ which gives $A \times 1 = A$ by **CS**.5 and the theorem is established.

Exercise. Prove that, for all objects A, $A+0 = A$.

We now define the notions of subset and element in our system.

Definition 21. Let $A \xrightarrow{x} B$ and $C \xrightarrow{y} B$ be two monomorphisms with codomain B. We then say that x is a *subset of* y, and we write "$x \subset y$" if and only if there exists $A \xrightarrow{z} C$ such that $x = zy$. We also say that "x precedes y" or "x is a subobject of y" to express this same notion.

Exercise. Prove that the z of Definition 6 is a monomorphism and uniquely determined.

THEOREM 11. *Let A be some object. Then $x \subset A$ if and only if x is a monomorphism with codomain A.*

Proof. The reader will prove Theorem 11 as an exercise.

Definition 22. We say that x *is an element of* y, and we write "$x \in y$" if and only if $A \xrightarrow{y} B$, $1 \xrightarrow{x} B$, y is mono, and there exists $1 \xrightarrow{h} A$ such that $hy = x$.

Exercise. Prove that the mapping h is uniquely determined (it is trivially monomorphic).

THEOREM 12. *Let A be some object. Then $x \in A$ if and only if $1 \xrightarrow{x} A$.*

Proof. The reader will prove Theorem 12 as an exercise.

We also read "$(x \in y)$" as "x is a *member* of y", the terms "element" and "member" being used interchangeably as in set theory.

Let us now give an intuitive justification for these definitions. When we introduced the concept of a monomorphism, we noted (see the exercise following Definition 6) that monomorphisms are 1-1 functions in the category of sets. The domain of any 1-1 function is thus isomorphic to the subset of the codomain that is the range of the function. In other words, a 1-1 function picks out a unique subset of the codomain. This justifies Theorem 11, which characterizes a subset of an object A to be a monomorphism with codomain A. Notice, it is the mapping (monomorphism) itself that is the subset, not the domain of the mapping.

Keeping this in mind, we can now see that, in Definition 21, x and y are both subsets of B and z is a subset of C. Intuitively, then, the set A is a subset of C by the inclusion map z. Thus, the subset of B represented by x is intuitively a subset of the subset of B represented by y; i.e. $x \subset y$.

To understand the intuitive motivation for the definition of membership, observe that any mapping into a set A from a one-element set 1 will pick out uniquely an element of A as the image of the single element in 1. Theorem 12 tells us that such mappings from 1 into A are elements of A.

Thus, intuitively in Definition 22, A is the domain of the inclusion map y, x is an element of B, and h is an element of the subset determined by y. The equation $hy = x$ tells us that the element x of B is the image of h under the 1-1 mapping y. Hence, x is an element of the subset y.

Notice that an element of y is always an element of $C(y)$. This is intuitively natural, since y is mono and thus a subset of $C(y)$.

These intuitive justifications have some appeal, but care must be taken not to think of the analogy with the usual membership relation as being too close. Notice, for instance, that two objects with only one element in common must be equal, since two maps are equal only if they have the same domain and codomain. Thus, the extensionality property so heavily used in set theory is not true of the membership relation that we have defined here. In Section 8.10 we will see how this problem can be met.

We now posit some axioms about membership in **CS**.

CS.6. If $A \xrightarrow{f} B$ and $A \xrightarrow{g} B$ and if, for all $1 \xrightarrow{x} A$, $xf = xg$, then $f = g$.

Two mappings are equal if they are defined on the same domain and have the same codomain, and if they are equal on elements. This is certainly intuitively true of functions in the category of sets. However, **CS.6** is not true in most categories in which it is definable (those with a terminal object). It is thus a significant, though simple, structural property of the category of sets.

Also, in terms of our system, **CS.6** can be thought of as furnishing yet another way of proving that two mappings are equal. If we can prove they are equal for elements, then they are equal.

CS.7. Let $A \xrightarrow{i_A} (A+B)$ and $B \xrightarrow{i_B} (A+B)$ be injections of the sum $(A+B)$. Then $\text{Mono}(i_A)$ and $\text{Mono}(i_B)$, and if $x \in (A+B)$, then either $(x \in i_A)$ or $(x \in i_B)$.

CS.7 requires that each of the injections of a sum be subsets of the sum and that any element of the sum be an element of one of the injections.

Since $(A+B)$ intuitively represents *disjoint* sum, it should also be true that the two injections i_A and i_B have no elements in common. This theorem will be forthcoming after we have added our next axiom. In fact, our final set of axioms allows us to weaken the statement of **CS.7** by proving $\text{Mono}(i_A)$ and $\text{Mono}(i_B)$. Our presentation has been somewhat simplified by explicitly incorporating $\text{Mono}(i_A)$ and $\text{Mono}(i_B)$ into the statement of **CS.7**.

Note that the definition of sums by injections corresponds in the category of sets to the intuitive fact that if we have defined mappings on each piece of a partition of a domain, then we have really defined a mapping on the whole domain.

CS.8. The two injections $1 \xrightarrow{i_1} (1+1)$ and $1 \xrightarrow{i_2} (1+1)$ (whose existence is posited by **CS.4**) are different and are the only elements of $(1+1)$.

Definition 23. 2 for $(1+1)$.

CS.8 tells us that 2 is a two-element set. Intuitively this has been so all along, but we could not have proved it. Indeed, the axioms **CS.1** through **CS.7** have a model in a one-element domain. This one element, call it 0, is both initial and terminal. $0 \times 0 = 0$, $0 + 0 = 0$, and so on. Now, with **CS.8**, this is no longer possible. We can now prove, for instance, that $0 \neq 1$.

THEOREM 13. $\vdash 1 \neq 0$.

Proof. If $1 = 0$, then $1 \xrightarrow{i_1} 2 = 0 \xrightarrow{i_1} 2$ and $1 \xrightarrow{i_2} 2 = 0 \xrightarrow{i_2} 2$. Since 0 is initial, $i_1 = i_2$ which contradicts **CS.8**.

We also prove (as promised) the theorem that states sums are really disjoint.

THEOREM 14. *For any two objects A and B, the two injections* $A \xrightarrow{i_A} (A+B)$ *and* $B \xrightarrow{i_B} (A+B)$ *have no elements in common.*

Proof. There are unique terminal maps $A \xrightarrow{x} 1$ and $B \xrightarrow{y} 1$ that induce a mapping $(A+B) \xrightarrow{(x+y)} 2.^{\dagger}$ Now, if there is an element $1 \xrightarrow{h} (A+B)$ which is an element of both injections, then, by Df. 22, there are mappings $1 \xrightarrow{a} A$ and $1 \xrightarrow{b} B$ such that $ai_A = h = bi_B$. Thus, the following diagram commutes:

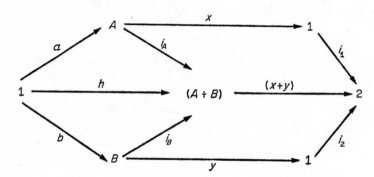

Now, $1 \xrightarrow{ax} 1 = 1 \xrightarrow{1} 1 = 1 \xrightarrow{by} 1$, since 1 is the only map from 1 to 1. This gives:

$$i_1 = 1i_1 = (ax)i_1 = a(xi_1) = a(i_A(x+y)) = (ai_A)(x+y) = h(x+y)$$
$$= (bi_B)(x+y) = b(i_B(x+y)) = b(yi_2) = (by)i_2 = 1i_2 = i_2.$$

But this is not possible, since $i_1 \neq i_2$ by **CS**.8. Thus there is no such h.

We can also prove that 0 is really the null (no-element) set.

THEOREM 15. *0 has no elements.*

Proof. If 0 has elements, then there is a mapping $1 \xrightarrow{x} 0$. But, by **CS**.1, there is exactly one mapping $0 \xrightarrow{y} 1$. Thus, $0 \xrightarrow{yx} 0 = 0 \xrightarrow{0} 0$, since there is only one mapping with domain and codomain 0. Similarly, $1 \xrightarrow{xy} 1 = 1 \xrightarrow{1} 1$ and so Iso(x). Thus $0 \cong 1$ which, by **CS**.5, gives $0 = 1$. This contradicts Theorem 13, and so our assumption that 0 has elements is false.

Intuitively 0 should be the *only* no-element set. Consequently, we should be able to prove that any nonzero set does have elements. This will follow from our next axiom, the axiom of choice.

CS.9. If $A \neq 0$ and $A \xrightarrow{f} B$, then there is a morphism $B \xrightarrow{g} A$ such that $fgf = f$.

Notice that this is one of the forms of the axiom of choice already discussed in Chapter 5, Section 5.4. The reader is referred to that section.

---

† $(x+y)$ is our sum map defined in the exercise following Definition 20.

THEOREM 16. *If $A \neq 0$, then A has elements.*

Proof. By **CS.1**, there is exactly one mapping $A \xrightarrow{f} 1$. Since $A \neq 0$, then $A \not\cong 0$ by **CS.5**. Thus, **CS.9** tells us there is a $1 \xrightarrow{g} A$ such that $fgf = f$. Thus $g \in A$ by definition and our theorem is established.

Our present collection of axioms permits us to prove a certain number of the usual properties of the set and element relation. For example:

THEOREM 17. *Let $C \xrightarrow{a} A$ and $B \xrightarrow{b} A$ be subsets of A. Then $a \subset b$ if and only if, for every $x \in A$, $x \in a$ implies $x \in b$.*

Proof. Suppose $a \subset b$ and $1 \xrightarrow{x} A$. If $x \in a$, then there is $1 \xrightarrow{h} C$ such that

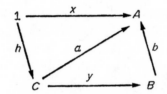

commutes, where y is the monomorphism that exists because $a \subset b$ (Definition 21). Since the diagram is commutative, we have immediately that $(hy)b = h(yb) = ha = x$, and so $x \in b$.

Conversely, suppose that, for all $x \in A$, $x \in a$ implies $x \in b$. We consider several cases. If $C = B = 0$, then $a = b$ by **CS.2** and the result follows. If $C \neq 0$, then, by Theorem 16, C has an element, say x. But $xa \in a$ and so, by hypothesis, $xa \in b$, which means there is an element $y \in B$. Thus, $B \neq 0$. This shows that in the case $C \neq 0$, $B = 0$ is impossible under our hypotheses. If $B \neq 0$, then by our axiom of choice and **CS.5**, there is a $A \xrightarrow{g} B$ such that $bgb = b$. We prove that $(ag)b = a$, which gives $a \subset b$ by Definition 21. If $C = 0$, then $C \xrightarrow{(ag)b} A$ and $C \xrightarrow{a} A$ are unique and so $a = (ag)b$.

Finally, if $C \neq 0$, let x be any element of C. Then $1 \xrightarrow{xa} A$ is an element of A and $xa \in a$ by Definition 22. Hence, $xa \in b$ by our hypothesis. Since $xa \in b$, there is a map $1 \xrightarrow{y} B$ such that $xa = yb$. We obtain the following diagram:

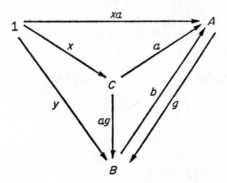

We have the following equations:

$$x(ag)b = (xa)gb = (yb)gb = y(bgb) = yb = xa.$$

Thus, for any element $x \in C$, $x(agb) = xa$ and the two maps agb and a are equal on elements. By **CS**.6 they are equal and our theorem is proved.

This is typical of many theorems (in which essential use is made of **CS**.6) that we shall prove. Notice also that the axiom of choice is involved in the proof of this theorem, which is a definition in ordinary set theory.

Exercise. Prove (i) $a \subset a$, and (ii) if $a \subset b$ and $b \subset c$, then $a \subset c$, where a, b, c are all subsets of a given object A. Notice that antisymmetry does not hold in general.

We should like to develop the theory of natural numbers in our system, but we have not yet considered how to do so. One obvious idea is to continue using sums, defining 3 as $(2+1)$, 4 as $(3+1)$, and so on. We already know that 0 has no elements, 1 has one element and 2 has two elements. We could proceed to prove that 3 has exactly three elements, 4 has four elements, and so on. But this manner of developing the natural numbers one by one is obviously unsatisfactory. We would have to have some way of "collecting" the objects 0, 1, 2, ... into a given set (object). However, there is no evident way of performing such a collecting procedure in **CS**. The answer to the dilemma is further postulation. We will now posit the existence of an object that intuitively represents the set of natural numbers N. This is an axiom of infinity for **CS**.

CS.10. There exists an object N together with mappings $1 \xrightarrow{o} N$, called the *zero map*, and $N \xrightarrow{s} N$, called the *successor map*, such that for all mappings $1 \xrightarrow{x_0} X$ and $X \xrightarrow{f} X$, where X is some object, there is one and only one map $N \xrightarrow{x} X$ such that $x_0 = ox$ and $xf = sx$.

N represents the set of natural numbers, conceived here as the finite *ordinal* numbers. The mapping $1 \xrightarrow{o} N$ is the *ordinal* number zero whereas our initial object 0 is the *cardinal* number zero. We thus have, in **CS**, a distinction between ordinal and cardinal even for finite numbers. Several definitions will help make this more precise.

Definition 24. By a *natural number* or a *finite ordinal* we mean a map $1 \xrightarrow{n} N$. The map $1 \xrightarrow{o} N$ of **CS**.10 is called *zero*, the *first ordinal*.

Definition 25. By the *successor* of a number n we mean the natural number $1 \xrightarrow{ns} N$ where s is the successor mapping of **CS**.10.

We have a hierarchy of cardinal numbers and a hierarchy of ordinal numbers. The cardinal numbers are the sequence 0, 1, 2, $(2+1)$, $((2+1)+1)$, and so forth. The ordinal numbers are the sequence o, os, oss, $osss$, and so on. The finite ordinals are all elements of our object N. It is for N that we can prove the Peano postulates in **CS**.

CS.10 is the property of simple recursion, which we proved to be true of the natural numbers ω in **ZF**. We have previously discussed the importance of this property of the natural numbers, and pointed out (see Chapter 3, p. 89) that it is useful in proving the uniqueness of the natural numbers in set theory. It is by no means obvious that the property of simple recursion, which will be our only axiom dealing specifically with the natural numbers, allows us to recover, in conjunction with our other axioms, the usual properties of the natural numbers. That we can obtain a satisfactory theory of the natural numbers in this way is one of Lawvere's insights.

In set theory, a sequence is a function whose domain is the set of natural numbers. Thus the map $N \xrightarrow{x} X$ defined from x_0 and f by simple recursion is a sequence. In Chapter 5, we took the example of defining powers a^n of arbitrary elements in a monoid as an application of simple recursion. We also indicated how addition and multiplication of natural numbers could be defined. Let us take more examples here of sequences defined by simple recursion.

Let X be the set of real numbers and f the function $y = x^2$. Pick some element, let us say 2, as x_0. Then our uniquely defined sequence from N to X is the sequence 2, 4, 16, ..., that is, the sequence is a function g of natural numbers such that $(0)g = 2$ and $(n+1)g = ((n)g)^2$. The function can be explicitly given as $(n)g = 2^{(2^n)}$, $n = 0, 1, 2$, and so on. This can be proved by mathematical induction. If we pick a different initial member, say 1, we get a different sequence, namely, 1, 1, 1, ... If we pick e, then we get $e, e^2, e^4, \ldots, e^{2^n}, \ldots$.

Not only can we prove the Peano postulates for N, we can prove the theorem of primitive recursion as well. We now turn to the heuristic motivation for our next axiom.

In ordinary set theory, the set of all subsets of a given set has a natural algebraic structure that is a partial order under proper inclusion. For any class of subsets of a given set, ordered by inclusion and closed under arbitrary unions, there is always a unique maximal subset in the class. If we think of the class of subsets as being determined by some property, then the maximal subset will be the largest one satisfying the property (it will contain as subsets all the others in the class).

In set theory, the existence of a maximal subset for a class of subsets ordered by inclusion and closed under arbitrary unions is a consequence of the union operation furnished by the axiom of sum set. However, there does not seem to be any way of proving the existence of such maximal subsets in **CS** without further postulation, since we do not have an exact analogue of sum set in **CS**. Thus, we posit the following axiom.

CS.11. Let $A \xrightarrow{f} B$ and $A \xrightarrow{g} B$ be two mappings, then there exists a mapping $E \xrightarrow{k} A$ such that $kf = kg$ and such that, for every mapping $X \xrightarrow{u} A$ such that $uf = ug$, there is a unique mapping $X \xrightarrow{z} E$ such that $u = zk$. We call k the *equalizer* of f and g.

Exercise. Make a diagram for the foregoing axiom. Prove that the equalizer is a monomorphism.

To understand the intuitive meaning of **CS.11**, let f and g be two functions with a common domain A and codomain B. Imagine the class of all subsets X of A with the property that $(X)f = (X)g$, i.e. that a subset is in the class if the image of that subset under the two mappings is the same. This class of subsets has a unique maximal element, the largest subset on which the two functions are equal. This is the equalizer. In our exercise we have proved it is a monomorphism and hence a subset of A. The remainder of **CS.11** implies that any other equalizing monomorphism with codomain A is a subset of the equalizer. Thus, the equalizer is the largest such subset.

0 and 1 are dual, sums and products are dual, and now we posit as an axiom the dual of **CS.11**.

CS.12. For $A \xrightarrow{f} B$ and $A \xrightarrow{g} B$, there is a mapping $B \xrightarrow{q} Q$ such that $fq = gq$ and such that, for any other mapping $B \xrightarrow{u} X$ for which $fu = gu$, there is a unique mapping $Q \xrightarrow{z} X$ such that $qz = u$. We call q the *coequalizer* of f and g.

It is immediate by our duality principle that coequalizers are epimorphisms.

Equalizers are special subsets whereas coequalizers are special quotient sets. Again this is under our intuitive interpretation in the category of sets. To see this, let us think of an equivalence relation generated on the set B by two functions $A \xrightarrow{f} B$ and $A \xrightarrow{g} B$ as follows: We begin with the reflexive, symmetric relation "x is related to y if and only if $x = y$ or there is some element w of A such that $(w)f = x$ and $(w)g = y$, or $(w)f = y$ and $(w)g = x$". We take the transitive closure of this relation which will then be an equivalence relation on B. It is easy to see that Q will be the set of equivalence classes under this relation and q will be the canonical map putting every element of B in its equivalence class. There may be many different equivalence relations on B that identify $(w)f$ and $(w)g$ for all w in A, but Q will be the finest one that does. Intuitively, such a quotient set would exist in the category of sets, and **CS.12** guarantees that it exists in our system **CS**.

We can now state our final axiom for **CS**.

We are attempting to characterize something like the category of sets and functions of **ZF**. What basic operations have we neglected? We have an axiom of infinity, which gives us the set of all finite ordinal numbers, i.e. an infinite set. What we clearly do not have in **CS** is any way of generating further infinite cardinal numbers. The usual way of generating such higher cardinalities is by means of the operation of power set. We need some such process in **CS**.

In order to motivate the form of our axiom, let us observe that in set theory the operation of power set is a special case of the operation of cardinal exponentiation. Given any two sets A and B, let B^A represent the set of all functions with domain A and codomain B. If A and B are finite, and if A has n elements and B has m elements, then B^A will have m^n elements. (If we set out to choose an arbitrary function, then each element of A has m possible images in B. Since there are n elements in A, this gives exactly $m \times m \times \cdots \times m$ to n factors, or m^n possible choices to form a function.) The set of all subsets of a given A will have cardinality 2^n where n is the cardinality of A. This is so because each subset of A is associated with a unique mapping into a fixed two-element set, this mapping being the characteristic function of the set. Generally, associating subsets with their characteristic functions obtains the operation of power set as a special case of exponentiation. Since power set allows us to generate higher cardinalities, the more general operation of exponentiation will as well. Our last axiom, **CS.13**, posits the existence of exponentiation.

CS.13. For any two objects A and B, there exists an object B^A and a mapping $(A \times B^A) \xrightarrow{e} B$, called an *evaluation map* of B^A, such that, for any object X and any mapping $(A \times X) \xrightarrow{f} B$, there is a unique mapping $X \xrightarrow{h} B^A$ such that $(A \times h)e = \langle p_A, p_X h \rangle e = f$.

The reader should make a diagram for **CS.13**.

For an intuitive understanding of **CS.13**, let us think of B^A as the set of all functions from A to B. Now, given an element a of A and a function g from A to B, we can associate with this ordered pair $\langle a, g \rangle$ the element $(a)g$ of B (the image of a under g). Hence, we have a mapping e from $(A \times B^A)$ to B which associates with every ordered pair $\langle a, g \rangle$, $a \in A$ and $g \in B^A$, the element $(a)g$ of B. This is the evaluation map e, and its rule is that of evaluating the function g at its argument a.

Now, given a function f from $(A\times X)$ to B, we define a function h from X to B^A in the following way: For each element $x \in X$, we consider the set of all ordered pairs $\langle a, x \rangle$ where $a \in A$. This induces a function g from A to B by the rule $(a)g = (\langle a, x \rangle)f$. We choose this function g to be the image of x under h; that is, $(x)h = g$. In this way there is associated with each element $x \in X$ an element $g \in B^A$. Clearly we have, for every $\langle a, x \rangle$ in $A\times X$, $((\langle a, x \rangle)A\times h)e = (\langle a, g \rangle)e = (a)g = (\langle a, x \rangle)f$. h is clearly unique.

In this way, **CS**.13 provides for exponentiation in our theory.

Definition 26. For any map $(A\times X) \xrightarrow{f} B$, f^* is the unique map $X \xrightarrow{h} B^A$ satisfying the conditions of **CS**.13.

Exercise. Prove that exponentiation as given by **CS**.13 is completely determined by two operations $e_{A, B}$, on pairs of objects, and f^* on morphisms f as above, and the following equations: $\langle p_A, p_X f^* \rangle e_{A, B} = f$ and $(\langle p_A, p_X h \rangle e_{A, B})^* = h$, where all these maps have the obvious domains and codomains.

In the future, whenever we want to avoid any possible ambiguity concerning an evaluation map, we will subscript the appropriate objects as in the above exercise. The equations of this exercise will also be frequently used in handling the composition of morphisms involving exponentiation. The operation $(-)^*$ is called *transposition*, and f^* is the *transpose* (or *exponential transpose*) of f.

Let us recall that the original motivation for category theory was to study the abstract structure of systems. In this light, the purpose of **CS** is to study the structure of the category of sets. In previous discussion we have already contrasted the approach of **CS**, which considers sets as cardinal numbers, and the approach of **ZF**. **ZF**, beginning with simple sets and iterating basic operations, builds a tower of structure by means of complicated membership relations on sets. **CS**.13 characterizes what is perhaps the most fundamental of all of these operations by means of purely structural properties of objects and mappings.

Having completed our statement of the axioms of **CS**, we now turn to a more complete development of set theory within it.

Under our intuitive explanation, the elements of B^A should be mappings from A to B. Can this be made precise in our theory? The answer is "yes".

THEOREM 18. *For all $A \xrightarrow{f} B$, there is an element $1 \xrightarrow{\ulcorner f \urcorner} B^A$, called the name of f, such that, for all $1 \xrightarrow{a} A$, $\langle a, \ulcorner f \urcorner \rangle e = af$. e is an evaluation map of B^A.*

Proof. $A\times 1 \xrightarrow{p_A} A \xrightarrow{f} B$ is a map from $A\times 1$ to B. By **CS**.13, there is a unique map $1 \xrightarrow{\ulcorner f \urcorner} B^A$ such that $\langle p_A, p_1 \ulcorner f \urcorner \rangle e = p_A f$. (Thus, $\ulcorner f \urcorner = (p_A f)^*$.) Where $a: 1 \longrightarrow A$, we obtain, using the properties of products:

$$\langle a, \ulcorner f \urcorner \rangle e = \langle a, 1 \rangle \langle p_A, p_1 \ulcorner f \urcorner \rangle e = \langle a, 1 \rangle p_A f = af$$

and the theorem is established.

Another consequence of our axiom of exponentiation is that we can now prove that products are distributive over sums. We omit the proof, which the interested reader may treat as an exercise.

We now turn to the proof of an important theorem that is useful in proving the Peano postulates. It is the theorem of primitive recursion that generalizes the simple recursion already guaranteed to us by the axiom of infinity of **CS**.

We want to show that, given any maps $g : A \longrightarrow B$ and $f : (A{\times}N){\times}B \longrightarrow B$, there exists a unique map $h : A{\times}N \longrightarrow B$ such that, for all $a : 1 \longrightarrow A$ and $n : 1 \longrightarrow N$, we have $\langle a, 0 \rangle h = ag$ and $\langle a, ns \rangle h = \langle\langle a, n \rangle, \langle a, n \rangle h \rangle f$. Such an h is said to be defined by primitive recursion from the data g and f. Given g and f as above, the mapping h is in fact wholly determined as the unique map $h : A{\times}N \longrightarrow B$ such that

(1) $\langle A, to \rangle h = g$, and

(2) $\langle p_A, p_N s \rangle h = \langle\langle p_A, p_N \rangle, h \rangle f = \langle A{\times}N, h \rangle f$,

where t is the unique terminal map $A \longrightarrow 1$. This follows immediately by evaluating the equation (1) on arbitrary elements $a : 1 \longrightarrow A$ and the equation (2) on arbitrary elements $\langle a, n \rangle : 1 \longrightarrow A{\times}N$. We begin by proving a lemma which will be a useful first step in establishing the existence of functions defined by primitive recursion.

LEMMA. *Given any mappings* $g : A \longrightarrow B$ *and* $f : B \longrightarrow B$, *there exists a unique* $h : A{\times}N \longrightarrow B$ *such that (i)* $\langle A, to \rangle h = g$ *and (ii)* $\langle p_A, p_N s \rangle h = hf$, *where* t *is the unique terminal map* $A \longrightarrow 1$.

Proof. We define $f^A = (e_{A, B}f)^* : B^A \longrightarrow B^A$ which, with $\ulcorner g \urcorner = (p_A g)^* : 1 \longrightarrow B^A$, allows us to construct, using **CS.10**, a unique $k : N \longrightarrow B^A$ such that (i)' $ok = \ulcorner g \urcorner$ and (ii)' $sk = kf^A$. In other words, the following diagram commutes:

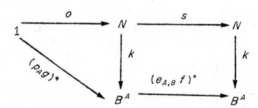

We now define $h = \langle p_A, p_N k \rangle e_{A, B} : A{\times}N \longrightarrow B$, and claim that h satisfies the conditions of the theorem. This is established by the following calculations where the properties of product maps and exponential transpose maps, as well as equations (i) and (ii)', are used:

$\langle A, to \rangle h = \langle A, to \rangle \langle p_A, p_N k \rangle e_{A, B} = \langle A, t \rangle \langle p_A, p_1 o \rangle \langle p_A, p_N k \rangle e_{A, B}$

$\qquad = \langle A, t \rangle \langle p_A, p_1 ok \rangle e_{A, B} = \langle A, t \rangle \langle p_A, p_1 (p_A g)^* \rangle e_{A, B} = \langle A, t \rangle p_A g = Ag = g$, verifying (i).

Also, $\langle p_A, p_N s \rangle h = \langle p_A, p_N s \rangle \langle p_A, p_N k \rangle e_{A, B} = \langle p_A, p_N sk \rangle e_{A, B} = \langle p_A, p_N k (e_A, Bf)^* \rangle e_{A, B}$

$\qquad = \langle p_A, p_N k \rangle \langle p_A, p_N (e_A, Bf)^* \rangle e_{A, B} = \langle p_A, p_N k \rangle e_{A, B} f = hf$, verifying (ii).

The uniqueness of h follows essentially from the uniqueness of k. It can also be verified by calculating on elements and appealing to **CS.6**. This establishes the theorem.

We now prove:

THEOREM 19. *Given the maps $g : A \longrightarrow B$ and $f : (A \times N) \times B \longrightarrow B$, there exists a unique map $h : A \times N \longrightarrow B$ such that the equations* (1) *and* (2) *above hold.*

Proof. We first define

$$\langle \langle A, to \rangle, g \rangle : A \longrightarrow (A \times N) \times B$$
$$\text{and} \quad \langle p_{A \times N} \langle p_A, p_{NS} \rangle, f \rangle : (A \times N) \times B \longrightarrow (A \times N) \times B.$$

Using the Lemma, we construct from these data a unique $k : A \times N \longrightarrow (A \times N) \times B$ such that

$$(1)' \ \langle A, to \rangle k = \langle \langle A, to \rangle, g \rangle \quad \text{and} \quad (2)' \ \langle p_A, p_{NS} \rangle k = k \langle p_{A \times N} \langle p_A, p_{NS} \rangle, f \rangle.$$

We define $h = k p_B : A \times N \longrightarrow B$, and assert that h satisfies the conditions of the theorem. Using (1)′, we have

$$\langle A, to \rangle h = \langle A, to \rangle k p_B = \langle \langle A, to \rangle, g \rangle p_B = g$$

which verifies (1). Using (2)′, we have

$$\langle p_A, p_{NS} \rangle h = \langle p_A, p_{NS} \rangle k p_B = k \langle p_{A \times N} \langle p_A, p_{NS} \rangle, f \rangle p_B = kf = k((A \times N) \times B)f$$
$$= k \langle p_{A \times N}, p_B \rangle f = \langle k p_{A \times N}, k p_B \rangle f = \langle k p_{A \times N}, h \rangle f.$$

This verifies (2) provided $k p_{A \times N} = A \times N$. We now establish that this last identity does hold. Consider the following diagram:

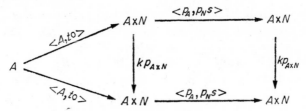

We establish, using (1)′ and (2)′, that it is commutative.

$$\langle A, to \rangle k p_{A \times N} = \langle \langle A, to \rangle, g \rangle p_{A \times N} = \langle A, to \rangle,$$
$$\text{and} \quad \langle p_A, p_{NS} \rangle k p_{A \times N} = k \langle p_{A \times N} \langle p_A, p_{NS} \rangle, f \rangle p_{A \times N} = k p_{A \times N} \langle p_A, p_{NS} \rangle.$$

Now, by the Lemma, there is only one map in the vertical positions of the rectangle making this diagram commute. Since the identity $A \times N$ obviously makes the diagram commute when it is substituted for $k p_{A \times N}$, then we must have $k p_{A \times N} = A \times N$.

This establishes that $\langle p_A, p_{NS} \rangle h = (A \times N, h)f$, i.e. equation (2), holds.

As before, the uniqueness of h follows from the uniqueness of k, as well as by elementwise calculation and appeal to **CS.6**. This completes the proof of the theorem.

The categorial proof of the theorem of primitive recursion was first established by Lawvere. A detailed proof was given in Hatcher [3], but one more complicated both in form and conception. The present, simplified proof is essentially due to Freyd [2]. Cf. also Goldblatt [1] where a proof is only sketched but where, unfortunately, there is an error in the construction of h.

For the sake of completeness and utility, we obtain a frequently used version of primitive recursion, without the parameter A, as a corollary to Theorem 19:

COROLLARY. *Given* $b : 1 \longrightarrow B$ *and* $f : N \times B \longrightarrow B$, *there exists a unique* $h : N \longrightarrow B$ *such that* $oh = b$ *and* $sh = \langle N, h \rangle f$.

Proof. We apply Theorem 19 with $A = 1$ to the data $b : 1 \longrightarrow B$ and

$$f' = \langle p_{1 \times N} p_N, p_B \rangle f : (1 \times N) \times B \longrightarrow B,$$

obtaining a unique $k : 1 \times N \longrightarrow B$ such that

$$\langle 1, o \rangle k = b \quad \text{and} \quad \langle p_1, p_{NS} \rangle k = \langle 1 \times N, k \rangle f'.$$

Letting $h = \langle t, N \rangle k$, where

$$t : N \longrightarrow 1, \quad \text{we have} \quad oh = o \langle t, N \rangle k = \langle 1, o \rangle k = b \quad \text{and}$$
$$sh = s \langle t, N \rangle k = \langle st, s \rangle k = \langle t, s \rangle k = \langle t, N \rangle \langle p_1, p_{NS} \rangle k$$
$$= \langle t, N \rangle \langle 1 \times N, k \rangle f' = \langle \langle t, N \rangle, h \rangle f'$$
$$= \langle \langle t, N \rangle, h \rangle \langle p_{1 \times N} p_N, p_B \rangle f = \langle N, h \rangle f$$

which establishes the theorem (the uniqueness of h follows from the uniqueness of k).

We now wish to prove Peano's postulates for the natural numbers $1 \longrightarrow N$. We already have that $o \in N$ and that the successor of any natural number is a natural number, for this follows from Definitions 24 and 25. To establish the remaining principles, we shall use the following theorem whose proof makes use of the corollary to the primitive recursion theorem.

THEOREM 20. *There is a map* $p : N \longrightarrow N$ *called* predecessor *such that* $op = o$ *and, for all* $n \in N$, $nsp = n$.

Proof. We apply the corollary of Theorem 19 with $B = N$, $o : 1 \longrightarrow N$, and $f = p_N^1 : N \times N \longrightarrow N$, this latter being the projection on the first factor. We thus obtain a unique map $h : N \longrightarrow N$ such that $oh = o$ and $sh = \langle N, h \rangle p_N^1 = N$. Thus, $p = h$ is the predecessor map.

Using the predecessor map, we can now prove:

THEOREM 21. *For all* $n \in N$, $o \neq ns$.

Proof. Assume $o = ns$ for some $n \in N$. Then $n = nsp = op = o$. Thus $o = os$. This means that our sequence of natural numbers, o, os, oss, etc., are all the same. We will draw a contradiction from this. Let $[i_1, i_2] : 2 \longrightarrow 2$ be the unique sum map making the following diagram commute:

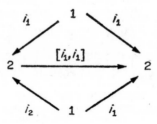

We now apply **CS**.10 to the simple recursion data $i_2 : 1 \longrightarrow 2$ and $[i_1, i_1] : 2 \longrightarrow 2$, defining a sequence $r : N \longrightarrow 2$ such that $or = i_2$ and $r[i_1, i_1] = sr$. We then have $i_2[i_1, i_1] = or [i_1, i_1] = osr = or = i_2$. But, by construction $i_2[i_1, i_1] = i_1$. Thus, we have $i_1 = i_2$, contradicting **CS**.8. This contradiction shows our original assumption of $o = ns$ for some n to be false, and the theorem is established.

By Theorem 21 o is not the successor of any natural number n. We now establish that the successor mapping is 1–1 on elements.

THEOREM 22. *If $ns = ms$, then $n = m$ where n and m are elements of N.*

Proof. If $ns = ms$, then $nsp = msp$, which gives

$$n = nsp = msp = m.$$

We have now obtained all the Peano postulates except the postulate of induction. In order to prove the postulate of induction, we need a new type of categorial construction, the "pullback".

THEOREM 23. *Given $A \xrightarrow{f} B$ and $C \xrightarrow{g} B$, there exist maps $G \xrightarrow{x} A$ and $G \xrightarrow{y} C$ for which $xf = yg$ and such that for any two maps $X \xrightarrow{x'} A$ and $X \xrightarrow{y'} C$ for which $x'f = y'g$, there is a unique mapping $X \xrightarrow{r} G$ such that $rx = x'$ and $ry = y'$.*

Proof. Given f and g as stated, consider the mappings

$$(A \times C) \xrightarrow{p_C} C \xrightarrow{g} B$$

and $(A \times C) \xrightarrow{p_A} A \xrightarrow{f} B$, and let $G \xrightarrow{k} (A \times C)$ be the equalizer of these two (**CS**.11). Let $x = kp_A$ and $y = kp_C$. We then have

$$xf = kp_A f = kp_C g = yg.$$

Moreover, if we have $x'f = y'g$ for some x' and y' with a common domain X, then $\langle x', y' \rangle$ also equalizes $p_C g$ and $p_A f$ since

$$\langle x', y' \rangle p_A f = x'f = y'g = \langle x', y' \rangle p_C g.$$

Thus, by the properties of an equalizer (**CS**.11), there is a unique map $X \xrightarrow{r} G$ such that $\langle x', y' \rangle = rk$. Thus

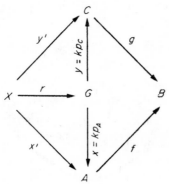

commutes, since $rx = rkp_A = \langle x', y' \rangle p_A = x'$ and

$$ry = rkp_C = \langle x', y' \rangle p_C = y'$$

and the proof is complete.

Definition 27. By the *pullback* of two maps $A \xrightarrow{f} B$ and $C \xrightarrow{g} B$ we mean the object G together with the maps x and y of Theorem 23.

Exercise. Prove that, for the pullback G of f and g, if Mono(f), then Mono(y), and also if Mono(g), then Mono(x). If Mono(g), then we call x the *inverse image of g under f* and write $x = f^{-1}[g]$. Similarly we have $y = g^{-1}[f]$ if Mono(f).

We now prove:

THEOREM 24. *If $a \subset N$ and if (i) $o \in a$ and (ii) $\forall n \in N, n \in a$ implies $ns \in a$, then $N \subset a$.*

Proof. By hypothesis $A \xrightarrow{a} N$ is Mono (let $D(a) = A$). Let G be the pullback of a and $N \xrightarrow{s} N$. We obtain the commutative diagram

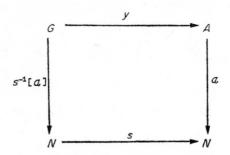

where $s^{-1}[a]$ is Mono by our preceding exercise. We first prove that $a \subset s^{-1}[a]$ by using Theorem 17. Let $1 \xrightarrow{n} N$ be any element of a. This means there is $1 \xrightarrow{r} A$ such that $ra = n$. We show that $n \in s^{-1}[a]$.

Since $n \in a$, our hypothesis (ii) tells us that $ns \in a$ which means there is a $1 \xrightarrow{w} A$ such that $ns = wa$. Hence $1 \xrightarrow{n} N$ and $1 \xrightarrow{w} A$ complete a pullback diagram. Consequently, there is a $1 \xrightarrow{x} G$ such that $xs^{-1}[a] = n$ and $xy = w$. Thus $n \in s^{-1}[a]$ and $a \subset s^{-1}[a]$ by Theorem 17, since n is any element of a.

Now $a \subset s^{-1}[a]$ means there is a map $A \xrightarrow{d} G$ such that $ds^{-1}[a] = a$. Let $t = dy$. Then $as = ds^{-1}[a]s = dya = ta$ by the properties of the pullback maps y and $s^{-1}[a]$. We have now arrived at the commutative diagram displayed at the top of page 279.

Invoking hypothesis (i), we have $o \in a$, which means there is $1 \xrightarrow{x_0} A$ such that $x_0a = o$. Using **CS.10** we define a sequence $N \xrightarrow{f} A$ from the data $1 \xrightarrow{x_0} A$ and $A \xrightarrow{t} A$. We have $of = x_0$ and $ft = sf$. Considering now $N \xrightarrow{fa} N$, we see that $ofa = x_0a = o$ and $sfa = fta = fas$. Thus, the second diagram on the opposite page commutes, and so $fa = N$ by the uniqueness of sequences defined by simple recursion.

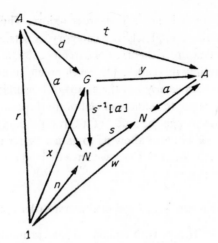

Now, for any $n \in N, n = nN = n(fa) = (nf)a$, and so $n \in a$ by definition. Appealing to Theorem 17, we conclude $N \subset a$ and our theorem is proved.

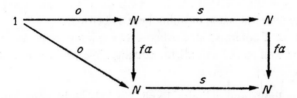

This completes the proof of the Peano postulates in **CS**. For more detailed indications of the development of analysis in **CS**, the reader should consult the basic papers Lawvere [1], Lawvere [2], and Lawvere [3].

8.6. Topos theory

It is natural to wonder about the adequacy of **CS** as a categorial formulation of set theory. What is clearly missing from **CS** is the strong "collecting up" principles of **ZF** like the axiom of replacement. In fact, **CS** quite clearly has a model in **ZF** among the sets of rank less than $\omega+\omega$: we take the set of all cardinals of rank less than $\omega+\omega$ and all the functions between these cardinals.

In Lawvere [1], Lawvere states and indicates the proof of the following metatheorem: Let M be any model for **CS** which, as a category, is complete. (By "complete" we mean that, in addition to satisfying the axioms of **CS**, M has the further property that sums and products exist in M on any indexing set whatever, instead of having only finite sums and products which already exist by **CS**.3 and **CS**.4.)[†] Then M is equivalent to the category $\mathcal{S}$ of sets and functions of **ZF**. By "equivalence" we mean the notion of equivalence between categories discussed in Section 8.4.

The additional hypothesis of completeness is necessary to guarantee the equivalence of any model with the category of sets. This condition of completeness cannot be incorporated in

---

[†] Generally speaking, a complete category is defined to be a category possessing equalizers and coequalizers, as well as infinite sums and products. But M already has equalizers and coequalizers if it is a model of **CS**.

our language **CS,** however, for it is not a first-order axiom. Thus, it appears that no first-order language will characterize completely the category of sets. However, since there exist denumerable models of any first-order theory, **CS** cannot really be singled out for any special criticism on this score. It certainly does allow the development of analysis and of a substantial portion of mathematics, and thus clearly has the status of a foundational system.

As it turns out, further attempts to refine and elaborate the category-theoretic approach to foundations are best handled through the study of a certain theory called *topos* which is more general than **CS.** Topos theory will also allow us to gain a better understanding of the structure of the category of sets than does **CS.** Once we have defined topos theory, we will establish clearly its relationship to the theory **CS.**

Definition 1. By a *topos* we mean any model of the first-order theory **T** to be defined below.

The theory **T** is a first-order theory with equality whose primitive symbols are the same as those of **C** with the addition of two new constant letters noted Ω and $\top$.

For this section, and indeed for the rest of the chapter, we will freely use the definitions, notions, and notations of the previous sections with the understanding that, unless otherwise stated explicitly, they have the same meaning here as before.

The proper axioms of **T** comprise all the first-order axioms for a category, the axioms **C.1**–**C.6** of Section 8.2, as well as the axioms for equality. We begin anew our numbering of the other proper axioms of **T.**

T.1. T has all finite limits and colimits. Explicitly, this is the conjunction of axioms **CS.1, CS.3, CS.11** (finite limits), and **CS.2, CS.4,** and **CS.12** (finite colimits).

T.2. T has exponentiation. This is the same as **CS.13.**

T.3. Ω is a subobject classifier. Explicitly, this is the following statement: $Ob(\Omega)$ and $1 \xrightarrow{\top} \Omega$ and, for every monomorphism $A \xrightarrow{m} B$, there exists a unique map $B \xrightarrow{c_m} \Omega$, called the *characteristic map* of m, such that the diagram

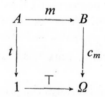

is a pullback where $A \xrightarrow{t} 1$ is the unique terminal map with domain A.

This completes the axioms for **T.**

We will not assume for **T** any equivalent of the skeletability axiom **CS.5.** We must thus suppose that "1" stands for some terminal object and "0" for some initial object. Terminal and initial objects are all respectively isomorphic to each other, so it will not matter which terminal or initial object is chosen. A similar remark applies for product and sum objects. For topos theory, we will be satisfied with determining objects up to isomorphism. We also adopt, as standard notation, "t_A" for any terminal map $A \longrightarrow 1$. We occasionally drop the subscript "A" if the domain is unambiguous.

The notion of topos defined by the above axioms is sometimes called an *elementary* topos to distinguish it from a *Grothendieck topos* which is somewhat more particular than an elementary one. We will have no need to consider Grothendieck toposes in this study.

The notion of an elementary topos first appeared in the literature in Lawvere [4] and in Tierney [1].

Let us first see that the category of sets and functions of **ZF** set theory is a topos. Since all the axioms of a topos except **T.3** are already axioms of **CS**, which has a model in **ZF**, we have only to interpret **T.3** in **ZF**. We let Ω be the two-element set $\{0, 1\}$ and we let $\top$ be the function f from $\{0\}$ to $\{0, 1\}$ such that $(0)f = 0$. As we have already seen in the previous section, a monomorphism is a subset. Thus, the unique map c_m is the characteristic map of m, giving the value 0 for elements of m and 1 elsewhere on B.

Any model of **CS** is already a topos where we interpret Ω to be $1+1$ and $\top$ to be the injection $1 \xrightarrow{i_1} 1+1$. We will prove this fact later on. We want first to develop a bit more the theory of toposes and to see what additional axioms are necessary in order to obtain a system equivalent to **CS**. It will soon become apparent that the axioms for **T** are deceptively simple and that **T.3**, in conjunction with the other axioms, is much stronger than might appear at first.

Axiom **T.3** allows us to define a unique characteristic map c_m for any monomorphism $A \xrightarrow{m} B$. On the other hand, given any map f from B to Ω, we can form the pullback of the maps $1 \xrightarrow{\top} \Omega$ and $B \xrightarrow{f} \Omega$. Notice that, since 1 is terminal, $\top$ is trivially a monomorphism. Thus, any map $A \xrightarrow{m} B$ defined by this pullback will be a monomorphism also (this is the result of the exercise following Definition 27 in Section 8.5), and f is thus its characteristic map. We refer to such an m as a *kernel* of f.

Since characteristic maps are unique, two different maps $B \xrightarrow{f} \Omega$ and $B \xrightarrow{g} \Omega$ give rise to different kernels. But the converse is not quite true. Two different kernels can give rise to the same characteristic map. The most that the universal property of pullbacks allows us to say is that two monomorphisms $A \xrightarrow{m} B$ and $C \xrightarrow{n} B$ having the same characteristic map, $c_m = c_n$, are *isomorphic* meaning that there is an isomorphism from A to C which carries m to n.

Exercise. Show that two monomorphisms $A \xrightarrow{m} B$ and $C \xrightarrow{n} B$ are isomorphic in the above sense if and only if $m \subset n$ and $n \subset m$. Observe that this latter condition immediately holds if m and n have the same characteristic map.

T.3 thus establishes a natural bijection between Hom (B, Ω) and the isomorphism classes of subobjects of B.

We illustrate a simple use of **T.3** with the following theorem:

THEOREM 1. *In any topos, if a monomorphism m is also epi, then it is iso.*

Proof. Given $A \xrightarrow{m} B$, we define the characteristic map $c_m : B \longrightarrow \Omega$. We thus have the commutative square

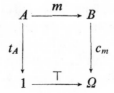

Clearly $mt_B\top = t_A\top = mc_m$. Thus, if m is epi, $t_B\top = c_m$. By the universal property of pullbacks, we have a map $f : B \longrightarrow A$ such that $fm = B$. But also $Am = m = mB = mfm$. Since m is mono, this yields $A = mf$, and m is thus iso.

A category satisfying the property of Theorem 1 is called *balanced*. We have already seen (cf. the exercise following Definition 8, Section 8.2) that this is not a trivial property. Theorem 1 thus says that all toposes are balanced. Here is another property true in all toposes:

THEOREM 2. *In any topos, $A\times 0 \cong 0$, for any object A.*

Proof. We prove that $A\times 0$ is initial. Given any object X, there is exactly one morphism $k : 0 \longrightarrow X^A$. Thus, the composite $\langle p_A, p_0k\rangle e_{A,X}$ is a map from $A\times 0$ to X. But this is the only map from $A\times 0$ to 0, for let $h : A\times 0 \longrightarrow X$ be any such. Then, by exponentiation, $h^* : 0 \longrightarrow X^A$ and $\langle p_A, p_0h^*\rangle e = h$. But 0 is initial, so $h^* = k$ which immediately gives $h = \langle p_A, p_0k\rangle e$.

We use this as a lemma to prove the following:

THEOREM 3. *In any topos, if $a : A \longrightarrow 0$, then $A \cong 0$.*

Proof. Let $f : A\times 0 \longrightarrow 0$ and $g : 0 \longrightarrow A\times 0$ be inverses of each other. Then $(\langle A, a\rangle f)(gp_A) = A$ and $(gp_A)(\langle A, a\rangle f) = 0$, the first by the properties of products and the second because 0 is initial.

COROLLARY. In any topos initial objects are *strict*, i.e. for every object A, there is at most one morphism $a : A \longrightarrow 0$ and any such a is iso.

Proof. If $a : A \longrightarrow 0$, then by Theorem 3 $A \cong 0$. Thus, A is also initial and a is therefore the unique morphism from A to 0. a is necessarily iso, the unique inverse of the unique initial map from 0 to A.

Our axioms for a topos are not independent. In fact, the axiom set can be significantly simplified. For example, we can construct sums $A+B$ from products, equalizers, and a terminal object by means of exponentiation. The study of these questions is best carried out using some of the sophisticated tools of functor theory, and we do not intend to go through the details here (for a fairly complete discussion of them, see Johnstone [1]).

To get an idea of the approach involved, observe that by the laws of exponents which hold in topos, $\Omega^{A+B} \cong \Omega^A\times\Omega^B$. The right-hand side involves only products and exponentiation. Moreover, any object X is the domain of a monomorphism $X \xrightarrow{\{\ \}} \Omega^X$ called *singleton* and defined by $\{\ \} = (c_{\langle X, X\rangle})^*$. Thus, $\{\ \}$ is the exponential transpose of the characteristic map of the diagonal $\langle X, X\rangle : X \longrightarrow X\times X$. It can be shown that $\{\ \}$ is a mono in any topos. In this way, $X = A+B$ is definable as the domain of a certain monomorphism $m : X \longrightarrow \Omega^A\times\Omega^B$.

Not only can sums be obtained from finite limits and exponentiation, but general exponentiation can be obtained by assuming exponentiation only on Ω: for every map $f : A\times B \longrightarrow \Omega$, there exists a unique map $f^* : B \longrightarrow \Omega^A$ such that $\langle p_A, p_Bf^*\rangle e_{A,\Omega} = f$. In the category of sets, Ω is 2 and so Ω^A is the power set of A. It turns out that Ω^A operates substantially as a power set in any topos, and so B^A is forthcoming as the domain of a monomorphism with codomain

$\Omega^{(A \times B)}$, i.e. as "the subset of functional relations of the set of all binary relations from A to B".

By similar kinds of constructions, coequalizers and the initial object 0 can also be obtained from products, equalizers, the terminal object, and exponentiation on Ω. The proper axioms for a topos can thus be taken to be the existence of a terminal object, products, equalizers, exponentiation on Ω, and T.3. This is a useful fact to keep in mind in verifying whether a given category constitutes a topos. It will also be useful for our study of the "internal logic" of toposes (Sections 8.11 and 8.12).

Before pushing any further with a deductive study of the theory T, let us return to a consideration of models of it. So far, the only models we have really constructed are in the category of sets. Here the object Ω is fairly trivial, namely the two-element set {0, 1}. If these were the only models of T, then topos theory would certainly not have had the immense success that it has known. It turns out that the object Ω can be of almost arbitrary complexity. Indeed, we will now describe a class of models for T which we call *simple models*. It is to the existence of such naturally defined classes of models that topos theory owes much of its success.

Let $\mathcal{C}$ be any small category. By a *presheaf* over $\mathcal{C}$, we mean any functor $F: \mathcal{C}^{op} \longrightarrow \mathcal{S}$, where $\mathcal{S}$ is the category of sets and functions of **ZF**. The set of all such presheaves is itself a category where the morphisms are simply the natural transformations. This category is noted $\mathcal{S}^{\mathcal{C}^{op}}$.

THEOREM 4. *For any small category $\mathcal{C}$, the category $\mathcal{S}^{\mathcal{C}^{op}}$ of presheaves over $\mathcal{C}$ is a topos.*

Proof. A somewhat painful but straightforward verification which we leave to the reader (see Goldblatt [1], pp. 204–210).

By a *simple model* of T we now understand any full subcategory of a category of presheaves which is a model of T. In particular, if the full subcategory in question is closed under the formation of all finite limits and colimits, is closed under exponentiation, and contains 0, 1, Ω, and $\top$, then it will be a simple model. Among the simple models are thus the *sheaves* over a topological space (in this case, the category $\mathcal{C}$ is the set of open sets ordered by inclusion).

The simple models of a topos are thus "parametrized sets". We obtain the sets themselves as a simple model by taking the base category $\mathcal{C}$ to be the category with exactly one (identity) morphism 1. The category of all finite sets is a topos where we take 1 for the base category and exclude those functors which map 1 to a infinite set.

Since a monoid M or a group G is a one-object category, the categories $\mathcal{S}^M$ and $\mathcal{S}^G$ each constitute a topos.[†] A functor F from M to $\mathcal{S}$ is an action of M on the set $(M)F$. Natural transformations of these functors are thus morphisms of actions, and the category $\mathcal{S}^M$ is therefore the category of all M-actions and their homomorphisms. If we let M be the free monoid over a finite alphabet A, then $\mathcal{S}^M$ is just the category of all automata having A as input alphabet, and their morphisms. If we restrict to the subcategory of finite-set-valued functors, then we obtain the set of all finite-state automata over the input alphabet A, and their morphisms.

Besides forming set-valued functor categories, there is another way of constructing categories which yields a topos when applied to a topos. For any category $\mathcal{C}$, and any object A of $\mathcal{C}$, we form the category $\mathcal{C}/A$, called "$\mathcal{C}$ over A", whose objects are morphisms of $\mathcal{C}$ with

---

[†] We make use here of the fact that the category M^{op} is just the opposite monoid of M and that $(M^{op})^{op} = M$.

codomain A and whose morphisms $(X \xrightarrow{x} A) \longrightarrow (Y \xrightarrow{y} A)$ are the

commutative triangles

$$
\begin{array}{ccc}
X & \xrightarrow{\;z\;} & Y \\
& {}_{x}\searrow \quad \swarrow_{y} & \\
& A &
\end{array}
$$

It is a fairly deep theorem that $\mathcal{C}/A$ is a topos whenever $\mathcal{C}$ is. Moreover, there is a naturally defined functor F from $\mathcal{C}$ to $\mathcal{C}/A$, given by $(X)F = p_A : X \times A \longrightarrow A$, and

$$
(X \xrightarrow{\;f\;} Y)F =
\begin{array}{ccc}
X \times A & \xrightarrow{f \times A} & Y \times A \\
{}_{p_A}\searrow & & \swarrow_{p_{A'}} \\
& A &
\end{array}
$$

which is commutative by virtue of the equations $(f \times A)p_{A}' = \langle p_X f, p_A \rangle p_{A}' = p_A$. It turns out that this functor is *logical*, i.e. that it preserves all the topos data (finite limits and colimits, exponentiation, Ω, and $\top$).

This construction also yields interesting toposes. For example, where A is any set, $\mathcal{S}/A$ is a topos whose Ω-object is $(2)F = p_A : 2 \times A \longrightarrow A$, and whose terminal object is $A : A \longrightarrow A$.

Let us note in passing that for $\mathcal{C} = \mathcal{S}$, $\mathcal{C}/A$ is a simple model up to an equivalence of categories. This follows from the observation that any set A is a discrete category. Hence, there is an obvious injective correspondence between functions $I \longrightarrow A$ (objects of $\mathcal{S}/A$) and functors $A \longrightarrow \mathcal{S}$ (objects of $\mathcal{S}^A$). This correspondence is given by associating with each $f : I \longrightarrow A$, the function $f^{-1} : A \longrightarrow \mathcal{P}(I)$. We leave as an exercise the verification that this correspondence extends naturally to a functorial one between $\mathcal{S}/A$ and $\mathcal{S}^A$, and that this correspondence is an equivalence of categories (see Goldblatt [1], p. 204).

We can thus see that part of the interest and power of topos theory derives from the fact that it generalizes set theory in an interesting way: a topos shares a number of important structural features with the category of sets and yet there is a vast class of naturally defined toposes which are significantly more general than the category of sets.

One of the ways that a topos is more general than set theory is that non-zero objects A do not necessarily admit mappings $1 \xrightarrow{x} A$. In Section 8.5 we have called such maps elements of A. Here we will call them *global elements* of A. Whereas in sets, the object 1 is "undecomposable", having as it does only the initial map and itself as subobjects, the terminal object in a topos can be more complex as is clear from the simple models where the base category is rich.[†]

Not only can non-zero objects A be empty of global elements, two morphisms $f : A \longrightarrow B$ and $g : A \longrightarrow B$ may differ even though they have the same value for all global elements of A. In a certain sense, toposes for which this is true do not have enough "points" (i.e. global elements) to determine the behavior of morphisms and to make them into genuine functions. As we will also see, the question of the existence of an adequate supply of global elements is closely related to the structure of the object Ω. We now begin a closer study of this question.

8.7. Global elements in toposes

In this section we continue the numbering of definitions and theorems begun in the last section.

[†] However, a global element is always mono since 1 is terminal.

First, let us observe that Ω has, besides $\top$, another distinguished global element. Because of Theorem 3 and its corollary, any initial map $k : 0 \longrightarrow A$ is a monomorphism. We therefore let $\bot : 1 \longrightarrow \Omega$ be the unique characteristic map of the monomorphism $0 \longrightarrow 1$.

Definition 2. A topos is *degenerate* if $0 \cong 1$.

THEOREM 5. *Any topos is non-degenerate if and only if* $\top \neq \bot$.

Proof. Clearly, if $0 \cong 1$, the 1 is initial and there is only one map, $\top = \bot$, from 1 to Ω. Conversely, if $\top = \bot$, then by the universal property of pullbacks, there exists a unique map $1 \longrightarrow 0$ such that $1 \longrightarrow 0 \longrightarrow 1 = 1$. Hence, $1 \cong 0$.

Exercise. In any degenerate topos, every object A is isomorphic to 0.

The result of the exercise justifies the terminology of Definition 2.

Definition 3. A topos is *well-pointed* if it is non-degenerate and if, further, 1 is a generator (i.e. axiom **CS.6** holds). We use **WT** to denote the theory obtained from **T** by adding the two axioms $0 \ncong 1$ and **CS.6**.

The notio,n and terminology of Definition 3 is due to Freyd [2]. He there proves, among other things that any small topos can be embedded in a product of well-pointed toposes. We will show here that all of the axioms of **CS** except the axioms of infinity (**CS.10**), choice (**CS.9**), and of course skeletability (**CS.5**) are theorems of **WT**. In fact, somewhat surprisingly, the lone assumption that 1 is a generator allows us to determine completely the structure of Ω.

From now to the end of the present section, all theorems are for the theory **WT** unless the contrary is explicitly stated.

THEOREM 6. *Every* $A \ncong 0$ *has global elements.*

Proof. The maps $t_A \top$ and $t_A \bot$ are different, for otherwise there exists a map $a : A \longrightarrow 0$ rendering the following diagram commutative:

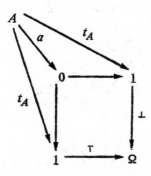

where the inner square is a pullback. Hence, by Theorem 3, $A \cong 0$ contradicting our hypothesis. Since 1 is a generator, there exists $x : 1 \longrightarrow A$ such that $xt_A \top \neq xt_A \bot$, and the theorem is established.

THEOREM 7. *If $A \not\cong 0$, then t_A is epi.*

Proof. By Theorem 6, A has a global element $x : 1 \longrightarrow A$. Since 1 is terminal, the composite $xt_A = 1$, i.e. x is a left inverse for t_A. It follows immediately that t_A is epi.

COROLLARY. *The only subobjects of 1 are the initial map $0 \longrightarrow 1$ and 1 itself.*

Proof. If $t_A : A \longrightarrow 1$ is mono, and $A \not\cong 0$, the t_A is also epi and therefore iso by Theorem 1. Hence $A \cong 1$.

THEOREM 8. *The only global elements of Ω are $\top$ and $\bot$.*

Proof. Let $x : 1 \longrightarrow \Omega$ be any global element of Ω and form the pullback

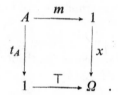

m is a kernel of x. If $A \cong 0$, then $x = \bot$ since $\bot$ is the unique characteristic map of $0 \longrightarrow 1$. If $A \not\cong 0$, then m is epi (by Theorem 7) and mono. By Theorem 1 m is this iso and $A \cong 1$ which yields immediately $x = \top$.

Ω is said to be *two-valued* in any topos in which Theorem 8 is true. We already have here a partial characterization of the structure of Ω in a well-pointed topos.

Definition 4. $\neg : \Omega \longrightarrow \Omega$ is the characteristic map of the mono $\bot : 1 \longrightarrow \Omega$.

THEOREM 9. $\neg\,\neg = \Omega$.

Proof. By definition, $\bot\,\neg = \top$. We show that also $\top\,\neg = \bot$. If $\top\,\neg = \top$, then by the universal property of pullbacks, there is a map $x : 1 \longrightarrow 1$ making the following diagram commute:

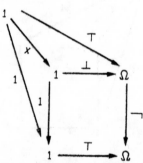

where the inner square is a pullback. But 1 is the only map from 1 to 1, so $\top = \bot$ contradicting nondegeneracy. But $\top\,\neg$ is a global element of Ω, and by Theorem 8 Ω has only two global elements, namely $\top$ and $\bot$. Since $\top \neg \neq \top$, $\top \neg = \bot$. Thus, $\neg\neg$ and Ω have the same value for all global elements of Ω. Since 1 is a generator, this immediately gives $\neg\neg = \Omega$.

Notice that it follows from the two-valuedness of Ω alone that $\neg\neg$ and Ω have the same value on global elements of Ω. But we need the fact that 1 is a generator to conclude that $\neg\neg = \Omega$. Indeed, there are toposes satisfying Theorem 8 but not Theorem 9. Hence, we have yet a further property of Ω in well-pointed toposes.

Any topos in which Theorem 9 is true is said to be *Boolean*.[†] We have thus established that a well-pointed topos is Boolean with a two-valued Ω object. The true force and usefulness of these facts will become clearer as we progress in our development of the theory **WT**.

We now establish a few useful properties of sums and their injections.

THEOREM 10. *The injections of a sum are monomorphisms.*

Proof. Let $A \xrightarrow{i_A} A+B \xleftarrow{i_B} B$ be given. If $B \cong 0$, then $A \xrightarrow{A} A \longleftarrow 0$ is the sum of A and 0, and each of the injections here is a monomorphism. If $B \ncong 0$, the B has a global element $x : 1 \longrightarrow B$. We therefore can define a unique sum map $[(t_A x), B] : A+B \longrightarrow B$. Thus, $i_B[(t_A x), B] = B$. Hence i_B has a right inverse and is therefore mono. By a symmetrical argument, the injection i_A is also mono.

THEOREM 11. *Sums are disjoint, i.e. no global element of a sum can be an element of both of the injections.*

Proof. Let $x : 1 \longrightarrow A+B$ be given. Define the sum map $[(t_A \top), (t_B \bot)] : A+B \longrightarrow \Omega$. If, now, there exist y and z such that $yi_A = x = zi_B$, then $x[(t_A \top), (t_B \bot)] = yt_A \top = \top = zt_B \bot = \bot$ which contradicts nondegeneracy. Thus, x cannot factor through both injections.

THEOREM 12. *Every global element of a sum must factor through one of the injections.*

Proof. Given the sum $A \xrightarrow{i_A} A+B \xleftarrow{i_B} B$, suppose there is some global element $m : 1 \longrightarrow A+B$ which does not factor through either of the injections i_A or i_B. Since m is a monomorphism it has a characteristic map c_m. We show first that c_m is the sum map $[(t_A \bot), (t_B \bot)] : A+B \longrightarrow \Omega$. Let $a : 1 \longrightarrow A$ be any global element of A. Then $ai_A c_m = \bot = at_A \bot$, for otherwise $ai_A c_m = \top$ which implies by the universal property of pullbacks that $ai_A = m$, i.e. that m factors through i_A. Since 1 is a generator, $i_A c_m = t_A \bot$. Arguing similarly for global elements of B, we obtain $i_B c_m = t_B \bot$ and so $c_m = [(t_A \bot), (t_B \bot)]$ as claimed. But $[(t_A \bot), (t_B \bot)] = t_{A+B} \bot$. Hence, $\top = mt_{A+B} \bot = \bot$, contradicting nondegeneracy. Thus, there is no global element which does not factor through at least one of the injections i_A and i_B, and the theorem is proved.

COROLLARY. **CS**.7.

Proof. This is precisely the conjunction of Theorem 10 and Theorem 12.

[†] There are also Boolean toposes for which the Ω object is not two-valued. Thus, the conditions represented by Theorems 8 and 9 are quite independent in the theory **T**.

Thus, we have already obtained one of the promised axioms of **CS** as a theorem of **WT**. The other one, **CS**.8, is immediately forthcoming as well.

THEOREM 13. (**CS**.8.) *The two injections of* $1+1$ *are different and are the only global element of* $1+1$.

Proof. Define the sum map $[\top, \perp]: 1+1 \longrightarrow \Omega$. If the two injections of $1+1$ are the same, then $\top = \perp$ contradicting nondegeneracy. By Theorem 12, every other global element $1 \longrightarrow 1+1$ must factor through one of the injections and is therefore equal to that injection through which it factors. Hence, the injections are the only two global elements of $1+1$.

In fact, the sum map defined in the proof of Theorem 13 is an isomorphism. This can be seen quite easily in the light of the next two theorems which characterize monos and epis respectively in terms of their action on global elements.

THEOREM 14. $m : A \longrightarrow B$ *is mono if and only if it is injective on global elements.*

Proof. Since a mono is right cancellable, it always takes distinct global elements to distinct global elements. Conversely, suppose m injective and let $fm = gm$ for two parallel maps

$$C \underset{g}{\overset{f}{\rightrightarrows}} A.$$

If $f \neq g$, then, since 1 is a generator, there exist $x : 1 \longrightarrow C$ such that $xf \neq xg$ and thus $xfm \neq xgm$ since m is injective. But $xfm = xgm$ since $fm = gm$. Thus, $f = g$ and m is mono.

A bit trickier is the proof of the following:

THEOREM 15. $q : A \longrightarrow B$ *is epi if and only if it is surjective on global elements, i.e. for every* $1 \xrightarrow{x} B$, *there exist* $1 \xrightarrow{y} A$ *such that* $yq = x$.

Proof. Suppose q surjective and let $qf = qg$ for a pair of parallel maps

$$B \underset{g}{\overset{f}{\rightrightarrows}} C.$$

For every global element $b : 1 \longrightarrow B$, there is a global element $a : 1 \longrightarrow A$ such that $aq = b$. Hence, $bf = aqf = aqg = bg$, for every map $b : 1 \longrightarrow B$. Since 1 is a generator, we conclude $f = g$ and q is therefore epi.

Conversely, suppose q epi but not surjective and let $b : 1 \longrightarrow B$ such that, for all $a : 1 \longrightarrow A$, $aq \neq b$. Since b is mono, it has a characteristic map c_b, and clearly $aqc_b = \perp = at_A \perp$ for all $a : 1 \longrightarrow A$. Since 1 is a generator, this gives $qc_b = t_A \perp$. But $t_A \perp = (qt_B) \perp$. Since q is epi, this yields $c_b = t_B \perp$. But then $\top = bc_b = bt_B \perp = \perp$ contradicting nondegeneracy. Hence q is surjective.

THEOREM 16. $1+1 \cong \Omega$.

Proof. The sum map $[\top, \perp]: 1+1 \longrightarrow \Omega$ is easily seen to be both injective and surjective and therefore iso.

With Theorem 16 we have now arrived at a complete determination of the structure of Ω. We use its Boolean character (Theorem 9) to construct, for each subobject m of B, a complement m'.

THEOREM 17. *For any subobject* $A \xrightarrow{m} B$ *of an object* B, *there exists a subobject* $A' \xrightarrow{m'} B$ *such that* $[m, m'] : A + A' \longrightarrow B$ *is an isomorphism.*

Proof. Letting c_m be the characteristic map of m, let $A' \xrightarrow{m'} B$ be a kernel map of $c_m \neg$. Then, $m'c_m = m'c_m \Omega = m'c_m \neg \neg = t_{A'} \top \vdash = t_{A'} \bot$. We have established the commutativity of the following diagram:

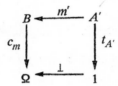

We now establish that this diagram is, in fact, a pullback.

Suppose that $X \xrightarrow{f} B$ such that $fc_m = t_X \bot$. Then $fc_m \neg = t_X \bot \neg = t_X \top$. There is therefore a unique map $g : X \longrightarrow A'$ such that $gm' = f$ and, of course, $gt_{A'} = t_X$. Hence the diagram is a pullback.

Let, now, $x : 1 \longrightarrow B$ be any global element of B. Then xc_m is a global element of Ω and hence either $\top$ or $\bot$. If x does not factor through m, then $cx_m = \bot$. But since we have established that our diagram is a pullback, there then must exist a unique $y : 1 \longrightarrow A'$ such that $ym' = x$, i.e. x factors through A'. If x factors through m and m', then $\top = xc_m = \bot$ which contradicts nondegeneracy.

We now use these facts to show that the sum map $[m, m'] : A + A' \longrightarrow B$ is an isomorphism. Every global element $b : 1 \longrightarrow B$ of B factors through m or m'. Suppose, for a given $b : 1 \longrightarrow B$, that there is $a : 1 \longrightarrow A$ such that $am = b$. Then $b = a(i_A[m, m']) = (ai_A)[m, m']$. If b factors through A', then there will likewise be an $a' : 1 \longrightarrow A'$ such that $b = (a'i_{A'})[m, m']$. Thus, the map $[m, m']$ is surjective and, by Theorem 15, epi.

Let, now, $x[m, m'] = y[m, m']$ where x and y are two global elements of $A + A'$. x and y must each factor through one and only one of the injections i_A or $i_{A'}$ (Theorem 12). They obviously must both factor through the same injection, for otherwise we will have $\top = x[m, m']c_m = y[m, m']c_m = \bot$. Suppose, then, that $x = zi_A$ and $y = wi_A$ (the argument is similar if they both factor through $i_{A'}$). Then, $x[m, m'] = (zi_A)[m, m'] = zm$, and $y[m, m'] = (wi_A)[m, m'] = wm$. Thus, $zm = wm$ which yields $z = w$ (since m is mono) and hence $x = y$. $[m, m']$ is therefore injective and consequently mono. $[m, m']$ is thus iso and the theorem is established.

We have now established as theorems of **WT** every axiom of **CS** except the axiom of infinity, **CS**.10, of choice, **CS**.9, and of course skeletability (**CS**.5). If we add these further axioms to **WT**, we will then have a theory which includes **CS**. In fact, it will be equivalent to **CS** as follows from the fact (to be shown later) that **CS** is a topos.

WT thus isolates and identifies a certain constructive fragment of **ZF** set theory. Indeed, any well-pointed topos $\mathcal{C}$ can be embedded into **ZF** in a simple and natural way: To every object we associate the set of its global elements and to every map f the function from the set of global elements of its domain to the set of global elements of its codomain induced by composition with f. The other operations and special objects of **WT** correspond to their obvious set-theoretic counterparts. That this correspondence is functorial is immediate. That it is an embedding follows from the fact that 1 is a generator in **T**. In a later section, we will exhibit a theory expressed in the language of **ZF** set theory which corresponds to that portion of **ZF** described by **WT**.

Let us note, in passing, another way of axiomatizing **WT**:

THEOREM 18. *Every nondegenerate topos in which Theorems 6 and 17 of the present section hold is well-pointed.*

Proof. We deduce that 1 is a generator in any such topos. Let parallel maps

$$A \underset{g}{\overset{f}{\rightrightarrows}} B \quad \text{be given.}$$

Suppose f and g are the same on all global elements of A. Let $k : E \longrightarrow A$ be the equalizer of f and g. As always for equalizers, k is mono. Moreover, every global element $a : 1 \longrightarrow A$ factors through k by the universal property of equalizers, since f and g are equal on such global elements. Letting $k' : E' \longrightarrow A$ be the complement of k (Theorem 17), we have immediately, for any global element $x : 1 \longrightarrow E'$, $(xk')f = (xk')g$ which implies that xk' must factor through k, i.e. there exists z such that $zk = xk'$. Now, the sum map $[k, k']$ is an isomorphism, and so $zi_E[k, k'] = zk = xk' = xi_E[k, k']$ immediately yields $zi_E = xi_{E'}$. But this latter identity says that the element $xi_{E'}$ of $E+E'$ factors through both injections, which contradicts nondegeneracy (see the proof of Theorem 11). Thus, E' cannot have any global elements and hence $E' \cong 0$ (Theorem 6). Thus, $A \cong E+0 \cong E$. In particular, $k : E \longrightarrow A$ is an isomorphism since $i_E : E \longrightarrow E+0$ is, and $k = i_E[k, k']$ is thus the composite of two isomorphisms.

Hence, the identity morphism A equalizes f and g, and $f = g$. The theorem is established.

Theorem 18 allows us to give an even more elegant form to our axioms for **CS** in the light of a theorem of Diaconescu (cf. Johnstone [1], p. 141). Diaconescu shows that complements exist (Theorem 17) in any topos in which the axiom of choice is true. The form of the axiom of choice used by Diaconescu is somewhat weaker than our **CS.9** and will be discussed in detail in the next section. Nevertheless, it is implied by **CS.9** in any topos. Since **CS.9** also implies Theorem 6 in any topos (cf. Theorem 16, Section 8.5), we have immediately that any nondegenerate topos satisfying **CS.9** satisfies as well every other axiom of **CS** except infinity and skeletability. Recalling now our simplified version of axioms for a topos, we can recover **CS** in the extremely succinct form as a nondegenerate topos satisfying (strong) choice (**CS.9**), infinity (**CS.10**), and, if desired, skeletability (**CS.5**).

Let us mention that some of the theorems we have proved in **WT** actually hold in **T** itself (Theorem 10, for example), but the more general proofs are sometimes involved. Still other theorems, while not holding in **T**, nevertheless are special cases in **WT** of more general theorems which do hold in **T**. For example, in any topos, the pullback of the two injections of a sum is an initial object 0. In **WT** this immediately specializes to Theorem 11.

Similarly, some of the separate results we have proved in **WT** are equivalent in **T**. For example, Theorems 9, 16, and 17 are equivalent in any topos, but the proof of this fact involves results (such as the preservation of sums and of epimorphisms under pullbacks) which are best established using results from functor theory (in particular the existence of certain adjoints).

We have systematically avoided proofs of theorems in general topos theory which involve appeal to the techniques of functor theory and which are therefore not elementary (i.e. first-order). Of course, first-order versions of the proofs of these theorems can be given, but they may be horribly complicated. An alternative method of attack is to use a language of type theory appropriate to general topos theory. We will consider the details of this approach in a later section.

We now take a closer look at the axiom of choice in topos theory.

8.8. Image factorizations and the axiom of choice

One of the basic and useful facts of the category of sets is that every morphism can be factored as an epimorphism followed by a monomorphism. Such factorizations exist, however, in much more general categories than sets, and they are quite useful. We now establish that image factorizations exist in any topos.

LEMMA. *In any topos, every monomorphism is an equalizer.*

Proof. Given a monomorphism $m: A \longrightarrow B$, m is the equalizer of $t_B \top$ and c_m.

In the proof of the following theorem, we make use of *pushouts* which are the precise dual of pullbacks: Given two maps $f: A \longrightarrow B$ and $g: A \longrightarrow C$, their pushout consists of two maps $f': B \longrightarrow X$ and $g': C \longrightarrow X$ such that $ff' = gg'$ and which are couniversal for this property. Just as pullbacks can be constructed as equalizers of products (Theorem 23, Section 8.5), pushouts can be constructed as coequalizers of sums. Thus pushouts exist in any topos.

THEOREM 1. *In any topos, every morphism $f: A \longrightarrow B$ can be factored as $f = qm$ where $q: A \longrightarrow C$ is epi and $m: C \longrightarrow B$ is mono. Moreover, m is the smallest subobject of B through which f factors.*

Proof. Let $m: C \longrightarrow B$ be the equalizer of the pushout maps $h: B \longrightarrow X$ and $g: B \longrightarrow X$ of f with itself. By the universal property of equalizers, there is thus a unique map $q: A \longrightarrow C$ such that $qm = f$. We need to establish that q is epi.

Let us first observe that m precedes any other monomorphism n through which f factors. For, any monomorphism $n: Y \longrightarrow B$ is an equalizer of some two maps $a: B \longrightarrow Z$ and $b: B \longrightarrow Z$. Thus, if $f = yn$ for some $y: A \longrightarrow Y$, the couniversal property of pushouts yields a unique map $d: X \longrightarrow Z$ such that $hd = a$ and $gd = b$. But by construction $mh = mg$ which immediately yields $ma = mb$. Since n is the equalizer of a and b, this yields a unique map $w: C \longrightarrow Y$ such that $wn = m$. In other words, m precedes n as claimed. It follows immediately that for any map $f: A \longrightarrow B$, the monomorphism $m: C \longrightarrow B$ constructed as above is unique up to isomorphism of subobjects of the codomain B. We call $f = qm$ the *image factorization* of f, and m the *image* of f, noted $\text{Im}(f)$.

To prove, now, that q is always epi in the image factorization, we observe that by the construction of $\mathrm{Im}(f)$, f is epi if and only if B is the image of f. Applying now the image construction to q in the image factorization $f = qm$, we immediately obtain that $\mathrm{Im}(q)$ is C and thus q is epi. This establishes the theorem.

The form of the axiom of choice most frequently used in topos theory can be stated as follows: Every epi splits, i.e. given any epimorphism $q : A \longrightarrow B$, there exists a map $k : B \longrightarrow A$ such that $kq = B$. k is thus a left inverse of q.

In the category of sets, this form of the axiom of choice has the following interpretation: Since g is epi, it is surjective. Thus, every element $b \in B$ determines a unique subset $q^{-1}(b) \subset A$, and $q^{-1}(B)$ is a disjoint collection of sets. k thus picks exactly one element $k(b)$ out of each class $q^{-1}(b)$.

Exercise. Show that, in any topos, the k above is a (not necessarily unique) monomorphism.

We refer to this version of the axiom of choice as *weak choice*, while the version **CS.9** will henceforth be called *strong choice*.

THEOREM 2. *In any topos, strong choice implies weak choice.*

Proof. Given an epimorphism $q : A \longrightarrow B$. If $A \cong 0$, then q is also mono (because 0 is strict initial) and thus iso. Thus, q has a full inverse k. If $A \ncong 0$, then, applying **CS.9**, we obtain $k : B \longrightarrow A$ such that $qkq = q$. But q is epi and therefore left cancellable. Thus, $kq = B$.

THEOREM 3. *In any well-pointed topos, weak choice implies strong choice.*

Proof. Let $f : A \longrightarrow B$, and let $f = qm$, $q : A \longrightarrow C$, be its image factorization. If $A \ncong 0$, then $C \ncong 0$ since 0 is strict initial. Thus, if $A \ncong 0$, C has a global element, say $d : 1 \longrightarrow C$. Let $m' : C' \longrightarrow B$ be the complement of the mono m, and $j : B \longrightarrow C+C'$ the isomorphism which is the inverse of $[m, m']$. Finally, we define $[C, t_{C'} d] : C+C' \longrightarrow C$ and we let $k : C \longrightarrow A$ be the left inverse of the epi q (weak choice). We claim that the composite $g = j[C, t_{C'} d]k$ is the choice map satisfying $fgf = f$ (strong choice). Calculating, we find:

$$fgf = (qm)\,g(qm) = qmj[C, t_{C'}d]kqm = qmj[C, t_{C'}d]m = qmj[Cm, t_{C'} dm]$$

$$= q(ic[m, m'])\,j[m, t_{C'} dm] = qic(C+C')\,[m, t_{C'} dm] = qic[m, t_{C'} dm] = qm = f$$

as claimed. This establishes the theorem.

Notice that we have used both Theorem 6 and Theorem 17 of the previous section in the above proof. In view of Theorem 18 of that section, we have made essential use of well-pointedness.

COROLLARY. *In any well-pointed topos, strong and weak choice are equivalent.*

Proof. Immediate from the above.

In the last section, we discussed the result of Diaconescu that the axiom of choice implies the existence of complements of subobjects. Weak choice is the form of the axiom of choice used by Diaconescu in this result. Thus, in any topos weak choice implies Theorem 17 of Section 8.7. We therefore obtain the following somewhat sharper version of the relationship between the two forms of the axiom of choice:

THEOREM 4. *In any topos, strong choice is equivalent to weak choice plus the proposition that every non-initial object has global elements (Theorem 6, Section 8.7).*

Proof. If the topos is degenerate, then both strong and weak choice are trivially true. If the topos is nondegenerate, weak choice implies Theorem 17 of Section 8.7 (Diaconescu) which, in conjunction with Theorem 6 of Section 8.7, implies well-pointedness (Theorem 18 of Section 8.7) from which the equivalence of the two versions of choice follows. Thus, weak choice and Theorem 6 of Section 8.7 imply strong choice. The converse is immediate.

There exist toposes which satisfy weak choice but not strong choice. A particularly simple example is the topos $\mathcal{S}/2$ (cf. Goldblatt [1], p. 295). Finally, we remark that both forms of the axiom of choice are independent in **WT** in view of Cohen's independence result for full **ZF** set theory of which **WT** is a proper part.

8.9. A last look at CS

Before taking a closer look at general topos theory, we want to tie up the loose ends in our analysis of the relationship between **T** and **CS**. *For the remainder of our discussion in this book, we shall always understand that **CS**.5 is excluded as an axiom from the theory **CS**,* but we shall not bother to give another name for this weaker form of **CS**.

We have seen that all of the axioms of **CS** are forthcoming in the theory **WT** plus choice plus infinity, and we have mentioned that in fact **CS** is equivalent to this theory. Thus, **WT** is a subtheory of **CS**. However, we cannot extract the subtheory **WT** from **CS** by any combination of **CS** axioms, and it is interesting to see why. The reason lies in the fact that **CS** makes extensive use of (strong) choice at very elementary levels. In particular, choice is necessary to prove T.3 in **CS**. But since choice does not hold in **WT**, we cannot hope to recover **WT** by any combination of **CS** axioms: if we exclude infinity and choice the system will be too weak, and if we include choice it will be too strong.[†] In particular, there is a model of **CS** without choice in the category of partially ordered sets, a model which is not a topos.

A typical example of the use of the axiom of choice at elementary levels of **CS** is Theorem 17 of Section 8.5. The axiom of choice was there used to prove a statement which is a definition in ordinary set theory. Let us see how this statement would be proved in **WT** without using the axiom of choice.

Given $a : C \longrightarrow A$ and $b : B \longrightarrow A$, each monomorphisms, suppose that every element of A which factors through a factors also through b. We want to establish that a factors through b (i.e. precedes b as a subobject of A). Now, for any element $x : 1 \longrightarrow C$, xa must factor through b, i.e. there exists $y: 1 \longrightarrow B$ such that $xa = yb$ which yields $xac_b = ybc_b = \top =$

[†] Choosing infinity but not choice obviously yields a system having some theorems not holding in **WT** while lacking some theorems of **WT**.

$xat_A \top$. Since 1 is a generator, $ac_b = at_A \top$, whence, by the universal property of pullbacks, there exists a unique $h : C \longrightarrow B$ such that $hb = a$, and a factors through b as claimed.

In this and similar ways, uses of choice in **CS** are replaced by uses of **T.3** in **T** and in **WT**. This allows us to appreciate the superiority of topos theory as a framework for formulating questions about set-like categories in a categorial language. Not only is the continual use of choice in **CS** intuitively repugnant, the need to invoke choice on such elementary levels makes it an integral part of the system. This means that no subsystem of **CS** which excludes choice will be very interesting, while finding models of subsystems which include choice may not be easy. On the other hand, the theory **T** comes with a host of naturally defined models which serve as guiding examples in our study of various extensions such as **WT**.

Let us now give the proof that **T.3** holds in **CS**. Specifically, we claim that the object 2 is a subobject classifier with the injection $i_1 : 1 \longrightarrow 2$ playing the role of $\top$. We thus need to show that every monomorphism m in **CS** has a characteristic map c_m relative to these data. We take up our study of **CS** where we left it at the end of Section 8.5.

Let us first notice that, since i_1 is mono, the pullback

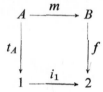

for any given map $f : B \longrightarrow 2$ determines a mono m with f as a characteristic map. Monos m which are defined in this way are called "special" until we can prove that all monos are special. We will accomplish this by constructing complements for arbitrary subobjects. We begin by constructing image factorizations in **CS**.

LEMMA. *In* **CS**, *every monomorphism is an equalizer.*

Proof. We show that any mono $m : A \longrightarrow B$, $A \not\cong 0$, is the equalizer of B and fm where $f : B \longrightarrow A$ is, by **CS**.9, a map such that $mfm = m$. By **CS**.11, B and fm have an equalizer h which is the largest subobject of B such that $hB = hfm$. Since $mfm = m = mB$, m precedes h. On the other hand, since $h(fm) = h$, h precedes m. Hence, m is isomorphic to the equalizer h of B and fm and is therefore an equalizer of them. Finally, any $m : 0 \longrightarrow B$ is the equalizer of the pair of maps $t_B i_1$ and $t_B i_2$ from B to 2.

THEOREM 1. *Every map* $f : A \longrightarrow B$ *in* **CS** *has an image factorization* $f = qm$ *where* $q : A \longrightarrow C$ *is epi, and* $m : C \longrightarrow B$, *is mono, the smallest subobject of* B *through which* f *factors.*

Proof. Using the above lemma, the proof is identical with the proof of Theorem 1, Section 8.8, which does not use **T.3** (except in the proof of the lemma which we have here proved using the axiom of choice and without using **T.3**). All the other constructions used in the proof are true in **CS**.

Let us also note that Theorems 14 and 15 of Section 8.7 are immediately forthcoming in **CS**. In particular, the axiom of choice immediately implies that epis are surjective. Strong choice also allows us to conclude immediately that any epi-mono is iso, thus that models of **CS** are balanced categories. These properties of **CS** will be useful in what follows.

We now show how to construct certain unions which we will need in the construction of complements for subobjects.

THEOREM 2. *Given any map* $k : I \longrightarrow 2^X$, *there exists a mono* $a : Y \longrightarrow X$ *which is the union of the* k_j, $j \in I$, *i.e. such that for* $x \in X$, $x \in a$ *if and only if there exists* $j \in I$ *such that* $\langle x, jk \rangle e_X = i_1$. $e_X : X \times 2X \longrightarrow 2$ *is the evaluation morphism* $e_{X, 2}$.

Proof. Let $\bar{k} = \langle p_X, p_I k \rangle e_X : X \times I \longrightarrow 2$. Let $h : A \longrightarrow X \times I$ be the equalizer of $\bar{k}$ and $t_{X \times I} i_1$. Now, the map $hp_X : A \longrightarrow X$ has an image factorization $hp_X = qa$. We show that the mono $a : Y \longrightarrow X$ of this factorization has the desired property. Since q is epi, every $x : 1 \longrightarrow X$ which factors through a, say $x = ya$, gives rise to a $z : 1 \longrightarrow A$ such that $zq = y$ (q is surjective). Hence, $x = ya = zqa = zhp_X$. Letting $j = zhp_I$, we have

$$\langle x, jk \rangle e_X = \langle x, j \rangle \langle p_X, p_I k \rangle e_X = \langle x, j \rangle \bar{k} = \langle x, zhp_I \rangle \bar{k} = \langle zhp_X, zhp_I \rangle \bar{k}$$
$$= zh \langle p_X, p_I \rangle \bar{k} = zh\bar{k} = zht_{X \times I} i_1 = i_1.$$

Conversely, if there is a $j \in I$ such that $i_1 = \langle x, jk \rangle e_X = \langle x, j \rangle \bar{k}$, then, since $i_1 = 1i_1 = (\langle x, j \rangle t_{X \times I}) i_1$, $\langle x, j \rangle$ equalizes $\bar{k}$ and $t_{X \times I} i_1$. Thus, $\langle x, j \rangle$ factors through h, i.e. $\langle x, j \rangle = wh$ for some unique $w : 1 \longrightarrow A$. Hence $(wq)a = whp_X = \langle x, j \rangle p_X = x$, i.e. $x \in a$. This establishes the theorem.

LEMMA. *Given any mono* $a : A \longrightarrow B$, *and any global element* $x : 1 \longrightarrow B$ *which does not factor through* a, *then there is a map* $c_x : B \longrightarrow 2$ *such that* $xc_x = i_1$ *while* $ac_x = t_A i_2$.

Proof. Since a is mono and x does not factor through a, the sum map $[a, x] : A + 1 \longrightarrow B$ is injective and therefore mono. Applying **CS**.9, we obtain a right inverse h for $[a, x]$, i.e. $[a, x]h = A + 1$. Let $i_{1'} : 1 \longrightarrow A + 1$ be the injection of 1 into $A + 1$. We define $c_x = h[t_A i_2, i_1]$. Then, $xc_x = (i_{1'}[a, x])c_x = i_{1'}([a, x]h) [t_A i_2, i_1] = i_{1'}[t_A i_2, i_1] = i_1$, and $ac_x = (i_A[a, x])c_x = {}_A[t_A i_2, i_1] = t_A i_2$, establishing the theorem.

THEOREM 3. *Complements to subobjects exist, i.e. for every mono* $A \xrightarrow{a} B$, *there exists a mono* $A' \xrightarrow{a'} B$ *such that* $[a, a'] : A + A' \longrightarrow B$ *is an isomorphism.*

Proof. The construction is a nice exercise in the use of exponents. We first construct the map $2^a : 2^B \longrightarrow 2^A$ as follows: Given the map $\langle p_A a, p_{2^B} \rangle e_B : A \times 2^B \longrightarrow 2$, where e_B is the evaluation map $e_{B, 2}$, we apply exponentiation (**CS**.13) to obtain 2^a such that $\langle p_A, p_{2^B} 2^a \rangle e_A = \langle p_A a, p_{2^B} \rangle e_B$ (thus $2^a = (\langle p_A a, p_{2^B} \rangle e_B)^*$), e_A abbreviating $e_{A, 2}$. Letting $j_0 : 1 \longrightarrow 2^A$ be the exponential transpose $(p_1 i_2)^*$ of $p_1 i_2 : A \times 1 \longrightarrow 1 \longrightarrow 2$, $k : K \longrightarrow 2^B$ is the equalizer of 2^a and $t_{2^B} j_0 : 2^B \longrightarrow 2^A$. We now apply Theorem 2 to k, obtaining its union $a' : A' \longrightarrow B$. We must now prove that $[a, a'] : A + A' \longrightarrow B$ is an isomorphism. We first show that this map is mono by establishing that it is injective.

Suppose that $y[a, a'] = z[a, a']$ for y, z elements of $A + A'$. Each of y and z must factor through one of the injections i_A or $i_{A'}$. Since a and a' are each mono, $y = z$ immediately follows if they both factor through the same injection. We show that this must be the case. Applying the conclusion of Theorem 2 to a', we have the fact that, for every $x \in B$, $x \in a'$ if and only if there is $j \in K$ with $\langle x, jk \rangle e_B = i_1$. Suppose, now, that $y = wi_{A'}$ and $z = si_A$. Composing with $[a, a']$ we obtain an $x = wa' = sa$. Since $x \in a'$, there is some $j : 1 \longrightarrow K$ such that $\langle x, jk \rangle e_B = i_1$. On the other hand, for the same j we have:

$$\langle x, jk \rangle e_B = \langle sa, jk \rangle e_B = \langle s, jk \rangle \langle p_A a, p_{2B} \rangle e_B = \langle s, jk \rangle \langle p_A, p_{2B} 2^a \rangle e_A = \langle s, jk 2^a \rangle e_A$$
$$= \langle s, (jkt_{2B}) j_0 \rangle e_A = \langle s, 1(p_1 i_2)^* \rangle e_A = \langle s, 1 \rangle \langle p_A, p_1(p_1 i_2)^* \rangle e_A$$
$$= \langle s, 1 \rangle p_1 i_2 = 1 i_2 = i_2.$$

Hence $i_1 = i_2$ contradicting **CS**.8. Thus, y and z must factor through the same injection and $[a, a']$ is therefore mono.

Finally, we show that $[a, a']$ is surjective and therefore epi. Given some $x : 1 \longrightarrow B$, if x does not factor through a, then, applying the lemma, there is c_x such that $xc_x = i_1$ while $ac_x = t_A i_2$. First, we remark that the name of c_x, i.e. the mapping $\ulcorner c_x \urcorner = (p_B c_x)^*$, equalizes $t_{2B} j_0$ and 2^a. We have immediately $\ulcorner c_x \urcorner t_{2B} j_0 = 1(p_1 i_2)^* = (p_1 i_2)^*$. Since this latter is the unique map such that $\langle p_A, p_1(p_1 i_2)^* \rangle e_A = p_1 i_2$, we show equality with $\ulcorner c_x \urcorner 2^a$ by evaluating $\langle p_A, p_1 \ulcorner c_x \urcorner 2^a \rangle e_A$. We have:

$$\langle p_A, p_1 \ulcorner c_x \urcorner 2^a \rangle e_A = \langle p_A, p_1 \ulcorner c_x \urcorner \rangle \langle p_A, p_{2B} 2^a \rangle e_A = \langle p_A, p_1 \ulcorner c_x \urcorner \rangle \langle p_A a, p_{2B} \rangle e_B$$
$$= \langle p_A a, p_1 \ulcorner c_x \urcorner \rangle e_B = \langle p_A a, p_1 \rangle \langle p_B, p_1 \ulcorner c_x \urcorner \rangle e_B$$
$$= \langle p_A a, p_1 \rangle (p_B c_x) = p_A a c_x = p_A t_A i_2 = t_{A \times 1} i_2 = p_1 i_2,$$

as claimed. Since $\ulcorner c_x \urcorner$ equalizes these maps, it factors through their equalizer $k : K \longrightarrow 2^B$. Thus, $\ulcorner c_x \urcorner = jk$ for some $j : 1 \longrightarrow K$. Now, $\langle x, \ulcorner c_x \urcorner \rangle e_B = xc_x = i_1$ by Theorem 18, Section 8.5. Thus, $\langle x, jk \rangle e_B = i_1$ for $j \in K$, which implies $x \in a'$ by the construction of a'.

We have shown that every $x : 1 \longrightarrow B$ which does not factor through a must factor through a' from which it immediately follows that $[a, a']$ is surjective and therefore epi. Since it is mono it is iso and the theorem is established.

We now apply this theorem immediately to conclude that every mono is special, i.e. has a characteristic map:

THEOREM 4. *Every monomorphism* $m : A \longrightarrow B$ *has a unique characteristic map* $c_m : B \longrightarrow 2$ *such that the diagram*

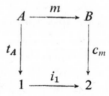

is a pullback.

Proof. Applying Theorem 3, we can assume without loss of generality that B is $A + A'$ with injections $A \xrightarrow{m} A + A' \xleftarrow{m'} A'$. We define c_m to be the sum map $[t_A i_1, t_{A'} i_2] : A + A' \longrightarrow 2$.

We have the following commutative diagram:

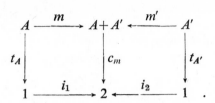

We want to establish that the left-hand square is a pullback (the right-hand one will, of course, also be a pullback). Let $f: X \longrightarrow A + A'$ be any map such that $fc_m = t_X i_1$. Now, every global element $x: 1 \longrightarrow X$ gives rise to a global element $xf: 1 \longrightarrow A + A'$ which must factor through exactly one of the injections m or m'. In fact, xf can only factor through m since otherwise we will have $i_1 = i_2$ contradicting **CS**.8. We distinguish two cases. In the case that X has no global elements, $X \cong 0$, and so there is a unique initial map $g: X \longrightarrow A$ such that $mg = f$ (since f is also an initial map). Otherwise, if X has global elements so does A, and we apply **CS**.9 to m, obtaining $h: A + A' \longrightarrow A$ such that $mhm = m$. We claim that fh is the unique map such that $(fh)m = f$. For every $x: 1 \longrightarrow X$ we have $xf = ym$ for some $y: 1 \longrightarrow A$. Thus, $xf = ym = y(mhm) = (ym)hm = xfhm$ which yields $f = (fh)m$ by **CS**.6. The uniqueness of fh follows from the fact that m is mono. For, if $gm = f$ for some g, then $gm = gmhm = fhm$ which yields $g = fh$ upon cancellation of m. This establishes that the left-hand square is indeed a pullback.

Finally, c_m is unique since any other map $B \longrightarrow 2$ making the left-hand square a pullback can only be the sum map $[t_A i_1, t_{A'} i_2]$ as follows from **CS**.6 and **CS**.7. This completes the proof.

With Theorem 4 we now have established the property **T**.3 for **CS**. We thus have the equivalence of **CS** with the theory **WT** plus infinity and (weak or strong) choice. The immense advantage of the topos-theoretic form of **CS**, as we have seen, is that applications of **T**.3 avoid elementary uses of choice.

Exercise (for masochists). Count the minimum number of distinct applications of the axiom of choice which we have used essentially in deriving Theorem 4 in **CS**. Do not forget use of lemmas which ultimately depend on choice, as well as redundancies which may occur through convenient appeal to unnecessarily strong principles based on choice.

From now on, we will always think of **CS** in the form **WT** plus infinity and choice. In the next section we obtain an even clearer understanding of **CS** by constructing a set-theoretic language and theory which is essentially equivalent to the **WT** "core" of **CS**.

8.10. ZF and WT

In Chapter 5 we have briefly considered the system **Z** of Zermelo set theory which is **ZF** without the axioms of infinity, replacement, and choice. We now consider an even weaker system $\mathbf{Z_0}$ which is obtained from **Z** by weakening the axiom of separation (the axiom **ZF**.2) in the following way:

ZF.2*. $(z)(x)(x \in \{y \,|\, y \in z \land A(y)\} \equiv x \in z \land A(x))$ where $A(y)$, besides satisfying the other restrictions of **ZF.2**, has all of its quantified variables restricted to sets. Thus, every occurrence of universal quantification in $A(y)$ is of the form $(x)(x \in t \supset \ldots)$ and every occurrence of existential quantification is of the form $(Ex)(x \in t \land \ldots)$ where, in each case, t is some term of **Z**. The other restrictions for **ZF.2** are carried over unchanged. We call **ZF.2*** the *weak separation scheme*.

One of the characteristics of the weak separation scheme is that it can be replaced by a finite number of axioms. Specifically, we can replace **ZF.2*** in Z_0 by axioms affirming the existence of the operation of cartesian product, of forming relative complements, of forming the domain of the relational part of any set, etc. (see Johnstone [1], p. 313). We do not enter into detail since the present form of the system is adequate for our purposes.

It is easy to see that the restricted separation scheme is enough to carry through in Z_0 all of the usual set-theoretical constructions of such things as ordered pairs, cartesian products, relations, etc. In particular, functions will be defined in the usual way as certain ordered triples of sets (i.e. domain, graph, and codomain). Thus, associated with any model $\langle D, g \rangle$ of Z_0 will be a category, namely the category whose objects are the elements of D and whose morphisms are the interpretations in $\langle D, g \rangle$ of the functions of Z_0. We have:

THEOREM 1. *The category associated with any model of Z_0 is a well-pointed topos.*

Proof. The proof of this amounts to observing that the axioms of Z_0 easily contain enough set-theoretical power to construct all of the topos-theoretic data of **WT**. The usual **ZF** construction of such things as coproducts, coequalizers, and characteristic functions for injective functions (monomorphisms) all go through unchanged in Z_0. That the topos so constructed is well pointed follows from extensionality and the relevant definitions.

We have stated Theorem 1 model-theoretically but it can be given a purely syntactic form. Using the idea of the construction of the category associated with any given model of Z_0, we can construct a syntactic translation of the formulas of **WT** to the formulas of Z_0 which preserves negation and which carries the axioms of **WT** to theorems of Z_0. This translation defines a syntactic interpretation of **WT** in Z_0 and amounts to a relative consistency proof of **WT** with respect to Z_0.

Going the other way is not so easy, however. Though the theory **WT** seems to have sufficient strength to do most of the constructions of Z_0, it is not at all clear how to deal with extensionality. In the system **CS** we have already observed that the membership relation defined with global elements and monomorphisms is not extensional. Thus, the naïve approach of interpreting the sets of Z_0 to be the objects of **WT** will not work. If we are to be successful in modelling the membership relation of Z_0 in **WT**, we will have to define our set-objects in a more sophisticated manner.

The basic observation, due primarily to Osius [1], is that, though the membership relation is defined globally on the whole universe of sets, any two sets x and y can be considered as elements of an englobing set z, reducing the question of the membership between x and y to that of determining the membership among elements of z. Thus, the global membership question can, in any given instance, be reduced to a "local" membership question concerning elements of a fixed set z.

In particular, if z is transitive, then x and y will also be subsets of z. But any set z is con-tained as a subset of a transitive set, namely its transitive closure t.[†] We can therefore "imitate" the membership relation of $\mathbf{Z_0}$ if we can give a categorial definition of membership between two monomorphisms with transitive codomain (representing two given elements of a transitive set). This is done as follows:

Let D be some transitive set with elements A and B. A and B are also subsets of D and thus domains of canonical monomorphisms $a : A \longrightarrow D$ and $b : B \longrightarrow D$. Since D is transitive, A and B are also elements of $\mathcal{P}(D)$. This latter fact can be translated into category language. The powerset of any transitive set is transitive. Thus, there is a canonical inclusion mono-morphism $m : D \longrightarrow \mathcal{P}(D)$ (not to be confused with the singleton map $\{\,\} : D \longrightarrow \mathcal{P}(D)$ which is also monomorphic but whose image cannot be D).[‡] Also, $\ulcorner c_a \urcorner : 1 \longrightarrow \mathcal{P}(D)$ picks out A as an element of $\mathcal{P}(D)$. This is expressed by the fact that there exists a unique map $\hat{a} : 1 \longrightarrow D$ such that $\hat{a}m = \ulcorner c_a \urcorner$. Similarly, there exists a unique $\hat{b}$ such that $\hat{b}m = \ulcorner c_b \urcorner$. Finally, we can express the notion "$a \in b$" by asserting the existence of a map $f : 1 \longrightarrow B$ such that $fbm = \ulcorner c_a \urcorner$. Notice that bm is a monomorphism and so we have reduced the general notion of membership, $a \in b$, to the original categorial one given in Definition 22 of Section 8.5.

It is immediately clear how to formulate this notion in the language of topos theory where $\mathcal{P}(D)$ is, of course, represented as Ω^D, and where $m : D \longrightarrow \Omega^D$ is monomorphic. The only remaining problem is how to characterize an object D of a topos as a transitive set. This is equivalent to determining the monomorphism $m : D \longrightarrow \Omega^D$ as a canonical inclusion map since there are, even in the category of sets, other monos from D to Ω^D. We now see how to do this, following Osius [1].

In any topos, a map $m : A \longrightarrow \Omega^A$ can be thought of as a binary relation on A. For, uniquely associated with m is the map $(A \times m)e : A \times A \longrightarrow \Omega$. Pulling back this map with $\top : 1 \longrightarrow \Omega$ gives a monomorphism $k : X \longrightarrow A \times A$. This represents m as a subobject of $A \times A$, thus as a binary relation on A. Returning, now, to the category of sets, a well-known theorem of Mostowski (see Mostowski [2]) gives necessary and sufficient conditions for any binary relation E on a set A to be isomorphic to the $\in$ relation on a transitive set B. These conditions are: (1) E must be extensional and (2) E must be well founded, i.e. every nonempty subset of E has an E-minimal element. We see how to represent these two conditions in the language of topos theory.

To say that the binary relation E on the set A is extensional means that, for any $x, y \in A$, if $\{z \mid zEx\} = \{z \mid zEy\}$, then $x = y$. But this simply says that the map $m : A \longrightarrow \mathcal{P}(A)$ determined by $E \subset A \times A$ is injective. Since, in the category of sets, injective maps are precisely the monomorphisms we can express condition (1) in a topos simply by requiring, as we already have, that $m : A \longrightarrow \mathcal{P}(A)$ be a monomorphism. As for condition (2), this is handled by the following theorem:

[†] To construct the transitive closure of a set, we need the full force of **ZF** since we will have to define a functional relation $F(x, y)$ whose domain is ω and then use **ZF**.9 to take an infinite union (see Jech [1], Lemma 9.1, pp. 70–71).

[‡] If any set D were equal to the set of singletons of its member, then every element of D would have another member of D as an element. There would therefore be no $\in$-minimal element of D contradicting the axiom of regularity.

THEOREM 2. *In the category of* Z_0 *sets, a monomorphism* $m : A \longrightarrow \mathcal{P}(A)$ *is well founded if and only if, for every* $g : \mathcal{P}(B) \longrightarrow B$, *there exists exactly one function* $f : A \longrightarrow B$ *such that the following diagram commutes:*

$$
\begin{array}{ccc}
A & \xrightarrow{\;m\;} & \mathcal{P}(A) \\
{\scriptstyle f}\big\downarrow & & \big\downarrow{\scriptstyle \Omega_f} \\
B & \xleftarrow{\;g\;} & \mathcal{P}(B)
\end{array}
$$

where $\Omega_f : \mathcal{P}(A) \longrightarrow \mathcal{P}(B)$ *is the covariant powerset functor which sends every subset* $X \subset A$ *to its image* $f(X) \in \mathcal{P}(B)$.

Proof. See Osius [1] (also, Goldblatt [1] and Johnstone [1]).

To complete the categorial version of transitivity, it remains only to define Ω_f in any topos. This is accomplished by the next definition.

Definition 1. Given any map $f : A \longrightarrow B$ in any topos, the map $\Omega_f : \Omega^A \longrightarrow \Omega^B$ is defined as follows: Starting with the maps $(f \times \Omega^A) : A \times \Omega^A \longrightarrow B \times \Omega^A$ and $e_A : A \times \Omega^A \longrightarrow \Omega$, we pull back e_A with $\top : 1 \longrightarrow \Omega$ to obtain a monomorphism $k : X \longrightarrow A \times \Omega^A$. The composite $k(f \times \Omega^A)$ has an image factorization qm where $m : ? \longrightarrow B \times \Omega^A$ is mono. The exponential transpose of the characteristic map of m, namely the map $(c_m)^*$, has domain Ω^A and codomain Ω^B. This is the map Ω_f.

Exercise. Show in detail that the construction of Definition 1 yields the standard covariant powerset functor in the category of sets.

Definition 2. In any topos, a *transitive set object* is a monomorphism $m : A \longrightarrow \Omega^A$ such that for every map $g : \Omega^B \longrightarrow B$, there exists a unique $f : A \longrightarrow B$ for which the following diagram commutes:

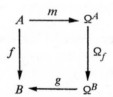

$$
\begin{array}{ccc}
A & \xrightarrow{\;m\;} & \Omega^A \\
{\scriptstyle f}\big\downarrow & & \big\downarrow{\scriptstyle \Omega_f} \\
B & \xleftarrow{\;g\;} & \Omega^B
\end{array}
$$

We now define an inclusion relation between transitive set objects.

Definition 3. In any topos, the transitive set object $a : A \longrightarrow \Omega^A$ *is included in* the transitive set object $b : B \longrightarrow \Omega^B$ if there exists a monomorphism $h : A \longrightarrow B$ such that $hb = a\Omega_h$. We write $a \sqsubset b$ when this is the case.

Definition 4. In any topos, a *set object* is a pair $\langle a, m \rangle$ such that $m : A \longrightarrow \Omega^A$ is a transitive set object and $a : X \longrightarrow A$ is a monomorphism.

Intuitively, a is an arbitrary set, an element (and subset) of the transitive set m. Of course, we have only required that a be mono, thus a subobject of A. While it is true in the category of sets that every element of a transitive set is a subset of it, the converse is not true. How, then, can we insure that the subobject a behaves as an element of m? The answer lies in the proper definition of membership between set objects. This is accomplished by the following definition.

Definition 5. In any topos, where $\langle a, m \rangle$ and $\langle b, n \rangle$ are set objects, we say that $\langle a, m \rangle$ *is a member of* $\langle b, n \rangle$, and write $\langle a, m \rangle$ R $\langle b, n \rangle$, if, for some transitive set object $d : D \longrightarrow \Omega^D$ which includes m and n via the inclusion monos x and y respectively, $\ulcorner c_{ax} \urcorner : 1 \longrightarrow \Omega^D$ is an element of the mono byd. Diagrammatically we have:

commutes for some $1 \longrightarrow Y$.

We are now ready to state our theorem concerning models of set theory in a well-pointed topos. For simplicity, we suppose that the systems $\mathbf{Z_0}$, $\mathbf{Z}$, and $\mathbf{ZF}$ are formulated with "$\in$" as the only primitive nonlogical symbol.

THEOREM 3. *Given any well-pointed topos $\mathcal{E}$, the collection of set objects $\langle a, m \rangle$ together with the membership relation R defined between them constitute a model of the theory $\mathbf{Z_0}$.*

Proof. See Osius [1], Goldblatt [1], and Johnstone [1]. Since $\mathbf{Z_0}$ has a presentation with a finite number of axioms, the proof consists in a direct verification that each of these axioms is satisfied. In particular, the axiom of extensionality is satisfied.

In verifying extensionality in the proof of Theorem 3, the following observation is helpful: It can be easily seen that two set objects $\langle a, m \rangle$ and $\langle b, n \rangle$ have the same elements only if they are both included in some transitive set object $d : D \longrightarrow \Omega^D$, with inclusion maps x and y respectively, such that ax and by are isomorphic as subobjects of D. It is immediately clear that, when this latter condition is fulfilled, $\langle a, m \rangle$ will be a member of any set-object of which $\langle b, n \rangle$ is a member.

COROLLARY. *There is a syntactic interpretation of $\mathbf{Z_0}$ in $\mathbf{WT}$.*

Proof. Just as in the case of Theorem 1, we imitate syntactically the model-theoretic interpretation given by Theorem 3. The only problem is that a set object cannot be defined as an ordered pair since ordered pairs do not exist in $\mathbf{WT}$. This is overcome by the following device. We represent the intuitively conceived set object $X \xrightarrow{a} A \xrightarrow{m} \Omega^A$ by the product map $\langle c_a, m \rangle : A \longrightarrow \Omega \times \Omega^A$. This allows us to formulate the notion "x is a set object" as a formula of $\mathbf{WT}$ with one free variable.

Definitions 1–5 have been stated for any topos, but it is clear that the collection of set objects with the membership relation will not usually form a model for $\mathbf{Z_0}$ if the topos is not well-pointed. In particular, as follows from our discussion above, the fact that 1 is a generator is crucial to proving that extensionality holds in the model of set objects and membership.

We have now established an equivalence of $\mathbf{WT}$ and $\mathbf{Z_0}$: each system is interpretable in the other. $\mathbf{Z_0}$ thus represents the $\mathbf{WT}$ "core" of $\mathbf{ZF}$ set theory. It is also clear that if we add the axioms of infinity or choice to $\mathbf{WT}$, then the model of $\mathbf{Z_0}$ in the set-objects of $\mathbf{WT}$ will also satisfy the $\mathbf{ZF}$ axioms of infinity and choice. Hence, the theory $\mathbf{Z_0}$ plus infinity and choice is the set theory that corresponds precisely to the system $\mathbf{CS}$ ($= \mathbf{WT}$ plus infinity and choice). Thus, the essential weakness of $\mathbf{CS}$ as a set theory, as we have already previously mentioned, is the lack of any strong collection principles like strong separation or replacement. However, Osius [1] shows how to express principles like replacement in the language of topos theory, thus obtaining an equivalent to $\mathbf{ZF}$ set theory in the language of $\mathbf{WT}$.

These results show that mathematics can be carried on in the language of topos theory and that the notion of set membership can be dispensed with as an irreducible, primitive notion of mathematics. Whether or not this should be done is the subject of philosophical debate. The advantages of the topos-theoretic approach to mathematics can be more clearly seen in the light of the general logic of toposes presented in the next section. Whether these advantages will ultimately be perceived as great enough to warrant a total displacement of set theory as a language for everyday mathematical activity is impossible to say at this time. The question is obviously complicated by the newness of category theory and the extra effort necessary to reformulate familiar notions in categorial terms. The paper Lawvere [5] contains strong arguments in favor of a thoroughgoing categorial approach.

8.11. The internal logic of toposes

What does it mean to say, as we often do, that the logic of mathematics is Boolean (or classical)? One answer is to observe that the lattice of subobjects (subsets) on any given set is Boolean: given any set X, its powerset $\mathcal{P}(X)$ is a (complete, atomic) Boolean lattice (algebra) whose partial order is the inclusion relation. This gives a precise sense to the notion that the category of sets is Boolean. Thus, insofar as mathematics is viewed as being carried on in set theory, its logic is also Boolean.

An alternative, but equivalent, way of characterizing the Booleanness of the category of sets is to observe that the set 2, with appropriately defined operations, is the smallest nontrivial Boolean algebra. Given any set X, the set of functions 2^X has a Boolean ring structure in the usual, naturally defined way (the operations are defined pointwise). The ring 2^X is isomorphic to $\mathcal{P}(X)$ via the bijection between subsets of X and their characteristic functions. Viewed in this way, the Booleanness of set theory results from the existence of a set of truth values (the set $2 = \{0, 1\}$) on which we can define a Boolean algebraic structure that can, in turn, be transferred to the subobject lattice of any mathematical object (set).

We now generalize this approach from the topos of sets to an arbitrary topos. The subobject classifer Ω is viewed as a truth-value object. We can define certain maps $\Omega \longrightarrow \Omega$ and $\Omega \times \Omega \longrightarrow \Omega$ which operate as truth functions and, by pulling back against $\top : 1 \longrightarrow \Omega$, we can induce a structure on the class of subobjects of any given object of the topos. We can also define a lattice structure directly on the poset of subobjects of a given object, and observe, as in the case of sets, that it is isomorphic to the structure induced by the operations defined on Ω. However, for a general topos the poset of subobjects of a given object is not Boolean. But it is always a Heyting algebra. Since Heyting algebras (sometimes called pseudo-Boolean algebras) are precisely the lattices which correspond to intuitionistic propositional logic, we say that the internal logic of a topos is intuitionistic.

Of course, we can always study toposes as mathematical objects viewed from the set-theoretical standpoint, thus using the usual Boolean logic. This is the external logic which can, in fact, be anything we choose it to be. But the internal logic of general topos theory is intrinsically intuitionistic in the sense defined above.

We now proceed to make all this precise.

Given any topos $\mathscr{E}$ and any object X of $\mathscr{E}$, let Sub(X) represent the class of monomorphisms with codomain X. Since any topos is well powered, we can think of Sub(X) as being any representable set of monos with codomain X. We have already defined the inclusion relation between subobjects of X (Definition 21, Section 8.5) and established that the relation of inclusion is antisymmetric (see the exercise preceding Theorem 1, Section 8.6). Since reflexivity and transitivity are immediate, Sub(X) is a poset. We now establish that Sub(X) is, in fact, a lattice.

THEOREM 1. *Given any object X in a topos, the pullback of two subobjects of X is their infimum, and the image of the unique map from their coproduct to X is their supremum.*

Proof. Given $a : A \longrightarrow X$ and $b : B \longrightarrow X$, each mono, form their pullback P where $x : P \longrightarrow A$, $y : P \longrightarrow B$. Since pullbacks preserve monomorphisms, x and y are each mono as is the composite map $xa \, (= yb)$. It follows immediately from the universal property of pullbacks that any other subobject of X which precedes a and b must also precede xa.

For the supremum we consider the map $[a, b] : A+B \longrightarrow X$ which has an image factorization $qm = [a, b]$. The mono part of this factorization, $m : Y \longrightarrow K$ (the image of $[a, b]$), must precede all subobjects of X preceded by a and b. To see this, observe that any mono $n : Z \longrightarrow X$ which is preceded by a and b becomes part of a factorization $dn = [a, b]$. Since the image m of $[a, b]$ precedes all such monos, $m \subset n$. Thus, m is the supremum of a and b in Sub(X).

This establishes that the Sub(X) is not only a poset but a lattice.

Before going further in examining the structure of Sub(X), let us see how these same operations can be induced on Sub(X) by operation maps defined on the truth value object Ω. We illustrate in detail with the easiest case, that of the intersection (infimum). We suppose we are working in an arbitrary topos.

The map $\langle \top, \top \rangle : 1 \longrightarrow \Omega \times \Omega$ is a monomorphism (every global element is). Its characteristic map $c_{\langle \top, \top \rangle} : \Omega \times \Omega \longrightarrow \Omega$ is called *conjunction* and noted $\wedge$. The map $\wedge$ induces, via pullbacks, a binary operation on subobjects of any object in the following way: Given monos $a : A \longrightarrow X$ and $b : B \longrightarrow X$, we form the product map $\langle c_a, c_a \rangle : X \longrightarrow \Omega \times \Omega$ and compose with $\wedge$ giving $\langle c_a, c_b \rangle \wedge : X \longrightarrow \Omega$. We claim that the kernel $k : K \longrightarrow X$ of this map is the infimum $a \cap b$ of a and b in Sub(X). Now, the infimum of a and b is the mono $d = xa = yb$ of the following pullback:

(1)

We need to establish that d and k are isomorphic subobjects of X.

Consider the following diagram:

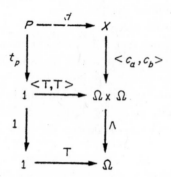

The bottom square is a pullback by definition. If we can show that the top square is a pullback, then the whole rectangle will be a pullback, as follows immediately from the definition of pullbacks (the reader is invited to check for himself this general fact about pullback diagrams). But, if the whole rectangle is pullback, then this means d is a kernel of $\langle c_a, c_b \rangle \wedge$ and thus isomorphic to k. To show that the top square is a pullback, let some object Z with $z : Z \longrightarrow X$ and $t_z : Z \longrightarrow 1$ such that $z \langle c_a, c_b \rangle = t_z \langle \top, \top \rangle$ be given. Then, calculating, we have: $\langle zc_a, zc_b \rangle = \langle t_z \top, t_z \top \rangle$ which implies $zc_a = t_z \top = zc_b$. Since the diagrams

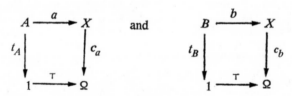

are pullbacks, there exist unique maps $f : Z \longrightarrow A$ and $g : Z \longrightarrow B$ such that $fa = z = gb$. But, since (1) is a pullback, there exists a unique $h : Z \longrightarrow P$ such that $hx = f$ and $hy = g$. Thus, $hd = hxa = fa = z$, and, of course, $ht_p = t_z$. Clearly, h is unique in this regard since d is mono. This establishes that the top square is a pullback, thus that k and d are isomorphic subobjects of X as claimed.

We now want to define the disjunction map $\vee : \Omega \times \Omega \longrightarrow \Omega$. We start with the two maps $\langle \Omega, t_\Omega \top \rangle : \Omega \longrightarrow \Omega \times \Omega$ and $\langle t_\Omega \top, \Omega \rangle : \Omega \longrightarrow \Omega \times \Omega$, and form the coproduct map $[\langle \Omega, t_\Omega \top \rangle, \langle t_\Omega \top, \Omega \rangle] : \Omega + \Omega \longrightarrow \Omega \times \Omega$. The image of this map is a monomorphism m with codomain $\Omega \times \Omega$. The characteristic map $c_m : \Omega \times \Omega \longrightarrow \Omega$ is the disjunction map $\vee$.

Just as with $\wedge$, $\vee$ induces a binary relation $\cup$ on subobjects of a given object X: given $a : A \longrightarrow X$ and $b : B \longrightarrow X$, $a \cup b$ is the kernel of $\langle c_a, c_b \rangle \vee$.

Exercise. Prove that, for any two monos $a : A \longrightarrow X$ and $b : B \longrightarrow X$, $a \cup b$ as defined above is the sup of a and b in Sub(X). You may assume that pullbacks of epimorphisms are epimorphisms and also that pullbacks preserve coproducts.

Finally, we use the previously defined negation map $\neg : \Omega \longrightarrow \Omega$ to define an operation of pseudo-complementation on Sub(X) : given any mono $a : A \longrightarrow X$, the kernel of $c_m\neg$ is the pseudo-complement of m, noted $m.\neg$

Verifying that the operations of inf, sup, and *p-c* define a Heyting-algebra structure on Sub(X) is somewhat tedious but involves no essential difficulties (see Freyd [2] for details as well as a precise definition of a Heyting algebra). A Heyting algebra, like a Boolean algebra, is distributive as a lattice. The main difference between the two structures is the action of pseudo-complementation in a Heyting algebra: whereas $p \leqslant p\neg\neg$ as for a Boolean algebra, $p\neg\neg \leqslant p$ does not hold in general. The collection of all open sets of any topological space forms a Heyting algebra under inclusion with inf and sup being the usual set-theoretic intersection and union. The pseudo-complement of an open set is defined to be the interior of its set-theoretic complement, which is an operation from open sets to open sets. Thus, if we take the real line as our space, the set $R-\{1\}$ is open and has R as its double pseudo-complement, but R is not contained in $R-\{1\}$.

We sometimes sum up the above results by saying that the propositional logic of a topos is intuitionistic. Clearly, a Boolean topos is one for which the internal logic is Boolean, i.e. for which the operation of pseudo-complementation is Boolean ($p\neg\neg \leqslant p$ always holds).

On the level of predicate logic, there is also a difference between the internal logic of toposes and classical logic. This derives from the fact that noninitial objects in some toposes may be empty of global elements (we have already encountered this problem in the course of formulating topos-theoretic axioms for set theory). In classical logic, this would correspond to having closed terms which do not always have a denotation (so-called free logic).

However, it has been realized by workers in topos theory that one obtains a set-like predicate logic for toposes if one quantifies not over global elements of an object X (i.e. maps $1 \longrightarrow X$) but rather over *local* elements (maps from $A \longrightarrow X$ where A is an arbitrary object of the topos). Moreover, because of the presence of exponentiation in a topos, the predicate logic is type-theoretic (higher-order).

Putting all this together, we may say that the internal logic of topos theory is higher-order, local, intuitionistic logic.

One of the central themes of this work has been that any precisely defined logical structure gives rise to an appropriate language (and theory within the language). This is also true of the logic of toposes. We will present a language and theory which has the same relationship to toposes as does the language and theory of types (see Chapter 4 of the present volume) to type hierarchies.[†] Various researchers have developed different (essentially equivalent) versions of such a language. The version we will present is based on the beautiful paper Boileau and Joyal [1], and is based ultimately on the presentation in Boileau [1].

[†] Thus, from the viewpoint of this language, topos theory is local, intuitionistic type theory.

8.12. The internal language of a topos

We begin by defining *type symbols over a set* X not containing the symbol "Ω". Type symbols will be called *sorts*.

Definition 1. Given the set X not containing the symbol "Ω", the set $S(X)$ of *sorts over* X or *type symbols* over X is the smallest set containing X and closed under the following operation: if $s_1, \ldots, s_n \in S(X)$, then so is the finite sequence $\Omega(s_1, \ldots, s_n)$ (we allow n to be 0, so $\Omega(\ ) \in S(X)$ for any X). The sort $\Omega(\ )$ will also be written simply as Ω. The elements of X are called *simple sorts*.

We now associate a class of languages with any set X of simple sorts. After describing the languages, we will establish explicitly their relationship to topos theory.

The symbols of any language L_X associated with the set X of simple sorts consists of the following: (1) A denumerable infinity of variables of each sort s. (2) A set of *function symbols*, each with a *source* and a *target*. The source of a function symbol is a fixed, finite sequence of sorts, and its target is a single sort. A *relation symbol* is a function symbol whose target is Ω. (3) The *logical symbols* "$\top$", "$\wedge$", "$=$", "$\in$", "$\{|\}$", "$($", "$)$", and "$,$". These are all the symbols of the language L_X. Clearly the language L_X is totally specified as soon as we select our set of function symbols. We suppose each of the sets described in clauses (1)–(3) to be disjoint from the others and disjoint from the set of sorts $S(X)$.

We now define the terms of L_X, defining at the same time their types.

(1) Every variable of sort s is a term of type s.
(2) If f is a function symbol whose source is $\langle s_1, \ldots, s_n \rangle$ and if $t_1, \ldots, t_n$ are terms of type $s_1, \ldots, s_n$ respectively, then $f(t_1, \ldots, t_n)$ is a term of type s, where s is the target sort of f.
(3) If t_1 and t_2 are terms having the same type, then $t_1 = t_2$ is a term of type Ω.
(4) If $t_1, \ldots, t_n$ are terms having types $s_1, \ldots, s_n$ respectively, $n \geqslant 1$, and if t is a term of type $\Omega(s_1, \ldots, s_n)$, then $(t_1, \ldots, t_n) \in t$ is a term of type Ω.
(5) If A is a term of type Ω and if $x_1, \ldots, x_n$ is a sequence of distinct variables of types $s_1, \ldots, s_n$ respectively, $n \geqslant 1$, then $\{(x_1, \ldots, x_n) \mid A\}$ is a term of type $\Omega(s_1, \ldots, s_n)$ in which the displayed variables are bound.
(6) If A and B are terms of type Ω, then $(A \wedge B)$ is a term of type Ω.
(7) $\top$ is a term of type Ω.
(8) Nothing is a term, has a type or is a bound variable except by the above.

We define a *formula* to be a term of type Ω.

Definition 2. Given any set $S(X)$ of sorts over X, any set H of function symbols (disjoint from $S(X)$), and any topos $\mathcal{E}$, we define an *interpretation of* $S(X) \cup H$ *in* $\mathcal{E}$ to be a function $| \ |$ from sort and function symbols to objects and morphisms respectively of $\mathcal{E}$ such that the following are satisfied:

(i) $|\Omega(s_1, \ldots, s_n)| = \Omega^{|s_1| \times \cdots \times |s_n|}$ and   (ii) $|f| : |s_1| \times \cdots \times |s_n| \longrightarrow |s|$,

where $f \in H$ has $\langle s_1, \ldots, s_n \rangle$, $s_i \in S(X)$, as source and $s \in S(X)$ as target. Notice that, by (i), $|\Omega| = \Omega$.

We want, now, to extend an interpretation $||$ of $S(X) \cup H$ in a topos $\mathcal{E}$ to an interpretation of any language L_X having the given set H of function symbols as its set of function symbols. For simplicity, we adopt, from now on, the definition of a topos in which only exponentiation on Ω and finite limits are primitive.

We have already noted that topos theory is a local rather than a global theory. For this reason our interpretation $|t|$ of each term of L_X in $\mathcal{E}$ will be relative to each given list of variables containing the free variables of t. We now state the clauses of the recursive definition which extends $||$ to every term of L_X relative to such lists.

(1) For any variable x, $|x|_o = |s|$ where s is the type of x.

Given, now, any term t of type s and any finite sequence of variables $x_1, \ldots, x_n$ containing every free variable of t, we define $|t|_{\vec{x_i}}$ relative to the given list as a morphism $|x_1|_o \times \cdots \times |x_n|_o \longrightarrow |s|$. When there is no danger of ambiguity, we will drop the subscript "$\vec{x_i}$" and continue to use just "$||$" as the name of the interpretation function.

(2) If t is a variable, then $|t|_{\vec{x_i}}$ is the canonical projection $|x_1|_o \times \cdots \times |x_n|_o \longrightarrow |t|_o$.

(3) If t is $f(t_1, \ldots, t_m)$, then the free variables of the terms t_j also occur in the list $x_1, \ldots, x_n$. Thus, for each j, $|t_j|$ is already defined as a morphism $|x_1|_o \times \cdots \times |x_n|_o \longrightarrow |s_j|$, where s_j is the type of t_j. Thus, $|t|$ is $\langle |t_1|, \ldots, |t_m| \rangle |f| : |x_1|_o \times \cdots \times |x_n|_o \longrightarrow |s|$ where s is the type of t.

(4) If t is the term $t_1 = t_2$, then $|t|$ is $\langle |t_1|, |t_2| \rangle \delta$ where $\delta : |s| \times |s| \longrightarrow \Omega$ is $c_{\langle |s|, |s| \rangle}$, s being the common type of t_1 and t_2.

(5) If t is $(t_1, \ldots, t_m) \in t_{m+1}$, then $|t|$ is $\langle |t_1|, \ldots, |t_m|, |t_{m+1}| \rangle e_{(|t_1| \times \cdots \times |t_m|), \Omega}$.

(6) If t is $\{(y_1, \ldots, y_m) | A\}$, where the variables y_j do not appear in the list x_i (we change bound variables if necessary), then $|A|_{\vec{x_i} \vec{y_j}}$ is already defined as a morphism

$$\underset{\cdot}{\times} |y_j|_o \underset{\cdot}{\times} |x_i|_o \longrightarrow \Omega.^{\dagger} \quad |t| \quad \text{is} \quad (|A|_{\vec{x_i} \vec{y_j}})^* : \underset{\cdot}{\times} |x_i|_o \longrightarrow \Omega^{\underset{\cdot}{\times} |y_j|_o}.$$

(7) $|(A \wedge B)| = \langle |A|, |B| \rangle \wedge$.

(8) $|\top| = t_{\overline{\times} |x_i|_o} \top : \overline{\times} |x_i|_o \longrightarrow \Omega$.

This completes the recursive definition of the interpretation of the language L_X in the topos $\mathcal{E}$. Notice that all of the essential data of a topos have been used in defining the interpretation. Notice also that the interpretation of a term of type s is a morphism with codomain $|s|$. The domain of this morphism depends on the types of the variables $x_1, \ldots, x_n$ relative to which the interpretation is defined. Thus, as we quantify over variables, we quantify over domains of the morphisms which interpret a given term, but not their codomain. This expresses the local character of topos theory.

What we need to do now is to relate the above-defined semantic notion to a deductive system with respect to which we can prove a completeness theorem. That it is possible to do this seems to have been first conceived by André Joyal.

Given a term A, we let $\sum(A)$ denote the set of all simple sorts of all free variables appearing in A. By a *sequent of* the language L_X we understand an expression of the form $A \underset{U}{\Longrightarrow} B$ where A and B are formulas of L_X and where U is a finite set of simple sorts containing $\sum(A) \cup \sum(B)$. We now define what it means for a sequent to be valid under an interpretation in a topos $\mathcal{E}$.

---

† We use "$\overline{\times}$" to indicate a finitely iterated application of "$\times$".

Definition 3. We say a sequent $A \xrightarrow[U]{} B$ of L_X is *valid relative to an interpretation* $| \ |$ of L_X, and write $| \ | \vDash_U A \xrightarrow[U]{} B$, if and only if there exists a sequence $x_1, \ldots, x_n$ of distinct variables containing all the free variables of A and B, $U = \{s_1, \ldots, s_n\} \cap X$ where s_i is the sort of x_i, and $|A|_{\vec{x}_i} = |(A \wedge B)|_{\vec{x}_i}$. We also say that $| \ |$ *satisfies* $A \xrightarrow[U]{} B$ when this condition holds.

Definition 4. By a *local higher-order theory*, we mean any set Γ of sequents. A sequent $A \xrightarrow[U]{} B$ is *valid in a local higher-order theory* Γ, and we write $\Gamma \vDash A \xrightarrow[U]{} B$, if and only if every interpretation in every topos which satisfies every sequent in Γ also satisfies $A \xrightarrow[U]{} B$.

We now proceed to define a deductive system which is complete with respect to the notion of validity defined above. The system consists of six axioms and six rules of inference given below. We use the following simplifying national convention:

$$A \Rightarrow B \quad \text{means} \quad A \xrightarrow[U]{} B \quad \text{with} \quad U = \textstyle\sum(A) \cup \sum(B).$$

Axioms

A.1. $A \Rightarrow A$.

A.2. $A \Rightarrow \top$.

A.3. $\top \Rightarrow y = y$, where y is any term.

A.4. $A(x) \wedge x = y \Rightarrow A(y)$, where y is a term free for the variable x in $A(x)$.

A.5. $A \Rightarrow (x_1, \ldots, x_n) \in \{(x_1, \ldots, x_n) \,|\, A\}$.

A.6. $(x_1, \ldots, x_n) \in \{(x_1, \ldots, x_n) \,|\, A\} \Rightarrow A$.

Rules of inference

R.1. $\dfrac{A \xrightarrow[U]{} B, \; U \subset V}{A \xrightarrow[V]{} B}$.

R.2. $\dfrac{A \xrightarrow[U]{} B, \; B \xrightarrow[U]{} C}{A \xrightarrow[U]{} C}$.

R.3. $\dfrac{A \xrightarrow[U]{} B, \; A \xrightarrow[U]{} C}{A \xrightarrow[U]{} B \wedge C}$.

R.4. $\dfrac{A \xrightarrow[U]{} B \wedge C}{A \xrightarrow[U]{} B, \; A \xrightarrow[U]{} C}$.

R.5. $\dfrac{A(x) \xrightarrow[U]{} B(x)}{A(y) \xrightarrow[W]{} B(y)}$, where y is a term of type s free for the variable x of type s in A and B and $W = (U - \{s\}) \cup \sum(A(y)) \cup \sum(B(y))$.

R.6. $\dfrac{A \wedge (x_1, \ldots, x_n) \in B_2 \xrightarrow[U]{} (x_1, \ldots, x_n) \in B_1, \quad A \wedge (x_1, \ldots, x_n) \in B_1 \xrightarrow[U]{} (x_1, \ldots, x_n) \in B_2}{A \xrightarrow[W]{} B_1 = B_2}$

where $x_1, \ldots, x_n$ are distinct variables none of which occur free in A, B_1, or B_2 and $W = (U - \{s_1, \ldots, s_n\}) \cup \sum(A) \cup \sum(B_1 = B_2)$, s_i being the type of x_i.

This completes the description of the deductive system for our languages L_X. Though the precise statement of the rules and axioms involves the usual somewhat complicated syntactical conditions, the conception and import of the rules and axioms should be intuitively quite clear. The first two axioms and the first four rules are propositional. Axioms 3 and 4 and rule 5 concern identity and substitution. Axioms 5 and 6 and rule 6 are topos-theoretic versions of comprehension (abstraction) and extensionality.

Definition 5. Where Γ is any local higher-order theory, we write $\Gamma \vdash A \underset{U}{\Longrightarrow} B$ to mean that the sequent $A \underset{U}{\Longrightarrow} B$ is deducible from the hypotheses Γ by means of the axioms and rules. We write $\Gamma \vdash A$ for $\Gamma \vdash \top \Rightarrow A$. Generally, we identify the formula A with the sequent $\top \Rightarrow A$ in discussions of provability.

We now formulate the soundness theorem for our deductive system:

THEOREM 1. *If $\Gamma \vdash A \underset{U}{\Longrightarrow} B$, then $\Gamma \vDash A \underset{U}{\Longrightarrow} B$.*

Proof. A direct calculation based on the definition of semantic validity shows that the axioms A.1 – A.6 are valid in any topos $\mathcal{E}$ under any interpretation $|\ |$, and that the rules of inference R.1 – R.6 preserve validity.

Notice that we have not, as yet, defined or discussed the usual quantifiers or some of the usual sentence connectives such as disjunction and negation. The reason is that they can all be defined in terms of the primitive notation already present in our system. Although it is not necessary to define them in order to deal with the deductive system, we nevertheless give the precise definitions below:

Definition 6. $\forall x A$ for $\{x \mid A\} = \top$, $A \supset B$ for $A \wedge B = A$, $A \vee B$ for $\forall C(((A \supset C) \wedge (B \supset C)) \supset C = \top)$, $\bot$ for $\forall A(A = \top)$, $\exists x B$ for $\forall A(\forall x(B \supset A) \supset A = \top)$, $\neg A$ for $A \supset \bot$.

Notice that the presence of propositional variables (variables of type Ω) in our language is essential for defining sentence connectives in terms of our primitive notions. Also essential to our semantic system is the internal definition of equality. It is this which allows us to interpret identity as a primitive symbol holding between all terms, including those of type Ω. The reader should again recall that the above-defined connectives obey the rules of intuitionistic rather than classical logic.

We now state our completeness theorem.

THEOREM 2. *Given any local higher-order theory Γ, there exists a topos $\text{Top}(\Gamma)$ and an interpretation $|\ |$ of the language L_X of Γ in $\text{Top}(\Gamma)$ which is universal, i.e. such that $\Gamma \vdash A \underset{U}{\Longrightarrow} B$ if and only if $|\ | \vDash A \underset{U}{\Longrightarrow} B$.*

Proof. We construct $\text{Top}(\Gamma)$ from the language itself. The objects are the formulas of L_X and a morphism $A \longrightarrow B$ is an equivalence class of Γ-functional formulas, i.e. the class of all those n-ary relational formulas having A as domain and B as codomain which are provably functional and provably equivalent in the theory Γ. We then check that the conclusion of the theorem holds in this category (see Boileau and Joyal [1]). We have, in effect, constructed the "Lindenbaum–

Tarski category" of a local higher-order theory much as we construct a Lindenbaum–Tarski (Boolean) algebra of a theory to prove the completeness theorem of first-order logic.

In particular, if we take Γ to be the empty theory, the set X of simple sorts and the set of function symbols to be empty (we denote the resulting language "L"), then the above construction of Top(Γ) yields the *free topos*. Applying Theorem 2, we see that a sequent will be valid in the free topos if and only if it is universally valid, i.e. valid in every topos. The free topos is an initial object (like the 0 within a topos itself) in the category of all toposes and logical morphisms (functors which preserve the topos structure). This means there is (up to equivalence) exactly one logical morphism from the free topos to any other topos. This follows immediately from the fact that there is essentially only one way to interpret the logical symbolism of L in any topos.

It may appear strange at first that the notion of consistency does not appear as a condition in the statement of Theorem 2. However, this point is clarified by the following observation: If a theory Γ is inconsistent, then every sequent is provable. In this case, the construction of Top(Γ) will simply yield a degenerate topos.

Theorems 1 and 2 show that we have completely captured the internal logic of topos theory, and it gives precise content to the conception of topos theory as local, intuitionistic type theory.

The language and deductive system for topos theory is very set-like in both its notation and properties. It is immediately clear that many theorems provable from the axioms of **ZF** set theory (and certainly from the axioms of type theory) can be carried through in our deductive system (i.e. in the empty theory). For, we have already seen from our previous study that extensionality and comprehension are the two fundamental principles of set-theoretical foundations, and we have (local, intuitionistic) versions of both these principles in our deductive system. Thus, all theorems of set theory provable in the language L of free topos theory are true in any topos when properly interpreted. Furthermore, we have a vast class of naturally defined toposes (the simple models of Section 8.6).

This last fact gives a "transfer principle" which enables us to transfer wholesale certain known theorems of set theory to toposes. Since the logic of topos theory is intuitionistic, it also gives precise content to the vague feeling of many mathematicians that "to prove something constructively is to prove more". Moreover, the transfer principle continues to hold for the appropriate class of toposes when we add other data to a topos such as a natural number object. Generally speaking, if a theorem is provable in set theory without using the law of excluded middle (the intuitionistic restriction) and without using the axiom of choice (which forces the topos to be Boolean and therefore structurally similar to the category of sets) it is true in any topos, thus in such categories as pre-sheaves over any small category and sheaves over any topological space.

This transfer principle has actually been used to obtain theorems of sheaf theory from theorems of set theory (see, for example, Reyes [1]).

Use of the internal languages L_X for toposes, and in particular the language L of the free topos, can replace appeal to principles of functor theory in establishing complicated properties of toposes. Workers in the field of topos theory differ considerably as to their preference for the linguistic approach or the functorial approach to studying toposes. On the one hand, the linguistic approach allows one to reason in a set-like language about toposes and thus to transfer certain thought patterns from set theory to topos theory. There is, however, the drawback that the linguistic approach does not give us much feeling for what is going on since the connection between the reasoning in the formal language and the toposes themselves is made via the fairly complicated interpretation function | | described above. Thus, one finds oneself constantly asking "Now what does this *really* mean?" On the other hand, the functorial and diagrammatic approach to topos theory has the advantage of allowing one to handle the toposes directly and to "see" schematically through the use of commutative diagrams and functors exactly why and how certain principles work. The heart of set-theoretical reasoning is the abstraction principle by which one simply thinks up the property one needs, writes it down, and then declares that there is a mathematical object (set) which satisfies the property. With the diagrammatic-functional approach, one must supply the link between concepts and objects since morphisms and functors must be explicitly defined. Functorial reasoning is therefore more explicit or "constructive" and contains more information than does set-theoretical reasoning, but one must pay the price of the extra effort necessary to obtain this extra information. At the time of this writing, it appears that those favoring the linguistic approach to topos theory are in the majority.

In trying to assess the relative merits of these approaches, it is probably important to realize that even the most advanced workers in category theory are mathematicians who, for the most part, were trained from the beginning in the set-theoretical approach to mathematics and only subsequently turned to category theory. This means that virtually everyone working in the field has to put forth the extra effort necessary to reformulate familiar notions in a new language. It will be interesting to see whether a new generation of workers trained from the outset in functorial thinking will find the diagrammatic language superior to the set-theoretic one and even abandon set theory altogether. The present writer, not naturally inclined to overcaution in making predictions, feels that some significant displacement of set theory by category theory as a language for mathematics will take place. In fact, in certain sectors it is already taking place and is quite advanced.

Many mathematicians and logicians have reacted to category theory by accusing it of fostering gratuitous abstraction and spurious generalization ("general abstract nonsense" as it has been called). I would like to go on record here as rejecting this evaluation of category theory. Whenever one is led to contemplate a new object of mathematical study, categorial thinking forces one to ask "Exactly how are these objects related to each other and to other objects having similar (or even dissimilar) kinds of structure?" Thus, in many ways, category theory is actually a counterbalance to gratuitous abstraction, for it helps us to find useful criteria for distinguishing between potentially fruitful concepts on the one hand and sterile ones on the other. A somewhat similar view is articulated in the Introduction to Johnstone [1].

8.13. Conclusions

Until rather recently, most mathematicians have carried on work in their special branch of mathematics without giving much thought to foundations. The most prevalent attitudes were that logicians had somewhere already solved the problem of giving a once-and-for-all foundation for mathematics or else that foundational problems were essentially irrelevant to mathematical practice anyway. The first idea has been undermined by the plethora of independence proofs in set theory starting with the work of Cohen. These results showed that there are many different, incompatible models of **ZF** set theory and thus tended to destroy set theory's (largely unwarranted) image as an absolute foundation for mathematics. To the affirmation "mathematics reduces to set theory", one has now to ask the question "To which set theory is mathematics reducible?"

The second notion, i.e. that foundational questions are irrelevant to mathematical practice, has also been considerably undermined. For example, an increasing number of problems occurring in mathematical practice have turned out to be reducible to or dependent on set-theoretical principles which are independent of the **ZF** axioms. In particular, a number of problems in the theory of abelian groups depend on whether one assumes the existence of inaccessible cardinals in set theory. The development of nonstandard analysis, which links foundational questions directly to the heart of applied mathematics, has also dramatized the importance of foundations for mathematical practice.

Indeed, the relevance of foundational questions for everyday mathematical practice is now beginning to be acknowledged by an ever broader segment of working mathematicians. There is also recent indication of a greater willingness on the part of mathematicians to become seriously involved in philosophy in their attempts to come to grips with the basis of their science (cf., for example, Goodman [1] and Snapper [1]).

Coupled with this loss of absolute faith in set theory has been the rise of attractive alternatives such as category theory. All of these developments seem to be consistent with the point of view expressed in the present study, as well as in Hatcher [3] and [6], that the search for a once-and-for-all foundation for mathematics is conceptually wrong.

It appears more and more clearly that what is truly foundational is not some arbitrary starting point (some list of axioms for a comprehensive system or other), but certain key, unifying notions common to many different aspects of mathematical practice. The comprehension (abstraction) scheme of set theory is certainly one of these foundational principles but not, as Frege, Dedekind, Russell, Carnap, and others had hoped, the only one. The notions of universality and naturality in category theory are clearly just as important, nor does anyone doubt that others will be forthcoming.

The developments of the last twenty years, and especially of the last twelve years since Hatcher [3] was written, have established beyond any serious doubt that foundational studies are a living branch of mathematics and not a sterile formalistic exercise in cataloguing axioms only to forget them in practice.

Selected Bibliography

BERNAYS, P. [1] *Axiomatic Set Theory*. Amsterdam, North-Holland Publishing Company, 1958.

BIRKHOFF, G. and LIPSON, J. D. [1] Heterogeneous algebras. *J. Combinatorial Theory* **8**: 115–132, 1970.

BIRKHOFF, G. and LIPSON, J. D. [2] Universal algebra and automata. *AMS Symposium on Pure Mathematics* **25**: 41–51, 1974.

BIRKHOFF, G. and MACLANE, S. [1] *A Survey of Modern Algebra*, Third edition. New York, The Macmillan Company, 1956.

BLANC, G. and DONNADIEU, M. R. [1] Axiomatisation de la catégorie des catégories. *Cahiers de topologie et géométrie différentielle* **17**: 135–170, 1976.

BLANC, G. and PRELLER, A. [1] Lawvere's basic theory of the category of categories. *J. Symbolic Logic* **40**: 14–18, 1975.

BOFFA, M. [1] The consistency problem for NF. *J. Symbolic Logic* **42**: 215–220, 1977.

BOILEAU, A. [1] *Types vs. Topos*. Université de Montréal, Thèse de Doctorat, 1976.

BOILEAU, A. and JOYAL, A. [1] La logique des topos. *J. Symbolic Logic* **46**. 6–16, 1981.

BOURBAKI, N. [1] Eléments de mathématique. *Théorie des ensembles*, Livre I. Paris, Hermann, 1960.

BROUWER, L. E. J. [1] De onbetrouwbaarheid der logische principes (The untrustworthiness of the principles of logic). *Tijd. Wijsbg.* **2**: 152–158, 1908.

CHANG, C. C. and KEISLER, H. J. [1] *Model Theory*. Amsterdam, North-Holland Publishing Company, 1973.

CHURCH, A. [1] A note on the Entscheidungsproblem. *J. Symbolic Logic* **1**: 40–41, 1936. (Correction, *ibid.*, pp. 101–102. Reprinted in Davis [2], pp. 110–115.)

CHURCH, A. [2] An unsolvable problem of elementary number theory. *Amer. J. Math.* **58**: 345–363, 1936. (Reprinted in Davis [2], pp. 89–107.)

CHURCH, A. [3] *Introduction to Mathematical Logic*. Princeton, Princeton University Press, 1956.

CHWISTEK, L. [1] Antynomje logiki formalnej. *Przeg. Filoz.* **24**: 164–171, 1921.

COHEN, P. [1] *Set Theory and the Continuum Hypothesis*. New York, W. A. Benjamin Inc., 1966.

CONWAY, J. H. [1] *On Numbers and Games*. London, Academic Press, 1976.

CORCORAN, J. and HERRING, J. [1] Notes on a semantic analysis of variable-binding term operators, *Logique et Analyse* **55**: 644–657, 1971.

CORCORAN, J., HATCHER, W. and HERRING, J. [1] Variable-binding term operators. *Z. math. Logik, Grundlag. Math.* **18**: 177–182, 1972.

DA COSTA, N. [1] On two systems of set theory. *Nederl. Akad. Wet. Proc.*, series A, **68**: 95–98, 1965.

DA COSTA, N. [2] Review of Corcoran, Hatcher and Herring [1], *Zentbl. Math.* **257**: 8–9, 1973.

DA COSTA, N. [3] A model-theoretical approach to variable binding term operators. *Mathematical Logic in Latin America* (ed. A. I. Arruda, R. Chaqui, and N. C. A. da Costa), Amsterdam, North-Holland Publishing Company, 1980, pp. 133–162.

DA COSTA, N. and DIAS, M. F. [1] Sur le système D* de théorie des ensembles. *Comptes Rendus Acad. Sci. Paris*, Série A, **282**: 5–7, 1976.

DAVIS, M. [1] *Computability and Undecidability*. New York, McGraw-Hill Book Company, 1958.

DAVIS, M. [2] *The Undecidable*. Hewlet, N.Y., Raven Press, 1965.

DEDEKIND, R. [1] In BEMAN, W. W. (Ed.) *Essays on the Theory of Numbers*. Chicago, University of Chicago Press, 1901.

DRUCK, I. F. and DA COSTA, N. [1] Sur les "vbtos" selon M. Hatcher. *Comptes Rendus Acad. Sci. Paris*, Série A, **281**: 741–743, 1975.

EHRENFEUCHT, A. [1] Theories having at least continuum many non-isomorphic models in each infinite power. *Notices Amer. Math. Soc.* **5**: 680, 1958.

EILENBERG, S, and MACLANE, S. [1] General theory of natural equivalences. *Trans. Amer. Math. Soc.* **58**: 231–294. 1945.

FOURMANN, M. P.. [1] The logic of topoi. *The Handbook of Mathematical Logic* (ed. J. Barwise). Amsterdam, North-Holland Publishing Company, 1977, pp. 1053–1090.

FRAENKEL, A. [1] Zu den Grundlagen der Cantor-Zermeloschen Mengenlehre. *Math. Ann.* **86**: 230–237, 1922.

FRAENKEL, A. [2] *Abstract Set Theory*. Amsterdam, North-Holland Publishing Company, 1961.

FRAENKEL, A., BAR-HILLEL, Y. and LEVY, A. [1] *Foundations of Set Theory*, Revised Edition. Amsterdam, North-Holland Publishing Company, 1973.

FREGE, G. [1] *Begriffschrift*. Halle, 1879. (Reprinted in van Heijenoort [1], pp. 1–82.)

FREGE, G. [2] *Die Grundlagen der Arithmetik*. Breslau, 1884 (English translation by J. L. Austin, Oxford, Blackwell Scientific Publications Ltd., 1950).

FREGE, G. [3] *Grundgesetze der Arithmetik*. Jena, Vol. I, 1893; Vol. II, 1903.

FREGE, G. [4] *The Basic Laws of Arithmetic*. Berkeley and Los Angeles, University of California Press, 1964. (Translation by M. Furth of selected parts of Frege [3].)

FREYD, P. [1] *Abelian Categories*. New York, Harper & Row, 1964.

FREYD, P. [2] Aspects of topoi. *Bull. Austral. Math. Soc.* **7**: 1–76 and 467–480, 1972.

GAGNON, M. [1] *Complétude d'une logique à opérateurs généralisés*. Université Laval, Thèse de Maîtrise, 1976.

GENTZEN, G. [1] Die Widerspruchsfreiheit der reinen Zahlentheorie. *Math. Ann.* **112**: 493–565, 1936.

GÖDEL, K. [1] Die Vollständigkeit der Axiome des logischen Funktionenkalkuls. *Monatsh. Math. Phys.* **37**: 349–360, 1930. (Reprinted in van Heijenoort [1], pp. 582–591.)

GÖDEL, K. [2] Über formal unentscheidbare Sätze der Principia Mathematica und verwandter Systeme I. *Monatsh. Math. Phys.* **68**: 173–198, 1931. (Reprinted in van Heijenoort [1], pp. 596–616.)

GÖDEL, K. [3] *On Undecidable Propositions of Formal Mathematical Systems*. Princeton, Princeton University Press, 1934. (Reprinted in Davis [2], pp. 39–74.)

GÖDEL, K. [4] *The Consistency of the Axiom of Choice and of the Generalized Continuum Hypothesis with the Axioms of Set Theory*. Princeton, Princeton University Press, 1940.

GOLDBLATT, R. [1] *Topoi, The Categorial Analysis of Logic*. Amsterdam, North-Holland Publishing Company, 1979.

GOODMAN, N. [1] Mathematics as an objective science. *The Amer. Math. Monthly* **86**: 540–551, 1979.

GRISHIN, V. [1] Consistency of a fragment of Quine's NF system. *Soviet Mathematics. Doklady* **10**: 1387–1390, 1969.

GRISHIN, V. [2] The equivalence of Quine's NF system to one of its fragments. *Nauchno-Tekhnicheskaya Informatsiya*, ser. 2, **1**: 22–24, 1972 (in Russian).

HALMOS, P. [1] *Naive Set Theory*. Princeton, D. Van Nostrand Company, 1960.

HATCHER, W. S. [1] Systèmes formels et catégories. *Comptes Rendus Acad. Sci. Paris* **260**: 3255–3528, 1965.

HATCHER, W. S. [2] Logical truth and logical implication. *J. Symbolic Logic* **31**: 561, 1966.

HATCHER, W. S. [3] *Foundations of Mathematics*. London and Philadelphia, W. B. Saunders & Co., 1968.

HATCHER, W. S. [4] Sur un système de da Costa. *Comptes Rendus Acad. Sci. Paris*, Série A, **268**: 1443–1446, 1969.

HATCHER, W. S. [5] Quasiprimitive subcategories. *Math. Ann.* **190**: 93–96, 1970.

HATCHER, W. S. [6] Foundations as a branch of mathematics. *J. Philosophical Logic* **1**: 349–358, 1972.

HATCHER, W. S. [7] Les identités dans les catégories. *Comptes Rendus Acad. Sci. Paris*, Série A, **275**: 495–496, 1972.

HATCHER, W. S. [8] A language for type-free algebra. *Z. math. Logik, Grundlag. Math.* **24**: 385–397, 1978.

HATCHER, W. S. and HODGSON, B. R. [1] Complexity bounds on proofs. *J. Symbolic Logic* **46**: 255–258, 1981.

HATCHER, W. S. and SHAFAAT, A. [1] Categorical languages for algebraic structures. *Z. math. Logik, Grundlag · Math.* **21**: 433–438, 1975.

HATCHER, W. S. and WHITNEY, S. [1] *Absolute Algebra*. Teubner Texte zur Mathematik, Teubner, Leipzig, 1978.

HENSEN, C. W. [1] Finite sets in Quine's New Foundations. *J. Symbolic Logic* **34**: 589–596, 1969.

HERRLICH, H. and STRECKER, G. [1] *Category Theory*. Boston, Allyn & Bacon, 1973.

HILBERT, D. [1] Neubegrundung der Mathematik. *Abhandl. Mathematischen Sem. Hamburg. Univ.* **1**: 151–165, 1922.

HILBERT, D. [2] Über das Unendliche. *Math. Ann.* **95**: 161–170, 1926. (Reprinted in van Heijenoort [1], pp. 367–392.)

HILBERT, D. [3] Die Grundlagen der Mathematik. *Abhandl. Mathematischen Sem. Hamburg. Univ.* **6**: 65–85, 1928. (Reprinted in van Heijenoort [1], pp. 464–479.)

HILBERT, D. and BERNAYS, P. [1] *Grundlagen der Mathematik*. Berlin, Vol. 1, 1934; Vol. 2, 1939.

ISBELL, J. [1] Structure of categories. *Bull. Amer. Math. Soc.* **72**: 619–655, 1966.

ISBELL, J. [2] Review of Lawvere [3], *Math. Reviews* **34**: 1354–1355 (#7332), 1967.

ISBELL, J. and WRIGHT, F. B. [1] Another equivalent form of the axiom of choice. *Proc. Amer. Math. Soc.* **17**: 174, 1966.

JECH, T. [1] *Set Theory*. New York, Academic Press, 1978.

JENSEN, R. [1] On the consistency of a slight (?) modification of Quine's New Foundations. *Synthese* **19**: 250–263, 1968–1969.

JOHNSTONE, P. T. [1] *Topos Theory*. London, Academic Press, 1977.

KELLEY, J. [1] *General Topology*. Princeton, D. Van Nostrand Company Inc., 1955.

KLEENE, S. [1] *Introduction to Metamathematics*. Princeton, D. Van Nostrand Company Inc., 1952.

KNUTH, D. [1] *Surreal Numbers*. Reading, Mass., Addison-Wesley, 1974.

LANDAU, E. [1] *Foundations of Analysis*. New York, Chelsea Publishing Company, 1951.

LANG, S. [1] *Algebra*. Reading, Mass., Addison-Wesley Publishing Company Inc., 1965.

LAWVERE, F. W. [1] An elementary theory of the category of sets. *Proc. Nat. Acad. Sci.* **52:** 1506–1511, 1964.

LAWVERE, F. W. [2] *An Elementary Theory of the Category of Sets*. University of Chicago, Mimeographed notes, 1965.

LAWVERE, F. W. [3] The category of categories as a foundation for mathematics. *Proceedings of the Conference on Categorical Algebra*. Berlin, Springer-Verlag, 1966, pp. 1–20.

LAWVERE, F. W. [4] Quantifiers and sheaves. *Actes du Congrès International des Mathématiciens*, Vol. 1, Nice, 1970, pp. 329–334.

LAWVERE, F. W. [5] Variable quantities and variable structures in topoi. *Algebra, Topology and Category Theory: a collection of papers in honor of Samuel Eilenberg* (ed. A. Heller and M. Tierney), New York, Academic Press, 1976, pp. 101–131.

LEVY, A. [1] *Basic Set Theory*. Berlin, Springer-Verlag, 1979.

LYNDON, R. [1] *Notes on Logic*. Princeton, D. Van Nostrand Company Inc., 1966.

MACLANE, S. [1] Duality for groups. *Bull. Amer. Math. Soc.* **56:** 485–516, 1950.

MACLANE, S. [2] Locally small categories and the foundations of set theory. *Infinitistic Methods, Proceedings of the Symposium on Foundations of Mathematics*. Warsaw, 1959, pp. 25–43.

MACLANE [3] *Categories for the Working Mathematician*. New York, Springer-Verlag, 1971.

MACLANE, S. and BIRKHOFF, G. [1] *Algebra*. New York, The Macmillan Company, 1967.

MENDELSON, E. [1] *Introduction to Mathematical Logic*. Princeton, D. Van Nostrand Company, 1964.

MITCHELL, W. [1] Boolean topoi and the theory of sets. *J. Pure Appl. Algebra* **2:** 261–274, 1972.

MORSE, A. [1] *A Theory of Sets*. New York, Academic Press, 1965.

MOSTOWSKI, A. [1] Über die Unabhängigkeit des Wohlordnungssatzes vom Ordnungsprinzip. *Fundamenta Mathematicae* **32:** 201–252, 1939.

MOSTOWSKI, A. [2] An undecidable arithmetical statement. *Fundamenta Mathematicae* **36** 143–164, 1949.

MOSTOWSKI, A. [3] Some impredicative definitions in the axiomatic set theory. *Fundamenta Mathematicae* **37:** 111–124, 1951. (Corrections, *ibid.*, **38:** 238, 1952.)

NOVAK, I. (GAL, I. N.) [1] A construction for models of consistent systems. *Fundamenta Mathematicae* **37:** 87–110, 1950.

OREY, S. [1] New Foundations and the axiom of counting. *Duke Mathematical J.* **31:** 655–660, 1964.

OSIUS, G. [1] Categorical set theory: a characterization of the category of sets. *J. Pure Appl. Algebra* **4:** 79–119, 1974.

PEANO, G. [1] *Arithmetices Principia, Nova Method Exposita*. Turin, 1889. (Reprinted in van Heijenoort [1], pp. 83–97.)

POINCARÉ, H. [1] Les mathématiques et la logique. *Révue de métaphysique et de morale* **13:** 815–835, 1905; **14:** 17–34, 294–317, 1906.

POINCARÉ, H. [2] La logique de l'infini. *Scientia* **12:** 1–16, 1912.

QUINE, W. V. [1] New foundations for mathematical logic. *Amer. Math. Monthly* **44:** 70–80, 1937.

QUINE, W. V. [2] *From a Logical Point of View*. Cambridge, Mass., Harvard University Press, 1953.

QUINE, W. V. [3] *Mathematical Logic*. Revised edition. Cambridge, Mass., Harvard University Press, 1955.

QUINE, W. V. [4] *Set Theory and Its Logic*. Cambridge, Mass., The Belknap Press of Harvard University Press, 1963.

RAMSEY, F. [1] *The Foundations of Mathematics and other Logical Essays*. New York, Harcourt, Brace & Co., 1931.

REYES, G. [1] Théorie des modèles et faisceaux. *Advances in Mathematics* **30:** 156–170, 1978.

ROBINSON, A. [1] *Introduction to Model Theory and to the Metamathematics of Algebra*. Amsterdam, North-Holland Publishing Company, 1963.

ROBINSON, A. [2] *Non-standard Analysis*. Amsterdam, North-Holland Publishing Company, 1966.

ROBINSON, R. M. [1] An essentially undecidable axiom system. *Proc. Int. Cong. Math.* **1:** 729–730, 1950.

ROGERS, H. [1] *Theory of Recursive Functions and Effective Computability*. New York, McGraw-Hill Book Company, 1967.

ROSSER, J. B. [1] Extensions of some theorems of Gödel and Church. *J. Symbolic Logic* **1:** 87–91, 1936.

ROSSER, J. B. [2] *Logic for Mathematicians*. New York, McGraw-Hill Book Company, 1953. Second, revised, edition, New York, Chelsea Publishing Company, 1978.

RUBIN, J. [1] *Set Theory for the Mathematician*. San Francisco, Holden-Day Inc., 1967.

RUSSELL, B. [1] Mathematical logic as based on the theory of types. *Amer. J. Math.* **30:** 222–262, 1908. (Reprinted in van Heijenoort [1], pp. 150–182.)

SCOTT, D. [1] A proof of the independence of the continuum hypothesis. *Math. Systems Theory* **1**: 89–111, 1967.

SHOENFIELD, J. [1] *Mathematical Logic*. Reading, Mass., Addison-Wesley Publishing Company, 1967.

SKOLEM, T. [1] Einige Bemerkungen zur axiomatischen Begründung der Mengenlehre. *Wiss. Vortrage*, 5. *Kongr. der Skandin. Math., Helsingfors, 1922*, Helsingfors, 1923, pp. 217–232.

SKOLEM, T. [2] Über die Nicht-charakterisierbarkeit der Zahlenreihe mittels endlich oder abzählbar unendlich vieler Aussagen mit ausschliesslich Zahlenvariablen. *Fundamenta Mathematicae* **23**: 150–161, 1934.

SMULLYAN, R. M. [1] *Theory of Formal Systems*. Princeton, Princeton University Press, 1961.

SNAPPER, E. [1] What is mathematics? *Amer. Math. Monthly* **86**: 551–557, 1979.

SONNER, J. [1] On the formal definition of categories. *Math. Zeit.* **80**: 163–176, 1962.

SPECKER, E. [1] The axiom of choice in Quine's New Foundations for mathematical logic. *Proc. Nat. Acad. Sci.* **39**: 972–975, 1953.

SPECKER, E. [2] Typical ambiguity. *Logic, Methodology and Philosophy of Science: Proceedings of the 1960 International Congress*. Stanford, Stanford University Press, 1962, pp. 116–124.

TARSKI, A. [1] *Logic, Semantics, Metamathematics*. New York, Oxford at the Clarendon Press, 1956.

TIERNEY, M. [1] Sheaf theory and the continuum hypothesis. *Toposes, Algebraic Geometry and Logic* (ed. F. W. Lawvere), *Springer Lecture Notes in Mathematics* **274**: 13–42, 1972.

VAN HEIJENOORT, J. [1] *From Frege to Gödel*. Cambridge, Mass., Harvard University Press, 1967.

VON NEUMANN, J. [1] Zur Einführung der transfiniten Zahlen. *Acta Lit. Sci. Reg. Univ. Hung. Fransisco-Josephinae, Sec. Sci. math.* **1**: 199–208, 1923. (Reprinted in van Heijenoort [1], pp. 346–354.)

VON NEUMANN, J. [2] Eine Axiomatisierung der Mengenlehre. *J. für Math.* **154**: 219–240, 1925. (Reprinted in van Heijenoort [1], pp. 393–413.)

VON NEUMANN, J. [3] Über die Definition durch transfinite Induction und verwandte Fragen der allgemeinen Mengenlehre. *Math. Ann.* **99**: 373–391, 1928.

WANG, H. [1] A formal system of logic. *J. Symbolic Logic* **15**: 25–32, 1950.

WANG, H. and MCNAUGHTON, R. [1] *Les Systèmes axiomatiques de la théorie des ensembles*. Paris, Gauthier-Villars, 1953.

WEYL, H. [1] *Das Kontinuum*. Leipzig, Veit, 1918.

WHITEHEAD, A. N. and RUSSELL, B. [1] *Principia Mathematica*. Cambridge, England, Cambridge University Press, Vol. 1, 1910; Vol. 2, 1912; Vol. 3, 1913. Second edition 1925 and 1927.

ZERMELO, E. [1] Beweis, dass jede Menge wohlgeordnet werden kann. *Math. Ann.* **59**: 514–516. (Reprinted in van Heijenoort [1], p. 139–141.)

ZERMELO, E. [2] Untersuchungen über die Grundlagen der Mengenlehre I. *Math. Ann.* **65**: 261–281, 1908. (Reprinted in van Heijenoort [1], pp. 199–215.)

Index

Abbreviation 14–16
Abstraction principle 77, 105, 312
  in class theory 173, 178
  in topos theory 309–311
  in type theory 114
  in ZF 136
Alphabet
  of a first-order system 19
  of a formal system 10
Alphabetic order 19, 109
Analysis, arithmetization of 69–70
Antecedent 3
Argument
  in English 7
  of a term 20
Argument number 20
Associated statement form 33
Automata as model of **T** 283
Axiomatizable 12
  recursively 203
Axiomatized 11
Axioms 10
  axiom of choice 127, 164, 167–170, 226–229, 249, 268, 292–294, 297, 302
  axiom of counting 235
  axiom of reducibility 113, 126
  independent 39
  logical 27, 28
  of category theory 243–244
  of **CS** 260–261, 267–268, 270–272
  of **F** 79
  of **MKM** 178
  of **ML** 234
  of **NBG** 173–175
  of **NF** 215–216
  of **NFU** 235
  of **P** 16
  of **T** 280
  of the predicate calculus 28
  of type theory 114, 120, 122, 125–126, 129, 130
  of **WT** 285
  of **Z** 154
  of **ZF** 138–141, 144, 154, 157, 168
  of Z_0 297–298
  proper 28

Berry's paradox 98
Biconditional *see* Connectives
Boolean algebra 6, 205, 302–305
Bound *see* Variables

Cantor's paradox 97, 217, 223
Cantor's theorem 97, 164, 217–218, 223
Cardinals
  in **NF** 222, 226–228
  in **ZF** 165
  inaccessible 175–177

Categorial 235
Categorical algebra 235 ff.
Category
  definition of 241, 243–244
  dual of 242, 246–248
  equivalent categories 258
  foundational problems related to category theory 253–259
  isomorphic categories 258
  large 255
  one-universe foundation for category theory 259
  skeleton of 249
  small 255
Classes, proper 172
Closure 90
  $\in$-transitive 299
  existential 25
  transitive 272
  universal 25
Compactness theorem 38
Completeness
  of a first-order theory 33
  of first-order logic 35
  of first-order logic with *vbto*s 61–63
  of the internal language of topos theory 308–310
Comprehension principle *see* Abstraction principle
Conclusion 3, 7
Conditional *see* Connectives
Conjunction *see* Connectives
Connectives
  adequate set of 8
  binary 2
  in first-order logic 19–20
  principal 5, 17
  sentential 1–3
Consequent 3
Consistency 32, 69, 190–191, 310
Constants 15
  constant letter 19
  dummy constant letter 40
Contradictions 32
  in set theory 70, 97–98
  logical and semantical 102
Constructivism 71, 98–100, 193
Conway's numbers 186–189

Decidable system 11
  recursively decidable system 206
Dedekind cut 101, 120–121, 127
Deduction
  c-deduction 41
  formal 10
  from hypotheses 10
  in first-order theory 28 ff.
  natural deduction 39 ff.
  natural deduction rules 43–44
Deduction theorem 35, 54

Deductive structure   *see* Deduction, formal
Definitions   14
   impredicative   100–101, 105, 119
   recursive   14, 65, 88, 162–163, 196–197,
      274–276
Dependency
   between formulas in a proof   12
   c-depends   41
   in natural deduction   43–44
Disjunction   *see* Connectives
Dwff   40

Elementarily equivalent   58
Equivalence
   elementary   58
   in a first-order theory   32
   logical   32
   tautological   6
Expressions   1
   formal   10
Extensionality   77
   in **NBG**   173
   in **NF** 215–216
   in **NFU** 235
   in topos theory   298–302
   in type theory   114
   in **ZF**   138–139

False   1
   logically   33
   tautologically   7
   under an interpretation   25
First-order theory   19 ff.
   categorical   67
   categorical in power   67
   complete   33
   consistent   32
   essentially incomplete   203
   extension of   37
   incomplete   33
   inconsistent   32
   with equality   57
   with *vbto*s   60
Formal language   10
   arithmetization of   194–196
   first-order   19
   Frege's notion of   72
Formal system   10
   first-order   19
   Frege's notion of   72
   see also Formal language
Formalism   193
Formulas   10
   closed   23
   deductively equivalent   12
   logically equivalent   32
   of a first-order system   20
   of the internal language of topos theory   306
   of type theory   109
   predicative   113
   prime 20, 103
   similar   28
   stratified   213–215
   subformula   22
Foundational system, criteria for   73–75
Functions
   composition of   237, 240–241

Functions
   defined as sets of pairs   159
   defined as triples of sets   242–243
Function letter   19

Global element   284
Gödel number   195
Gödel's incompleteness theorems   193–194

Higher-order theory   131, 134, 190, 194, 305, 308
Hypothesis   3
   continuum hypothesis   170–171
   in formal system   10
   in natural deduction system   43–44

Implication
   in first-order logic   32
   logical   32
   tautological   5
Implies   *see* Implication
Inconsistent   32
Inference   5
   formal   10
   immediate   10
Inferred   5
   by a rule   5
   in a formal system   10
Interpretation   13
   model   32
   of a first-order system   24
   of one theory in another   166
   of the internal language of topos theory   306–309
   of type theory   106–107
Intuitionism   *see* Constructivism
Intuitionistic logic   305, 309

Justification of a line in a proof   11
   in natural deduction   43–44
   in the predicate calculus   30

Language
   expression of   1
   first-order   19
   formal   10
   metalanguage   14
   object language   14
   of type theory   107–109
   see also Formal language *and* Formulas
Liar paradox   98, 102
Local theory   315, 317–311
Logical signs   19
Logicism   73, 123–124, 236

Models   10
   elementarily equivalent   58
   independence models   39
   isomorphic   67
   normal   57
   of arithmetic   65, 89, 207–212
   of first-order theories   32
   of **NBG**   172, 175–177, 185
   of one theory in another   165–167
   of **T**   281, 283
   of type theory   105–106, 128, 133
   of **WT** in Z_0 298
   of **ZF**   136–137, 175–177, 185
   of Z_0 in **WT**   301–302

Models
   simple   283
   standard   65, 133–134, 204
   used to prove consistency   33, 190
Modus ponens   5, 17, 28
   in natural deduction   43–44
   preserves validity   17, 29
MP   31

Natural numbers   63–67, 178–180, 191–192, 270
   see also Peano postulates
Negation   *see* Connectives
Number systems   178–189

Order-type symbols   124
Ordered pair   128
Ordinals   146–148

Paradoxes   *see* Contradictions
Peano postulates   64, 88–90, 119, 121–122, 143, 152–154, 220, 228, 270, 276–279
Predicate calculus   22, 28, 29
   of order ω 131
   of order n 131
Predicate letter   19
Predicative type theory   105 ff.
Predicativity restriction   114
Premiss   7
Primitive notation   15
Primitive recursion   88, 162–163, 196–197, 274–276
Primitive recursion rule proved   274–276
Primitive rule of inference   10
Proof   *see* Deduction
Proposition   *see* Sentence
Provable formula   *see* Theorem

Quantification   21
   existential   22
   of predicate letters   103
   universal   22
   see also Variables
Ramified type theory   113, 124–126
Rank   137, 176–177, 185
Recursive functions   196–198
Recursive set   197
Recursively enumerable set   199
Refutable   7
Rosser's theorem   199
Rule of inference   5, 10
   in natural deduction   43–45
   of predicate calculus   28
   of statement calculus   17
   preserves validity   17, 29
   primitive   10
Russell's paradox   73, 96

Satisfaction   24
Semantic   29
Sentence   1
   atomic   4
   compound   4
   in a first-order system   23
   open   21
   undecidable   200
   see also Statement calculus
Sentential calculus   *see* Statement calculus
Simple recursion   89, 160–162, 196
   as an axiom for the natural numbers   270

Statement   1
   see also Statement calculus
Statement calculus   1
   as a formal system   13 ff.
Substitution
   in sentences   6
   of terms for variables   28, 112, 114
Syntactic   29

Taut   31
Tautology   5
   equivalence determined by   6
   in predicate calculus   28
   in system **P**   17
   refutation of   7
Terms
   abstracts   105, 109
   formed form *vbto*s   60
   free for a variable   28
   of a first-order system   19
   order of   111
   predicative abstracts   111
Theorem
   Gödel's incompleteness theorem   194–195
   in a first-order theory   29
   in a formal system   10
   in system of natural deduction   44
   metatheorem   17
   of infinity   90–95, 226–229
Topos   279 ff.
   defined   280
   elementary   281
   free topos   310
   Grothendieck   281
True   1
   in all interpretations   26
   in an interpretation   25
   logically   5, 26
   tautologically   5
Truth function   8
Truth-functional   1
Truth set   61
Truth tables   2
Truth values   2
Type hierarchy   107
   simplified   128
Type symbol   106
   order of   107
   order-type symbol   112
Type theory   103 ff., 305–311
Typed relation   106
Typical ambiguity   123, 214, 229–230
   complete typical ambiguity   231–233

UG   31
Undecidable system   11
   essentially recursively undecidable   206
   recursively undecidable   206
Universal generalization   28

Valid
   argument   7
   in an interpretation of the internal language of topos theory   308
   logically valid   26
   universally valid   26, 310

Variables
  alphabetic order o   19
  bound   23, 60, 316
  canonical order of   109
  free   23
  in a first-order system   19
  scope of   23
  syntactical   16
  typographical order of   109

*Vbto*s   60, 110
Vicious circle   101–102

Well formed
  in sentential calculus   4
  see also Formulas
Well-formed formula   *see* Formulas
Wff   *see* Formulas